Signal Processing Strategies

Titles in this Series

Advances in Neural Engineering
Signal Processing Strategies

Advances in Neural Engineering
Brain-Computer Interfaces

Series Editors

Ayman S. El-Baz, Distinguished Professor, University of Louisville, Kentucky, United States, and the University of Louisville at Alamein International University (UofL-AIU), New Alamein City, Egypt

Jasjit S. Suri, Chairperson of AtheroPoint, Roseville, CA, USA.

Advances in Neural Engineering

Signal Processing Strategies

Edited by

Ayman S. El-Baz

Jasjit S. Suri

ACADEMIC PRESS
An imprint of Elsevier

ELSEVIER

Academic Press is an imprint of Elsevier
125 London Wall, London EC2Y 5AS, United Kingdom
525 B Street, Suite 1650, San Diego, CA 92101, United States
50 Hampshire Street, 5th Floor, Cambridge, MA 02139, United States

Notices
Knowledge and best practice in this field are constantly changing. As new research and experience broaden our understanding, changes in research methods, professional practices, or medical treatment may become necessary.

Practitioners and researchers must always rely on their own experience and knowledge in evaluating and using any information, methods, compounds, or experiments described herein. In using such information or methods they should be mindful of their own safety and the safety of others, including parties for whom they have a professional responsibility.

To the fullest extent of the law, neither the Publisher nor the authors, contributors, or editors, assume any liability for any injury and/or damage to persons or property as a matter of products liability, negligence or otherwise, or from any use or operation of any methods, products, instructions, or ideas contained in the material herein.

ISBN: 978-0-323-95437-2

For information on all Academic Press publications visit our website at
https://www.elsevier.com/books-and-journals

Publisher: Mara Conner
Acquisitions Editor: Chris Katsaropoulos
Editorial Project Manager: Emily Thomson
Production Project Manager: Sujithkumar Chandran
Cover Designer: Vicky Pearson Esser

Typeset by TNQ Technologies

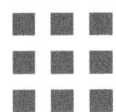

Contents

Contributors   xi

1. Framework for segmentation, optimization, and recognition of multivariate brain tumors   1

Hossam Magdy Balaha and Asmaa El-Sayed Hassan

  1. Introduction   1

  2. Literature review   4

  3. Background   6

  4. Methodology   14

  5. Experiments and discussions   21

  6. Conclusion, limitation, and future work   27

  References   28

2. The neural circuitry of PTSD—An RDOC approach   33

Anthony K. Allam, M. Benjamin Larkin, Ashwin Viswanathan, Sameer A. Sheth and Garrett P. Banks

  1. Introduction   33

  2. Pathophysiology of PTSD   35

  3. Neurotechnological strategies towards PTSD treatment   39

  4. Conclusion   41

  References   42

3. CNN-based artifact recognition from independent components of EEG signals   49

Matteo Polsinelli and Giuseppe Placidi

  1. Introduction   49

  2. ICA-based artifact removal pipeline   51

3. Report on a case of study method   59

4. Conclusions   65

References   66

4. **Deep multimodal representation learning for noninvasive neural speech decoding**   71

Ciaran Cooney, Raffaella Folli and Damien Coyle

1. Introduction   71

2. Multimodal decoding of overt and imagined speech   73

3. Results   81

4. Discussion   84

5. Conclusion   86

References   86

5. **Neural signals processing using deep learning for diagnosis of cognitive disorders**   91

Hamid Jahani and Ali Asghar Safaei

1. Introduction   91

2. Deep learning-aided medical diagnosis systems   92

3. Case study: Diagnosis of ADHD   108

4. Discussion   115

5. Conclusion   116

References   117

6. **Brain tumor recognition using semisupervised generative adversarial network**   119

Jyotismita Chaki

1. Introduction   119

2. Proposed methodology   125

3. Experimentations and results   128

4. Analysis   131

5. Conclusions and future scopes   132

References   133

7. Multivariate adaptive signal decomposition techniques
   and their applications to EEG signal processing:
   An introduction   137

Kritiprasanna Das, Achinta Mondal, Nabasmita Phukan and Ram Bilas Pachori

1. Introduction   137

2. Multivariate time series   138

3. Multivariate adaptive decomposition   140

4. Application of multivariate adaptive decomposition to EEG
   signal processing   154

5. Conclusion   158

References   158

8. Split learning for human activity recognition   163

Sandra Pavleska, Valentin Rakovic, Daniel Denkovski and Hristijan Gjoreski

1. Introduction   163

2. Related work   164

3. Data and preprocessing   165

4. Methodology   166

5. Experimental setup   168

6. Results and discussion   170

7. Conclusion   173

Acknowledgment   174

References   174

9. Machine learning approaches for epilepsy analysis in current
    clinical trials   175
    Ishan Ayus and Biswajit Jena

    1. Introduction   175
    2. Background studies   177
    3. Machine learning techniques for epilepsy analysis   184
    4. Challenges and considerations in clinical trials   186
    5. Case studies: Machine learning in current clinical trials   188
    6. Future directions and implications   189
    7. Conclusion   191
    References   192

10. Brainwave and head motion control of a smart home for
    disabled people   195
    Minoru Dhananjaya Jayakody Arachchige and Marwan Nafea

    1. Introduction   195
    2. Proposed methodology   198
    3. Results and analysis   204
    4. Conclusion   213
    Acknowledgment   213
    References   213

11. Independent component analysis methods for motor
    imagery-based brain-computer interfaces   217
    Paulo A.A.L. Viana, Sarah N.C. Leite and Romis Attux

    1. Introduction   217
    2. Methods   219
    3. Results   228

4. EEGNet   242

5. Conclusion   246

Acknowledgments   248

References   248

12. Advancing neural engineering: Hierarchical control strategies with human-centered focus for hand prosthetics   251

Tanaya Das and Dhruba Jyoti Sut

1. Introduction   251

2. Hierarchical control strategies in hand prosthesis   255

3. Sensory feedback in hierarchical control   266

4. Challenges and future directions   273

5. Conclusion   274

Acknowledgment   274

References   274

13. Advances in non-invasive EEG-based brain-computer interfaces: Signal acquisition, processing, emerging approaches, and applications   281

Shiu Kumar and Alok Sharma

1. Introduction   281

2. Signal acquisition techniques   283

3. Electroencephalography (EEG) and signal processing techniques   285

4. Existing approaches, packages, datasets, and applications of EEG signal processing   292

5. Conclusion and future perspectives   306

References   307

Index   311

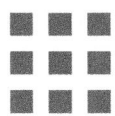

Contributors

Anthony K. Allam   Department of Neurosurgery, Baylor College of Medicine, Houston, TX, United States

Romis Attux   School of Electrical and Computer Engineering, University of Campinas (UNICAMP), Campinas, Brazil

Ishan Ayus   Department of Computer Science and Engineering, Institute of Technical Education and Research, SOA Deemed to be University, Bhubaneswar, Odisha, India

Hossam Magdy Balaha   Bioengineering Department, J.B. Speed School of Engineering, University of Louisville, Louisville, KY, United States; Computers and Control Systems Department, Faculty of Engineering, Mansoura University, Mansoura, Egypt

Garrett P. Banks   Department of Neurosurgery, Baylor College of Medicine, Houston, TX, United States

Jyotismita Chaki   School of Computer Science and Engineering, Vellore Institute of Technology, Vellore, Tamil Nadu, India

Ciaran Cooney   Intelligent Systems Research Centre, Ulster University, Derry, United Kingdom

Damien Coyle   Intelligent Systems Research Centre, Ulster University, Derry, United Kingdom; The Bath Institute for the Augmented Human, University of Bath, Bath, United Kingdom

Kritiprasanna Das   Department of Electrical Engineering, Indian Institute of Technology Indore, Indore, Madhya Pradesh, India

Tanaya Das   Independent Researcher, Digboi, Assam, India; School of Biomedical Engineering, The University of Sydney, Sydney, NSW, Australia

Daniel Denkovski   Ss. Cyril and Methodius University in Skopje, Faculty of Electrical Engineering and Information Technologies Skopje, Macedonia

Raffaella Folli   School of Communication and Media, Ulster University, Belfast, United Kingdom

Hristijan Gjoreski   Ss. Cyril and Methodius University in Skopje, Faculty of Electrical Engineering and Information Technologies Skopje, Macedonia

Asmaa El-Sayed Hassan   Mathematics and Engineering Physics Department, Mansoura University, Faculty of Engineering, Mansoura, Egypt

Hamid Jahani   Department of Data Science, Faculty of Interdisciplinary Science and Technology, Tarbiat Modares University, Tehran, Iran

Minoru Dhananjaya Jayakody Arachchige   Department of Automatic Control and Systems Engineering, Faculty of Engineering, University of Sheffield, Sheffield, United Kingdom

Biswajit Jena   Department of Computer Science and Engineering, Institute of Technical Education and Research, SOA Deemed to be University, Bhubaneswar, Odisha, India

Shiu Kumar   School of Electrical and Electronics Engineering, Fiji National University, Suva, Fiji

M. Benjamin Larkin   Department of Neurosurgery, Baylor College of Medicine, Houston, TX, United States

Sarah N.C. Leite   Division of Electronics Engineering, Aeronautics Institute of Technology (ITA), São José dos Campos, Brazil

Achinta Mondal   Department of Electrical Engineering, Indian Institute of Technology Indore, Indore, Madhya Pradesh, India

Marwan Nafea   Department of Electrical and Electronic Engineering, Faculty of Science and Engineering, University of Nottingham Malaysia, Selangor, Malaysia

Ram Bilas Pachori   Department of Electrical Engineering, Indian Institute of Technology Indore, Indore, Madhya Pradesh, India

Sandra Pavleska   Ss. Cyril and Methodius University in Skopje, Faculty of Electrical Engineering and Information Technologies Skopje, Macedonia

Nabasmita Phukan   Department of Electrical Engineering, Indian Institute of Technology Indore, Indore, Madhya Pradesh, India

Giuseppe Placidi   Deparment of Life, Health & Environmental Sciences, University of L'Aquila, L'Aquila, Italy

Matteo Polsinelli   Department of Computer Science, University of Salerno, Salerno, Italy

Valentin Rakovic   Ss. Cyril and Methodius University in Skopje, Faculty of Electrical Engineering and Information Technologies Skopje, Macedonia

Ali Asghar Safaei   Department of Data Science, Faculty of Interdisciplinary Science and Technology, Tarbiat Modares University, Tehran, Iran; Department of Medical Informatics, Faculty of Medical Sciences, Tarbiat Modares University, Tehran, Iran

Alok Sharma   Laboratory for Medical Science Mathematics, RIKEN Center for Integrative Medical Sciences, Yokohama, Japan; Institute for Integrated and Intelligent Systems, Griffith University, Brisbane, QLD, Australia

Sameer A. Sheth   Department of Neurosurgery, Baylor College of Medicine, Houston, TX, United States

Dhruba Jyoti Sut   Department of Mechnaical Engineering, SRM Institute of Science and Technology, Chennai, Tamil Nadu, India

Paulo A.A.L. Viana   School of Electrical and Computer Engineering, University of Campinas (UNICAMP), Campinas, Brazil; AI R&D Lab, Samsung R&D Institute Brazil, Campinas, Brazil

Ashwin Viswanathan   Department of Neurosurgery, Baylor College of Medicine, Houston, TX, United States

1

Framework for segmentation, optimization, and recognition of multivariate brain tumors

Hossam Magdy Balaha[1,3] and Asmaa El-Sayed Hassan[2]

[1]BIOENGINEERING DEPARTMENT, J.B. SPEED SCHOOL OF ENGINEERING, UNIVERSITY OF LOUISVILLE, LOUISVILLE, KY, UNITED STATES; [2]MATHEMATICS AND ENGINEERING PHYSICS DEPARTMENT, MANSOURA UNIVERSITY, FACULTY OF ENGINEERING, MANSOURA, EGYPT; [3]COMPUTERS AND CONTROL SYSTEMS DEPARTMENT, FACULTY OF ENGINEERING, MANSOURA UNIVERSITY, MANSOURA, EGYPT

1. Introduction

Cancer is characterized by uncontrolled and abnormal cell growth and division within the body. Brain tumors (BTs), which manifest as masses of abnormal cell growth within brain tissue, represent one of the most lethal forms of cancer, although they are relatively uncommon [1]. Brain tumors do not discriminate based on age, gender, or ethnicity, affecting individuals across various demographics. According to the World Health Organization (WHO), the global 5-year prevalence of brain tumors stands at 837,152 [2]. In 2020, more than 308,102 people received a primary BT diagnosis, encompassing over 120 different tumor types. Tragically, in 2020, approximately 251,329 individuals lost their lives due to primary malignant brain tumors. The median age of diagnosis for all primary brain tumors is 60 years, with BT being the most common type of cancer among children aged 0 to 14 from 2013 to 2017. Furthermore, BT is the leading cause of cancer-related deaths in this age group [3]. Fig. 1.1, shown below, presents a heatmap depicting the proportions of brain tumor cases in 2020 [2]. Lithuania, Greece, and Bosnia and Herzegovina emerge as the top three countries with the highest incidence rates. The current section provides an overview of brain tumor grades, symptoms, segmentation, medical imaging techniques, and treatment modalities.

Brain tumors can be categorized as either primary or metastatic, depending on their origin. Primary brain tumors originate within the brain tissue, whereas metastatic tumors develop in another part of the body before spreading to the brain. Glioma is a specific type of brain tumor that arises from glial cells. The term "glioma" encompasses various glioma types, ranging from low-grade gliomas like oligodendrogliomas and

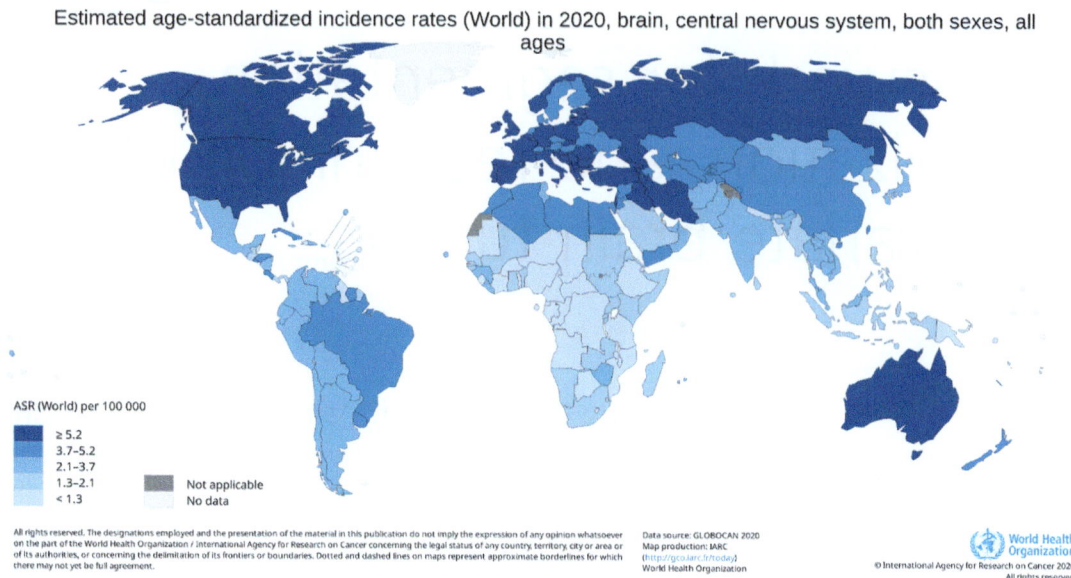

FIGURE 1.1 The 5-year estimated prevalence rate according to the world health organization (WHO) in 2023.

astrocytomas to high-grade glioblastoma multiforme (stage IV), which is the most aggressive and widespread form of primary malignant brain tumor [4]. The treatment of gliomas typically involves a combination of surgery, radiation therapy, chemotherapy, and targeted therapy [5]. Risk factors for developing a brain tumor include gender (with men being more susceptible than women), exposure to environmental factors at work or home (such as solvents, pesticides, rubber, oil products, or vinyl chloride), and a family history of brain tumors. Genetic conditions or factors are associated with only 5% of brain tumors, while exposure to infections, viruses, allergens, electromagnetic fields, and ionizing radiation may also contribute to the risk [3].

Brain tumor symptoms can be categorized as either generic or specific. Generic symptoms, also known as common symptoms, occur when the tumor exerts pressure on the brain and spinal cord. Specific symptoms arise when the tumor impacts the function of a specific area of the brain [6]. Early diagnosis of gliomas plays a crucial role in determining treatment options. Medical imaging methods such as positron emission tomography, computed tomography, single-photon emission computed tomography (SPECT), magnetic resonance imaging (MRI), and magnetic resonance spectroscopy are employed to aid in diagnosis and provide valuable information about the size, location, shape, and metabolism of brain tumors. While a combination of modalities typically provides the most detailed information, MRI is considered the standard technique due to its excellent soft-tissue contrast and widespread availability.

MRI, a noninvasive imaging method, utilizes radio-frequency signals in a strong magnetic field to generate images of target tissues. Varying the duration and frequency of stimulation during image acquisition produces different types of MRI sequences, each offering unique tissue contrast for structural information and tumor diagnosis [7].

During an MRI scan, approximately 150 slices of 2D images are generated to represent the brain's volume in 3D. When these slices from various sequences are combined for diagnostics, the data becomes complex and comprehensive.

Before initiating any treatment, it is essential to segment the tumor to protect healthy tissues while targeting tumor cells. Brain tumor segmentation involves defining and distinguishing tumor tissues, including the necrotic core, active cells, and the edema area, from normal brain tissues such as white matter (WM), gray matter (GM), and cerebrospinal fluid. Current clinical practice often involves manual segmentation, a time-consuming task. To achieve objective and efficient segmentation, powerful automatic segmentation methods have been developed in recent years [8].

Various machine learning (ML) and deep learning (DL) methods and approaches are employed for BT classification and segmentation. These include support vector machines (SVMs) [9] and fuzzy C-means [10] among classical ML methods. Traditional ML methods often require manual feature extraction [11], and classification and segmentation methods are applied using these features. DL, a popular branch of ML, captures complex relationships effectively [12]. It encompasses different algorithms and architectures that can automatically extract features [13], such as deep neural networks [14], recurrent neural networks [15], deep belief networks [16], convolutional neural networks (CNNs) [15], and autoencoders [13]. Following this, the present chapter is dedicated to the categorization and segmentation of BTs. Specifically, we employ U-Net CNN models for segmentation and pretrained CNN models for classification purposes. Furthermore, the optimization of CNN hyperparameters is achieved through the utilization of the Sparrow Search Algorithm (SpaSA) to attain cutting-edge performance metrics.

Numerous methodologies for detecting and segmenting brain tumors have been explored within the scientific literature. However, it is noteworthy that automated techniques often do not match the performance of manual methods involving human intervention. This disparity arises due to several factors, including the absence of prior user experience and the necessity for initialization. Although automated methods for brain tumor diagnosis typically outperform computer-assisted manual approaches in terms of accuracy and efficiency, their development poses substantial challenges encompassing performance and computational speed. Consequently, the principal objective of this research is to introduce an automated learning-based strategy for the classification and segmentation of brain tumors, incorporating the use of a metaheuristic optimizer. *The contributions of this chapter* can be succinctly summarized as follows:

- Tackling the task of brain tumor segmentation employing U-Net, U-Net++, Attention U-Net, and V-Net models.
- Introducing a hybrid model that combines DL techniques with the SpaSA for both the learning process and hyperparameter optimization.
- Providing a comprehensive report on state-of-the-art (SOTA) performance metrics and conducting a comparative assessment in relation to existing studies and methodologies in the field.

The remainder of this chapter is organized as follows: Section 2 reviews related studies in the field. Section 3 provides background information and covers various aspects including segmentation method, convolutional neural network (CNN), parameters optimization, transfer learning, data augmentation, metaheuristic optimization, and performance metrics. Section 4 outlines the methodology, which includes discussions on dataset acquisition, preprocessing, segmentation phase, learning and optimization, and an overall pseudocode description. Section 5 is dedicated to presenting the experiments, results, and subsequent discussions. Section 6 concludes the paper, highlighting chapter limitations and outlining potential avenues for future research.

2. Literature review

Research in the field of tumor segmentation and classification remains ongoing, with numerous automated techniques, approaches, and algorithms being developed to assist in computer-aided diagnosis [8,17]. Traditional ML algorithms encompass several phases, including preprocessing, feature extraction, feature reduction, and classification. Among these, feature extraction holds particular significance as the accuracy of detection relies on these extracted features. Feature extraction can be broadly categorized into two types: Low-level features (global features): These encompass fundamental statistics such as mean, standard deviation, and skewness, along with second-order statistics like wavelet transform, Gabor features, shapes, gray-level cooccurrence matrix (GLCM), and intensity and texture features. High-level features (local features): This category includes techniques such as Fisher vector (FV), Bag-of-Words (BoW), and scale-invariant feature transformation (SIFT). BoW, for instance, has been employed by numerous researchers for tasks such as grading tissue density in mammograms [18], X-ray image acquisition, retrieval, and grading at the pathology and organ levels [19], as well as content-based brain tumor retrieval [20].

Selvaraj et al. [21] applied the least-squares SVM, utilizing first- and second-order statistics, to create a binary classifier for distinguishing normal and abnormal brain MRI images. John [22] employed a combination of discrete wavelet transform and GLCM methods to identify and classify BTs. Due to the similarity in characteristics among most BT cases, the utility of low-level features and their representation was limited.

Ullah et al. [23] utilized discrete wavelet transform for extracting three-level decomposition details and approximate coefficients. They further employed color moments (CMs) to reduce these coefficients and employed artificial feedback neural networks to differentiate between normal and abnormal brain MRI images. Cheng et al. [24] employed a FV to capture BT features, extracting statistical characteristics from FV, BoW, and SIFT. Papageorgiou [25] employed fuzzy cognitive maps to categorize high-grade and low-grade gliomas, achieving an accuracy of 93.22% for high-grade tumors and 90.26% for low-grade brain tumors. Rajan and Sundar [26] introduced an automatic tumor detection and segmentation method based on an energy-efficient hybrid approach, consisting of seven phases, and reported an accuracy of 98%. However, it's worth noting that their model was found to be time-consuming.

Significantly, there have been two prominent limitations within the realm of feature extraction. Firstly, the focus has traditionally been on either low-level or high-level features exclusively. Secondly, classical ML methods heavily rely on manually extracted features, which demand robust prior knowledge and are susceptible to human errors. Thus, there arises a need for an approach that seamlessly integrates both low-level and high-level features without necessitating manual feature extraction. Recent advancements have witnessed the introduction of various techniques for tumor region recognition and segmentation in MRI images, as proposed by numerous researchers. DL methods have gained substantial traction in the classification of brain MRI images [9,27]. Bridging the semantic gap between the low-level and high-level visual information derived from human expertise and MRI imaging machines presents a significant challenge in MRI image classification. To address this gap, CNNs are employed as feature extractors to capture pertinent features for classification tasks.

In Deepak and Ameer [28], brain MRI image features were extracted using a pretrained GoogleNet model to recognize and classify three distinct brain tumor types, achieving an impressive 98% accuracy rate. Similarly, Çinar and Yildirim [29] explored various models, including GoogleNet, AlexNet, DenseNet201, InceptionV3, and ResNet50, to categorize brain MRI images and attain commendable accuracies. Among these pretrained models, they achieved the highest accuracy of 97.2%. Khawaldeh et al. [30] introduced a CNN model to discern normal brain MRI images from abnormal ones and differentiate between high-grade and low-grade gliomas. They adopted a modified AlexNet architecture, achieving an accuracy of 91%. Saxena et al. [31] leveraged InceptionV3, ResNet50, and VGG16 models, coupled with transfer learning (TL) techniques, for brain tumor classification. The ResNet50 model delivered the best accuracy of 95%. Finally, Díaz-Pernas et al. [32] presented a multipath CNN architecture designed for automatic BT segmentation and classification into categories such as meningioma, glioma, and pituitary tumor. They evaluated the model using a publicly available T1-weighted contrast-enhanced MRI dataset, achieving an accuracy of 97.3%.

In the development of fully automated methods for BT segmentation, the incorporation of user interaction and feature engineering assumes critical importance [33]. This endeavor involves the fusion of existing knowledge with artificial intelligence. Liu et al. [34] introduced a computer-aided system tailored for the automatic diagnosis of brain tumors. Their method was composed of four integral components: preprocessing and segmentation techniques, feature extraction, and the final categorization. They fine-tuned CNN hyperparameters employing a novel adaptation of the SpaSA for classification. The reported results underscored the success of their approach, achieving superior levels of accuracy, sensitivity, and specificity, with values of 94.77%, 97.15%, and 67.16%, respectively. Notably, these results outperformed three contemporary SOTA techniques.

In a distinct research endeavor, Liu et al. [35] championed an effective strategy aimed at segmenting brain tumors after the preprocessing of MRI images. Their approach harnessed the GLCM and discrete wavelet transform to extract crucial features. The utilization of an optimized CNN, guided by a well-balanced SpaSA, facilitated image classification. The reported results showcased the achievement of a peak accuracy of 93.65%, complemented by a specificity of 65.07%.

Saouli et al. [36] introduced the innovative concept of ensemble learning, merging two neural network-based models and progressively synthesizing their outcomes. They put forth three fully automated methods (2CNet, 3CNet, and EnsembleNet) for BT segmentation, all employing deep CNNs. In the work of Havaei [37], an extension of Axel et al.'s prior research was presented, featuring a CNN architecture that incorporated two distinct pathways, each integrating a unique 2D patch comprising four MRI sequences as channels. Additionally, the output from the first pathway was employed as an additional input for the second network. Furthermore, a postprocessing step was applied, contingent on the connected component, to eliminate extraneous regions adjacent to the skull.

The detection and segmentation of BTs pose substantial challenges within the domain of medical imaging applications. This chapter addresses these crucial tasks by introducing various CNN architectures designed for BT classification and segmentation. Subsequently, an evaluation of these architectures is conducted, followed by the application of a TL and SpaSA approach to optimize training parameters and hyperparameters. The chapter reports the best-performing architectures based on a series of experiments utilizing diverse performance metrics.

3. Background

In this section, we provide the reader with essential background information and foundational concepts that underpin the proposed approach. The subsequent methodology section relies on these fundamental elements and is organized into the following key components:

- Segmentation methods.
- Convolutional neural network (CNN).
- Parameters optimization.
- Transfer learning.
- Data augmentation.
- Metaheuristic optimization.
- Performance metrics.

3.1 Segmentation

BT segmentation methods can be classified into three primary categories based on user interaction levels: (1) fully automatic, (2) semiautomatic, and (3) manual techniques [38]. In manual segmentation, specialized clinicians or radiologists annotate regions of interest (ROI) or voxels of interest, often in a slice-by-slice or 3D fashion [39]. Semiautomatic methods involve user interaction for three main purposes: (1) initialization, (2) intervention or feedback, and (3) evaluation [40].

Automated segmentation techniques, on the other hand, operate autonomously and can be broadly divided into two categories: (1) learning-based and (2) nonlearning-based methods. These approaches offer consistent and reproducible segmentations, resulting in systematic errors rather than random variability commonly observed in manual or semiautomatic methods. Learning-based methods encompass generative and discriminative segmentation strategies [41]. Generative models rely on prior knowledge of healthy tissue voxels and tumor shapes, often utilizing atlases constructed from multiple healthy brains. These models frequently employ locally extracted image data, such as pixel-based measurements, neighborhood histograms, texture features, and region of interest analysis [42]. Some examples include Naive Bayes, hidden Markov models, autoencoders, and generative adversarial networks.

Conversely, discriminative models rely on local information and encompass techniques like SVMs, fuzzy C-means, k-nearest neighbor, deep neural networks (e.g., AlexNet, VGG, ResNet), and decision forests. Among these, random forest (RF) has demonstrated exceptional precision in BT segmentation. More recently, DL-based methods have gained prominence due to their efficacy in detecting image patterns, often outperforming classical discriminant models. Particularly noteworthy are CNN-based methods, which excel at extracting 2D or 3D patches from MRI images for central pixel classification [41].

Nonlearning-based methods are application-specific, leveraging disease characteristics and image data to perform segmentation tasks. They typically require the development of bespoke methodologies for different segmentation objectives [42].

3.2 Convolutional neural network

Neural networks are complex structures composed of an input layer, multiple hidden nonlinear layers, and an output layer. While they possess immense computational power, their extensive connectivity and countless trainable parameters render them unsuitable for resource-constrained devices, especially when applied to image and video data. Additionally, to prevent overfitting and ensure optimal performance, a substantial dataset is essential [43,44].

In the realm of DL, the CNN has emerged as a specialized tool designed for processing data characterized by grid-like structures, such as images. Inspired by the intricacies of the animal visual cortex [45,46], CNNs are engineered to autonomously acquire and adaptively abstract spatial hierarchies, spanning from fundamental to advanced patterns. These networks have proven remarkably effective in diverse computer vision applications, including enhancing autonomous vehicle safety, enabling autonomous ships, and bolstering disease detection efforts [47−49].

The architectural blueprint of a CNN consists of three core layer types: (1) convolution layers, (2) pooling layers, and (3) fully connected layers. Convolution and pooling layers shoulder the crucial task of feature extraction, while the fully connected layer orchestrates the mapping of these derived features to the ultimate network output [50].

At the heart of this architecture lies the convolution layer, a pivotal element that seamlessly integrates linear and nonlinear operations (i.e., convolution and activation functions). Here, kernels, represented as numerical arrays or tensors, are wielded to craft intricate feature maps. Each kernel engages in an element-wise interaction with the input tensor, computing an output value for the corresponding position in the output tensor. This process can be repeated with multiple kernels, each dedicated to encapsulating diverse characteristics of the input tensors, ultimately generating an arbitrary array of feature maps. During the training phase, these kernels undergo adaptation, as they are imbued with learnable parameters. To orchestrate this intricate dance, various hyperparameters, including kernel size, padding, the number of kernels, and stride, must be meticulously defined prior to initiating training. Within the convolution layer, K filters (kernels) take center stage, with each filter adopting a specific shape outlined by Eq. (1.1):

$$\text{Shape(ConvLayer)} = N \times N \times R \tag{1.1}$$

where R corresponds to the number of channels, and N denotes the filter size, encompassing both height and width dimensions. It's imperative that the number of classes, C, is equal to or greater than R to ensure the seamless functioning of the convolution layer. The end result is the production of K feature maps, each sized $M - N + 1$, as filters traverse the entire image, meticulously processing it through pixel by pixel [51].

The introduction of nonlinearity into the convolution outputs is facilitated by activation functions. While numerous activation functions exist, the Rectified Linear Unit (ReLU) remains a preferred choice in practice. ReLU acts as a piecewise linear function, promptly outputting the input if it's positive and defaulting to zero otherwise [52].

Pooling layers contribute an invaluable dimensionality reduction capability. They perform down-sampling operations to scale down the in-plane dimensions of feature maps. This not only curtails the number of learnable parameters but also imparts the network with translation invariance to minor shifts and deformations. While pooling operations encompass hyperparameters such as stride, filter size, and padding, max-pooling stands as the most widespread technique. Max-pooling operations typically employ an $N \times N$ filter size, with N often ranging from 2 to 5, effectively reducing dimensions and generating a fresh feature map [53].

Finally, the convolution output or the last feature map from the pooling layers undergoes a critical transformation into a one-dimensional array (vector). This vector is then channeled into one or more fully connected layers, which serve as the bridge connecting input nodes to output nodes via learnable weights. These fully connected layers orchestrate the mapping of the meticulously extracted features to the final network outputs. Typically, the last fully connected layer boasts a set of output nodes mirroring the number of classes to be classified. Nonlinear functions, like ReLU, are thoughtfully applied to each layer within this network segment [54].

3.3   Parameters optimization

Optimization techniques and algorithms play a crucial role in the quest to maximize (or minimize) the error function, which serves as the yardstick for measuring the disparity between model predictions and the target Y values. Among these techniques, the gradient descent algorithm stands as a fundamental tool, employed to iteratively update model parameters to minimize the error function [55−57].

Adam, short for Adaptive Moment Optimization Algorithm, introduced by Kingma and Ba in 2014, marries the principles of the RMSProp optimization method [58] with the heuristics of Momentum algorithms [59]. These two optimization approaches operate in distinct ways: RMSProp curbs the search process concerning oscillations, while Momentum accelerates the search toward the minima. Beyond Adam, there exists a plethora of other weight optimization algorithms, each with its unique characteristics and strengths. Some notable examples include Nadam, Adagrad, AdaDelta, AdaMax, and FTRL [41]. These optimization methods offer a diverse toolkit for fine-tuning ML models, catering to a wide range of applications and challenges.

3.4   Transfer learning

TL has become an invaluable tool, especially when tackling the challenge of limited training data [60]. It aims to bridge the gap between distinct domains by relaxing the stringent assumption that training and test data must share identical distributions and independence [61,62]. TL has demonstrated superior effectiveness and efficiency compared to traditional learning methods due to its incorporation of pretrained networks [63].

TL encompasses three primary settings: (1) inductive, (2) transduction, and (3) unsupervised TL. While prior research has predominantly concentrated on the first two settings, each of them can be categorized into four classes based on "what to transfer" during the learning process. These categories consist of the (1) instance-transfer approach, (2) feature-representation-transfer approach, (3) parameter transfer approach, and (4) relational-knowledge-transfer approach [64−66].

TL capitalizes on pretrained models, resulting in significant enhancements in performance and the expeditious advancement of various domains [67]. A multitude of pretrained CNN-based models are readily accessible for deployment, encompassing VGG16, ResNet, MobileNet, Xception, NASNet, and DenseNet. These pretrained models function as invaluable resources for a wide spectrum of applications, streamlining the transfer of knowledge and expertise across diverse domains.

3.5   Data augmentation

Data augmentation is a strategic approach within the realm of data manipulation that effectively addresses the problem of overfitting. It is applied to datasets to expand their scope, leveraging the benefits of large datasets to enhance model performance. This technique encompasses a set of methods aimed at enhancing both the quality and size of training datasets, thereby facilitating the development of more robust DL models [68,69]. In the context of the current chapter, various image augmentation techniques have been employed.

Rotation involves rotating images in either a clockwise or counterclockwise direction, typically within a range of 1 degree–359 degrees. In some cases, slight rotations (e.g., 1 degree–20 degrees or −1 degree to −20 degrees) can be beneficial, although it should be noted that as the degree of rotation increases, data labels may no longer be preserved. Horizontal axis flipping is more commonly used than vertical axis flipping. It represents a straightforward augmentation technique. Shearing introduces shifts to specific parts of an image, transforming its orientation. Random brightness augmentation is used to introduce variations in brightness levels across images. This helps DL models generalize better, especially when dealing with images captured under different lighting conditions.

Further, images can be shifted in various directions, including up, down, right, and left, to mitigate positional variations in the data. The remaining space can be filled with a constant value (e.g., 0s or 255s) or even Gaussian or random noise. Cropping involves extracting a central block from each image. This technique can be used to adjust images with disparate width and height dimensions. Random cropping is an alternative that yields effects similar to translation, although translation preserves spatial dimensions while cropping reduces input size. The zooming operation generates images at different zoom levels to diversify the dataset. It randomly zooms images in or out, adding a few pixels around the image to enlarge it.

3.6 Metaheuristic optimization

Hyperparameter optimization techniques encompass several methods, including trial-and-error, grid search, heuristic, and metaheuristic optimization algorithms. While trial-and-error is a less effective approach that fails to explore the entire hyperparameter space [70,71], grid search covers a broader range but is time-consuming, often taking months to complete. On the other hand, heuristic optimization aims to solve problems more rapidly than conventional methods, with algorithms like the brain storm optimization offering efficient solutions [72].

Metaheuristic optimization employs algorithms designed to tackle optimization problems under complex constraints. Many real-world optimization problems are highly nonlinear and multimodal, often involving conflicting objectives or even lacking an optimal solution altogether. Obtaining optimal or suboptimal solutions for such problems is a challenging task [73].

In essence, optimization can be viewed as a problem of minimization or maximization. For example, the function $f(x) = x^2$ has a minimum value of 0 at $x = 0$ across its entire domain $-\infty < x < \infty$. Generally, if the function is simple enough, the first derivative $f'(x) = 0$ can be applied to compute the probable locations, and the second derivative $f''(x)$ can be utilized to confirm if the solution is a minimum (or a maximum). However, for complex functions, particularly those that are nonlinear, multimodal, and multivariate, finding the minimum or maximum is not straightforward. Derivative information, often used in optimization, can be challenging to obtain for functions with discontinuities or complex structures, presenting difficulties for traditional methods like hill-climbing.

3.6.1 Sparrow search algorithm

In 2020, Xue and Shen introduced a novel suite of intelligent algorithms inspired by the foraging behavior of sparrows [74]. These algorithms stand out for their simplicity, minimal need for parameterization, and ease of implementation. Within this algorithmic framework, the sparrow population is categorized into two distinct roles during the food-searching process: (1) discoverers and (2) followers, each adhering to unique behavioral strategies.

Discoverers typically represent approximately 20% of the total population and function as guides for the remaining individuals. Their primary role is to lead the group in the exploration of food resources, and they possess a wide-ranging exploration capability. The formula governing the update of discoverers' positions is detailed in Eq. (1.2).

$$X_{i,j}^{t+1} = \begin{cases} X_{i,j}^t \times \exp\left(\frac{-h}{\alpha \times M_h}\right), & \text{if } (R\_2 < \text{ST}) \\ X_{i,j}^t + Q \times L, & \text{Otherwise} \end{cases} \tag{1.2}$$

Here, (h) corresponds to the current iteration number, (M_h) stands for the maximum number of iterations, $(X_{i,j})$ denotes the present position of the (i^{th}) sparrow in the (j^{th}) dimension, (α) is a randomly generated number within the $[0, 1]$ range, (Q) is a random variable drawn from a normal distribution, (L) is a $(1 \times D)$ matrix consisting solely of 1s, and (R_2) as well as (ST) are parameters representing warning and safety levels, where $(R2)$ varies between 0 and 1, while (ST) ranges from 0.5 to 1.

When $(R_2 < \text{ST})$, it indicates a safe environmental condition for the community, enabling discoverers to explore a wide spectrum of food resources, as no natural threats are detected. Conversely, when $(R_2 \geq \text{ST})$, it signifies that the group has identified a potential predator, prompting an alarm. In response, all individuals within the population adopt antipredation behaviors, with discoverers guiding the followers toward a secure location.

Followers closely observe the discoverers to access high-quality food resources. Some followers even take on the role of supervising the discoverers, particularly those who exhibit a high predation rate for food. This behavior enhances their nutritional intake. The equations governing the positional updates for followers are elucidated in Eq. (1.3).

$$X_{i,j}^{t+1} = \begin{cases} Q \times \exp\left(\frac{X_{\text{worst}}^t - X_{i,j}^t}{i^2}\right), & \text{if } (i > 0.5 \times n) \\ X_P^{t+1} + |X_{i,j}^t - X_P^{t+1}| \times A^+ \times L, & \text{Otherwise} \end{cases} \tag{1.3}$$

where X_P is the currently optimal discoverer position, and X_{worst} indicates the current worst position. A is a $1 \times D$ matrix, where an element is only -1 or 1, with $A^+ = A^T \times \left(A \times A^T\right)^{-1}$. When $(i > 0.5 \times n)$, indicating that the sparrow population is aware of the potential dangers associated with antipredation behaviors, the mathematical formula is applied to update their positions accordingly. In the context of Eq. (1.4), it dictates the positional adjustments for followers.

$$X_{i,j}^{t+1} = \begin{cases} X_{\text{best}}^t + \beta \times |X_{i,j}^t - X_{\text{best}}^t|, & \left(f_i \neq f_g\right) \\ X_{i,j}^t + K \times \left(\dfrac{|X_{i,j}^t - X_{\text{worst}}^t|}{(f_i - f_w) + \epsilon}\right), & \text{Otherwise} \end{cases} \tag{1.4}$$

Here, (X_{best}) represents the current global optimum, while (β) serves as the control step-size parameter, following a normal distribution with a mean of 0 and a variance of 1. (K), which ranges between -1 and 1, is another random variable signifying the direction of movement, with sparrows having control over the step size. (f_i) denotes the current sparrow's individual fitness value. Within the present search domain, $\left(f_g\right)$ and (f_w) are indicative of the optimal and worst fitness values, respectively. To prevent division by zero, (ϵ) is employed as the smallest real number. When $\left(f_i \neq f_g\right)$, this suggests that the current sparrow is positioned at the population's boundary and may be vulnerable to predator attacks, necessitating a modification of its position. Conversely, when individual sparrows perceive potential danger and need to remain in proximity to others to evade threats, the positional updates are adjusted accordingly.

3.7 Performance metrics

Designing or selecting an appropriate effectiveness measure for object classification and segmentation poses a challenging task. To gauge the model's predictive prowess, various performance metrics are computed [75–77]. The choice of assessment metric should align with the specific task's requirements, whether it involves diagnostic or interventional purposes. For instance, some tasks demand real-time operations, such as those in surgical and interventional procedures, while diagnostic tasks can be conducted offline. The significance attached to different performance metrics may vary when selecting the optimal segmentation approach in such cases.

Overlap-based metrics rely on four types of overlaps, namely true positive (TP), false positive (FP), false negative (FN), and true negative (TN), between the predicted output and the actual values. Accuracy, as defined in Eq. (1.5), is one such metric.

$$\text{Accuracy} = \frac{\text{TP} + \text{TN}}{\text{TP} + \text{TN} + \text{FP} + \text{FN}} \tag{1.5}$$

The true positive rate (TPR), also referred to as recall or sensitivity, evaluates the model's ability to correctly predict positive instances of each available category.

Similarly, the true negative rate (TNR), known as specificity, assesses the model's capacity to make accurate negative predictions for each available category. Their formulas are provided in Eqs. (1.6) and (1.7).

$$\text{Recall} = \text{Sensitivity} = \text{TPR} = \frac{\text{TP}}{\text{TP} + \text{FN}} \tag{1.6}$$

$$\text{Specificity} = \text{TNR} = \frac{\text{TN}}{\text{TN} + \text{FP}} \tag{1.7}$$

Additionally, two related metrics, the false negative rate (FNR) and the false positive rate (FPR) or fallout, are defined in Eqs. (1.8) and (1.9).

$$\text{Fallout} = \text{FPR} = \frac{\text{FP}}{\text{FP} + \text{TN}} = 1 - \text{TNR} \tag{1.8}$$

$$\text{FNR} = \frac{\text{FN}}{\text{FN} + \text{TP}} = 1 - \text{TPR} \tag{1.9}$$

Another significant metric is precision, also known as positive predictive value (PPV), which can be computed using Eq. (1.10).

$$\text{Precision} = \text{PPV} = \frac{\text{TP}}{\text{TP} + \text{FP}} \tag{1.10}$$

The Dice coefficient, also referred to as the F1-score and overlap index, is commonly used for validating medical images. Besides direct comparison between the output and true values, it is often employed to measure repeatability. It serves as a repeatability metric, checking the consistency of manual annotations by repeatedly annotating the same image and calculating the overlap between repeated pairs. The Dice coefficient is defined in Eq. (1.11).

$$\text{Dice} = \frac{2 \times \text{TP}}{2 \times \text{TP} + \text{FP} + \text{FN}} \tag{1.11}$$

The Jaccard index (JAC), also known as Intersection over Union (IoU), represents the intersection of two sets divided by their union \cite{jaccard1912distribution}. Eq. (1.12) provides the formula for calculating the Jaccard index.

$$\text{JAC} = \text{IoU} = \frac{\text{TP}}{\text{TP} + \text{FP} + \text{FN}} \tag{1.12}$$

Probabilistic-based metrics involve the computation of statistical functions based on the voxels in the overlap area. Gerig et al. [78] developed the probabilistic distance as a measure of the distance between fuzzy sets. The receiver operating characteristic (ROC) curve is a graph illustrating the relationship between TPR and FPR. Initially proposed by Hanley and McNeil [79] as a measure of diagnostic radiology accuracy, the area under the ROC curve (AUC) has been examined for validating ML algorithms by Bradley [80]. In the case of comparing output and true values (a single measurement), the AUC, defined according to Ref. [81], is determined as the trapezoidal area bounded by the measurement point and the lines TPR = 0 and FPR = 1, as calculated in Eq. (1.13).

$$\text{AUC} = 1 - \frac{\text{FPR} + \text{FNR}}{2} = 1 - 0.5 \times \left(\frac{\text{FP}}{\text{FP} + \text{TN}} + \frac{\text{FN}}{\text{FN} + \text{TP}} \right) \tag{1.13}$$

4. Methodology

In brief, the initial stage involves the acquisition layer, where incoming images are received. Subsequently, in the preprocessing phase, operations such as dataset balancing, augmentation, and scaling are applied to enhance the data. Following preprocessing, the U-Net models are employed for image segmentation. Finally, the learning and optimization phase is executed. Upon completion, various statistics, posttrained models, and graphical representations are generated. The subsequent subsections will provide detailed discussions of these phases.

4.1 Dataset acquisition and preprocessing

4.1.1 Datasets acquisition

In this research, six publicly available datasets were utilized, each serving a specific purpose. For the convenience of readers, a comprehensive overview of these datasets is provided, encompassing details such as the number of classes, image quantity, dimensions, format, and data source.

Brain Tumor Classification (MRI) dataset comprises 3264 images, featuring various dimensions. It is categorized into four distinct classes. Interested users can access and utilize this dataset via the following links: Kaggle (https://www.kaggle.com/sartajbhuvaji/brain-tumor-classification-mri) or GitHub (https://github.com/sartajbhuvaji/brain-tumor-classification-dataset). Br35H: Brain Tumor Detection 2020, with 3060 images of diverse sizes, this dataset is categorized into two classes. It can be obtained for research purposes from Kaggle (https://www.kaggle.com/ahmedhamada0/brain-tumor-detection). Brian Tumor Dataset, comprising 4600 images, this dataset primarily consists of images sized at (512, 512, 3) pixels, although other dimensions like (225, 225, 3) and (630, 630, 3) are also present. Images are predominantly in ".jpg" format, but ".tiff" and ".png" formats can also be found. It is categorized into two classes and is accessible for research from Kaggle (https://www.kaggle.com/preetviradiya/brian-tumor-dataset).

Brain Tumor Detection MRI, featuring 3060 images with varying dimensions, this dataset is divided into two classes. Researchers interested in using this dataset can find it in Kaggle (https://www.kaggle.com/abhranta/brain-tumor-detection-mri). Brain MRI Images for Brain Tumor Detection dataset comprises 506 images of assorted sizes and is categorized into two classes. It is available for research purposes from Kaggle (https://www.kaggle.com/jjprotube/brain-mri-images-for-brain-tumor-detection). Brain Tumor

Segmentation, specifically employed for segmentation experiments, this dataset encompasses 3929 images, all sized at 256×256 pixels. Researchers can access and utilize this dataset for their work from Kaggle (https://www.kaggle.com/muhammadusmansaeed/brain-tumor-segmentation). For a concise overview of these datasets, please refer to Table 1.1.

4.1.2 Datasets preprocessing

In the preprocessing phase, data balancing techniques are employed when the number of images per category is unequal [82]. These techniques aim to balance the categories and are implemented using data augmentation methods. The specific augmentation ranges used in this process include rotation: images are rotated by up to 25°, shift: both width and height are shifted by up to 15\%, shearing: shearing is applied with an amplitude of 15\%, flipping: horizontal and vertical flipping is performed, brightness adjustment: the brightness of images is adjusted within the range $[0.8, 1.2]$. Data augmentation is also utilized during the learning and optimization phase to diversify the dataset and mitigate the risk of overfitting.

Data scaling is a crucial preprocessing step for DL neural networks. It transforms the data to fit within a specific scale, such as $[0, 100]$ or $[0, 1]$. Several scaling techniques are employed in the current chapter. The normalization technique rescales the dataset to a new range of $[0, 1]$. It is achieved by dividing each data point by the maximum value in the dataset. The formula for normalization is given by Eq. (1.14), where the maximum pixel value for gray or RGB images is commonly 255.

$$\text{output} = \frac{\text{input}}{\max(\text{input})} \tag{1.14}$$

Standardization centers the data by subtracting the mean and scaling it by the standard deviation. This transformation results in a mean of 0 and a standard deviation

Table 1.1 The summarization of the used datasets.

Dataset	No. of classes	No. of images	Size of image	Image format	Source (link)
Brain tumor classification (MRI)	4	3264	Different sizes	".jpg"	https://www.kaggle.com/sartajbhuvaji/braintumor-classifcation-mri
Br35H: Brain tumor detection	2	3060	Different sizes	".jpg"	https://www.kaggle.com/ahmedhamada0/braintumor-detection
Brian tumor dataset	2	4600	512×512	".jpg," ".tif," and ".png"	https://www.kaggle.com/preetviradiya/briantumor-dataset
Brain tumor detection MRI	2	3060	Different sizes	".jpg"	https://www.kaggle.com/abhranta/brain-tumordetection-mri
Brain MRI images for brain tumor detection	2	506	Different sizes	".jpg," ".png," and ".jpeg"	https://www.kaggle.com/jjprotube/brain-mriimages-for-brain-tumor-detection
Brain tumor segmentation		3929	256×256	".tif"	https://www.kaggle.com/muhammadusmansaeed/brain-tumor-segmentation

of 1. It is especially useful when dealing with data points of varying scales. Standardization is expressed in Eq. (1.15).

$$\text{output} = \frac{\text{input} - \text{mean}}{\text{std}} \qquad (1.15)$$

Min-Max scaling rescales the feature range to fit within $[0,1] or [-1,1]$, depending on the data's nature. It is calculated using Eq. (1.16).

$$\text{output} = \frac{\text{input} - \text{min(input)}}{\text{max(input)} - \text{min(input)}} \qquad (1.16)$$

Max-Abs scaling involves finding the absolute maximum value in the dataset and dividing all values in the column by that maximum value. This ensures that all values fall within the range of -1 to 1. However, this technique is sensitive to outliers. Max-Abs scaling is represented in Eq. (1.17).

$$\text{output} = \frac{\text{input}}{|\text{max(input)}|} \qquad (1.17)$$

These data scaling methods serve to prepare the data for training DL models effectively.

4.2   Segmentation phase

The segmentation phase of the chapter focuses on isolating the tumor region within medical brain images. This phase employs four different variations of the U-Net architecture tailored for image segmentation. The U-Net models utilized in this research include U-Net, UNet++, Attention U-Net, and V-Net. These U-Net variants are leveraged to perform the crucial task of segmenting brain tumor regions within medical images, each offering unique advantages for accurate and efficient segmentation.

4.2.1   The U-Net model
U-Net [83] is an architecture designed for semantic segmentation, characterized by its symmetric U-shaped structure. It consists of an encoder and a decoder. The encoder, also known as the contracting path, captures context information, while the decoder, or expanding path, facilitates precise localization. The encoder follows the conventional convolutional network design to transform input data into a lower-dimensional space. It consists of modular convolution blocks. The decoder, on the other hand, aims to increase spatial dimensions by refining the encoder's feature maps.

4.2.2   The U-Net++ model
U-Net++ [84] is an extension of the U-Net architecture, incorporating deep supervision. It features a series of dense and nested skip connections that link subnetworks in the encoder and decoder. These redesigned paths aim to reduce the semantic gap between encoder and decoder subnetworks. U-Net++ retains the ability to capture fine-grained details while often outperforming the original U-Net in segmentation tasks.

4.2.3   The attention U-Net model

Similar to U-Net, the Attention U-Net [85] comprises an expansion path on the right and a contraction path on the left. At each level, it introduces skip connections equipped with attention gates. These gates are integrated into the standard U-Net architecture to emphasize salient features conveyed through the skip connections. The gating signal aggregates information from various image scales, resulting in improved performance and enhanced attention weight resolution.

4.2.4   The V-Net model

The V-Net [86] architecture closely resembles the widely used U-Net model, albeit with some distinctions. In the V-Net design, the left part is divided into phases operating at varying resolutions. Each stage incorporates one to three convolution layers. Nonlinearities are introduced to process the input and generate the residual function, which is then used in the convolution layers and added to the output of the respective phase's convolution layer. The V-Net network guarantees convergence, unlike non-residual learning architectures like U-Net.

4.3   Learning and optimization

To achieve SOTA performance in DL training, various hyperparameters need careful optimization [69,86–88]. The current study focuses on optimizing several critical hyperparameters, including batch size, dropout, pretraining tensorflow model learn ratio, optimizer parameters (weights), dataset scaling technique, data augmentation, rotation range, width shift range, height shift range, shear range, zoom range, horizontal flipping, vertical flipping, brightness range. Notably, the last eight hyperparameters are optimized when data augmentation is applied (as indicated in the sixth hyperparameter). Therefore, a minimum of six hyperparameters requires optimization. Using a grid search approach for optimization results in a computational complexity of $O(n^6)$.

To address this optimization challenge, the authors employ the SpaSA. This metaheuristic optimization algorithm is chosen for its superiority over other methods, such as the grey wolf optimizer, gravitational search algorithm, and particle swarm optimization, in terms of precision, accuracy, stability, robustness, fast convergence, and efficiency. SpaSA exhibits high performance across diverse search spaces and effectively mitigates issues related to local optima. The optimization process in SpaSA unfolds in several steps: initialization, objective function calculation, population sorting, selection, and population updating.

Initially, all sparrow populations and their associated parameters are randomly initialized within specified ranges (refer to Table 1.2). Then, all objective function is evaluated for each set of hyperparameters. The sparrow populations are sorted based on their fitness values. Sparrows are selected for reproduction based on their fitness, with better-performing sparrows having a higher chance of being chosen. New populations are generated by combining and modifying the parameters of selected sparrows. Detailed discussion about each phase and its equations is discussed in the following subsections. After a series of iterations, SpaSA identifies the best global optimal location and fitness value. This comprehensive optimization approach effectively tackles the challenge of hyperparameter tuning, ultimately leading to improved DL model performance.

Table 1.2 The used experiments configurations.

Configuration	Specifications
Dataset sources	Presented in Table 1.1
Number of classes	2 and 4
Classes	("HasTumor," "NoTumor") and ("GliomaTumor," "MeningiomaTumor," "NoTumor," and "PituitaryTumor")
Dataset size	3748 for 2-class and 11189 for 4-class images
Image size	$(100 \times 100 \times 3)$ for classification and $(256 \times 256 \times 3)$ for segmentation
Hyperparameter optimizer	Sparrow search algorithm (SpaSA)
Train split ratio	85%–15%
Shuffle dataset	True
SpaSA population size	10
SpaSA number of iterations	10
Number of Epochs	5
Output activation function	SoftMax
Early stopping patience	5
Pretrained parameter initializer	ImageNet
Pretrained models	VGG16, VGG19, MobileNet, MobileNetV2, MobileNetV3Small, MobileNetV3Large, EfcientNetB0, EfcientNetB1, EfcientNetB2, EfcientNetB3, EfcientNetB4, and EfcientNetB5
Loss	Categorical crossentropy, categorical hinge, KLDivergence, poisson, squared hinge, and hinge
Parameter optimizer	Adam, NAdam, AdaGrad, AdaDelta, AdaMax, RMSProp, SGD, Ftrl, SGD Nesterov, RMSProp centered, and adam AMSGrad
Dropout range	[0, 0.6] [0, 0.6]
Batch size	4 to 48 with a step of 4
Pretrained model learn ratio	1 to 100 with a step of 1
Scaling techniques	Normalize, standard, min max, and max abs
Apply data augmentation	[True, false]
Rotation range	$0°$ to $45°$ with a step of $1°$
Width shift range	[0, 0.25]
Height shift range	[0, 0.25]
Shear range	[0, 0.25]
Zoom range	[0, 0.25]
Horizontal flip range	[True, false]
Vertical flip range	[True, false]
Brightness range	[0.5, 2.0]
Programming language	Python
Python packages	Tensorfow, Keras, NumPy, OpenCV, Scikit-Learn, SciPy, Pandas, and Matplotlib
Learning and optimization environment	Google Colab (Intel(R) Xeon(R) CPU @ 2.00 GHz, Tesla T4 16 GB GPU with CUDA v.11.2, and 12 GB RAM)

4.3.1 Initialization

The initial population of sparrows and their corresponding parameters are randomly generated using a random method in the SpaSA. This random generation process can be defined as follows, in Eq. (1.18):

$$\text{Position}_{ij} = \text{LB}_j + \left(\text{UB}_j - \text{LB}_j\right) \times \text{random}(1, D) \qquad (1.18)$$

Here, Position$_{i,j}$ represents the position of the ith sparrow in the jth search space. The index i refers to the solution index, while j pertains to the dimension index. In the context of the current study, D is set to 14, which corresponds to the number of hyperparameters being optimized. This random initialization process ensures diversity in the initial population of sparrows, setting the stage for the subsequent optimization steps in SpaSA.

4.3.2   Objective function calculation

The objective function is a crucial component in the SpaSA as it assesses each sparrow's performance by assigning it a corresponding score. In this maximization problem, a higher score indicates a better-performing sparrow. To simplify this process, the objective function can be viewed as a black box, taking a solution as input and producing a score (i.e., accuracy in this context) as output.

Internally, the objective function follows a series of steps. First, it extracts the elements from the given solution, which in this case consist of 14 hyperparameters, as mentioned earlier. These hyperparameters are then used to initialize a pretrained CNN model, such as VGG16. The model utilizes these hyperparameters to commence the training and validation processes. Subsequently, it evaluates its performance on the entire dataset, generating a set of performance metrics. Finally, the accuracy is computed and returned as the output from the objective function. This accuracy score serves as the basis for evaluating the quality of each sparrow's solution.

4.3.3   Population sorting

Once the objective function has been applied to each sparrow in the population, their performance is evaluated, and they are then ranked or sorted in descending order based on their objective function values. This ranking allows for identifying and prioritizing the sparrows with the highest performance scores, which are indicative of better solutions.

4.3.4   Selection

In the updating process, the current best individual, denoted as X^t_{best}, and the worst individual, denoted as X^t_{worst}, are selected based on their fitness values. These individuals play crucial roles in influencing the optimization process.

4.3.5   Population updating

In the SpaSA, individuals with the best fitness values have a higher priority in the search process, taking on the role of producers and overseeing the entire population's movement. The updating of sparrow locations for producers is crucial and can be achieved using Eq. (1.2). Additionally, the positions of followers are updated using Eq. (1.3). It is assumed that only a fraction of the entire sparrow population, typically between 10% and 20%, is aware of potential dangers. The initial positions of the sparrows are randomly generated within the population using Eq. (1.4). This approach helps introduce diversity into the population and encourages exploration of the search space.

4.4   The overall pseudocode

The steps are performed iteratively for a specified number of iterations. Algorithm 1 provides a summary of the proposed learning and optimization approach.

Algorithm 1: The suggested learning and hyperparameters optimization pesudocode

Input model, dataset, configs   // Model name, Dataset, and Experimental configurations

Output Output: best, bestScore   //The best overall score and solution (i.e., combination)

1   SR = configs[SR]   // Extract the split ratio (SR).

2   M = configs[M]   // Extract the maximum number of iterations (M).

3   n = configs[n]   // Extract the maximum number of sparorws (n).

4   trainX, validationX, testX,trainY, validationY, testY = SplitDataset

5   model = CreateCNNModel(model)   // Create the initial pre-trained CNN model with the ImageNet pre-trained weights.

6   population = GenerateInitialPopulation(configs) // Generate (i.e., create) the initial population.

   // Executer the learning SpaSA hyperparameters optimization process for M terations.

// Executer the learning SpaSA hyperparameters optimization process for M iterations.

7   m = 1   // Initialize the iterations' counter where $m \leq M$.

8   while ($i \leq M$) do

//Calculate the scores for the population.

9   i = 1   // Initialize the sparrow' counter where $i \leq n$.

10 scoresList = []   // Initialize the scores list.

11 while ($i \leq n$) do

12 score = CalculateScore (model, population[i],trainX, trainY, validationX, validationY, con f igs) // Calculate the the score (i.e., accuracy) for the current solution.

13 Append(score, scoresList)   // Calculate the score into the scores list.

14 $i = i + 1$ // Increment the sparrow' counter.

// Sort the population scores.

15 Sort(population, scoresList)   // Sort the scores list in descending order.

16 best, worst, optimal, bestScore, worstScore = Extract(population, scoresList) // Extract the best, optimal, and worst solutions; and best and worst scores.

//Start the updating process using Equations 4, 5, and 5.

17 i = 1   // Initialize a counter.

18 while ($i \leq configs[PD]$) do

19 if ($configs[R2] < configs[ST]$) then

20 $population[i] = population[i] \times \exp^{\frac{-configs[h]}{configs[\alpha] \times M}}$

21 else

22 $population[i] = population[i] + configs[Q] \times configs[L]$

23 i = i + 1   // Increment the counter.

24 i = 1   // Initialize a counter.

25 while ($i \leq (n - configs[PD])$) do

26 if ($i > 0.5 \times n$) then

27 $population[i] = configs[Q] \times \exp^{\frac{worst - population[i]}{i^2}}$

28 else

29 $population[i] = optimal + |population[i] - optimal| \times configs[A]^{+} \times configs[L]$

30 $i = i + 1$ // Increment the counter.

31 $i = 1$ // Initialize a counter.

32 while ($i \leq configs[SD]$) do

33 if ($scoresList[i] \neq bestScore$) then

34 $population[i] = best + configs[\beta] \times |population[i] - best|$

35 else

36 $population[i] = population[i] + configs[K] \times \left(\frac{|population[i] - worst|}{(scoresList[i] - worstScore) + \varepsilon} \right)$

37 $i = i + 1$ // Increment the counter.

38 return best, bestScore   // Return the best score and solution

5. Experiments and discussions

The experiments conducted in this chapter can be categorized into two main groups: segmentation experiments and optimization, learning, and classification experiments.

5.1 Experiments configurations

In this study, Python was employed for scripting, and the Google Colab platform with GPU support served as the primary environment for learning and optimization. The essential Python packages used include TensorFlow, Keras, NumPy, OpenCV, Pandas, and Matplotlib, as outlined in a prior work [89]. The dataset was split into training and validation sets, which accounted for 85% of the data, while the remaining 15% was reserved for testing. Random shuffling was applied to the dataset. Image resizing was performed to dimensions of $(100 \times 100 \times 3)$ for classification tasks and $(256 \times 256 \times 3)$ for segmentation tasks in RGB. For further reference, Table 1.2 provides a summary of the experimental configurations.

5.2 Segmentation experiments

This subsection delves into the segmentation experiments conducted in the chapter, employing various architectural models. The models evaluated include U-Net with the VGG16 backbone, UNet++ with the VGG16 backbone, Attention U-Net with the VGG16 backbone, and V-Net. A summary of the results obtained from these segmentation experiments is presented in Table 1.3.

Based on the findings in Table 1.3, the U-Net model emerges as the top performer across various metrics, including loss, accuracy, specificity, IoU, and dice. Conversely, the attention U-Net model outperforms the others when it comes to the AUC value.

5.3 Random hyperparameters experiments

In this subsection, we delve into experiments that revolve around learning without hyperparameter optimization. These experiments harness the power of various pre-trained CNN models, including VGG16, VGG19, MobileNet, MobileNetV2, MobileNetV3Small, MobileNetV3Large, EfficientNetB0, EfficientNetB1, EfficientNetB2, EfficientNetB3, EfficientNetB4, and EfficientNetB5. The optimization process is

Table 1.3 The segmentation experiments and results.

Model	Loss	Accuracy	Specificity	AUC	IoU	Dice
U-Net	0.007	0.9973	0.9993	0.9935	0.9978	0.9980
U-Net++	0.019	0.9934	0.9984	0.9744	0.9950	0.9953
Attention U-Net	0.019	0. 9929	0.9935	0.9976	0.9913	0.9914
V-Net	0.29	0.9911	0.9978	0.9418	0.9934	0.9936

facilitated by the SpaSA metaheuristic optimizer. The number of epochs is standardized at 5, and both the SpaSA iterations and population size are set to 10 each. The metrics tracked encompass loss, accuracy, recall, specificity, AUC, sensitivity, IOU coefficient, Dice coefficient, cosine similarity, TP, TN, FP, and FN.

5.3.1   Two-classes dataset experiments

Table 1.4 provides a comprehensive summary of the outcomes obtained from the CNN experiments conducted on the 2-class dataset without hyperparameter optimization. Remarkably, the VGG19 pretrained model emerged as the top-performer, achieving an outstanding overall accuracy of 96.70%.

5.3.2   Four-classes dataset experiments

Table 1.5 offers a succinct overview of the findings from the CNN experiments conducted on the 4-class dataset without hyperparameter optimization. Notably, the EfficientNetB1 pretrained model stands out as the highest-performing model, achieving an impressive overall accuracy of 94.85%.

5.4   Learning and optimization experiments

This subsection discusses the experiments related to learning and optimization, utilizing the specified pretrained CNN models, which include VGG16, VGG19, MobileNet, MobileNetV2, MobileNetV3Small, MobileNetV3Large, EfficientNetB0, EfficientNetB1, EfficientNetB2, EfficientNetB3, EfficientNetB4, and EfficientNetB5, in conjunction with the SpaSA metaheuristic optimizer. The experimental setup involves configuring the number of epochs to 5, setting both the SpaSA iterations and the population size to 10. The metrics collected for evaluation encompass loss, accuracy, recall, specificity, AUC, sensitivity, IOU coefficient, Dice coefficient, cosine similarity, TP, TN, FP, and FN.

5.4.1   Two-classes dataset experiment

Table 1.6 provides an overview of the best-reported outcomes stemming from the conducted CNN experiments using the 2-class dataset. Notably, the MobileNetV3Large pretrained model achieved the highest reported overall accuracy, reaching an impressive 0.9999. The average accuracy across these experiments is 0.9992. The rounded average values for TP, TN, FP, and FN are 11,167, 11,167, 8, and 8, respectively. Augmentation was recommended in 8 of the experiments, while the Kullback–Leibler Divergence (KLDivergence) loss function was favored in six experiments. Furthermore, the AdaMax parameter optimizer and Max-Abs scaling technique received recommendations from five experiments.

5.4.2   Four-classes dataset experiment

Table 1.7 provides an overview of the best-reported outcomes resulting from the conducted CNN experiments utilizing the 4-class dataset. Impressively, the EfcientNetB2

Table 1.4 Two-classes dataset experiments without hyperparameters optimization.

Model	VGG16	VGG19	MobileNet	MobileNetV2	Mobile NetV3Small	Mobile NetV3Large	EfficientNetB0	EffecientNetB1	EffecientNetB2	Effecient NetB3	Effecient NetB4	Efficient NetB5
Accuracy	0.4745	0.9670	0.9270	0.6989	0.8526	0.7890	0.7794	0.9188	0.9168	0.6716	0.6071	0.6045
Precision	0.4745	0.9670	0.9270	0.6989	0.8526	0.7890	0.7794	0.9188	0.9168	0.6716	0.6071	0.6045
Recall	0.4745	0.9670	0.9270	0.6989	0.8526	0.7890	0.7794	0.9185	0.9168	0.6716	0.6071	0.6045
Specificity	0.4745	0.9670	0.9270	0.6989	0.8526	0.7890	0.7794	0.9188	0.9168	0.6716	0.6071	0.6045
AUC	0.4745	0.9757	0.9386	0.7614	0.8531	0.8611	0.8436	0.9628	0.9265	0.7630	0.6882	0.6588
Sensitivity	0.4745	0.9670	0.9270	0.6989	0.8526	0.7890	0.7794	0.9188	0.9168	0.6716	0.6071	0.6045
IoU	0.6496	0.9763	0.9472	0.6972	0.9016	0.7239	0.7912	0.9117	0.9446	0.7308	0.7165	0.6295
Dise	0.6496	0.9770	0.9488	0.7469	0.9017	0.7751	0.8204	0.9256	0.9450	0.7678	0.7381	0.6919
Cosine similarity	0.4645	0.9678	0.9281	0.7728	0.8527	0.8268	0.8191	0.9313	0.9186	0.7582	0.6618	0.7329
TP	5295	10,815	10,353	7816	9536	8812	8717	10,276	10,250	7508	6787	6756
TN	5295	10,815	10,353	7816	9536	8812	8717	10,276	10,250	7508	6787	6756
FP	5865	369	815	815	1648	2356	2467	908	930	3672	4393	4404
FN	5865	369	815	815	1648	2356	2467	908	930	3672	4393	4404

Table 1.5 Four-classes dataset experiments without hyperparameters optimization.

Model	VGG16	VGG19	MobileNet	MobileNetV2	Mobile NetV3Small	Mobile NetV3Large	Efcient NetB0	Efcient NetB1	Efcient NetB2	Efcient NetB3	Efcient NetB4	Efcient NetB5
Accuracy	0.2286	0.5844	0.4386	0.6488	0.4621	0.7003	0.8934	0.9485	0.9212	0.7842	0.6319	0.7831
Precision	0.2286	0.7247	0.484	0.6992	0.5786	0.7724	0.9038	0.9494	0.9258	0.7854	0.669	0.9038
Recall	0.2286	0.2503	0.3774	0.5809	0.3976	0.5169	0.8908	0.941	0.9151	0.7791	0.5192	0.6261
Specificity	0.7429	0.97	0.8657	0.9156	0.9022	0.9625	0.9682	0.9838	0.9756	0.929	0.9283	0.9784
AUC	0.4857	0.8167	0.7054	0.8675	0.7278	0.9005	0.9732	0.9922	0.9807	0.9287	0.8723	0.9459
Sensitivity	0.2286	0.2503	0.3774	0.5809	0.3976	0.5169	0.8908	0.941	0.9151	0.7791	0.5192	0.6261
IoU	0.4857	0.5414	0.5658	0.6743	0.5682	0.6385	0.9168	0.9378	0.9173	0.8377	0.649	0.671
Dice	0.4857	0.5952	0.6022	0.7124	0.6093	0.6882	0.9235	0.9474	0.9285	0.8475	0.6917	0.7219
Cosine similarity	0.2286	0.6507	53.77/100	0.7127	0.5714	0.7496	0.9066	0.9535	0.9301	0.7996	0.7112	0.8088
TP	855	937	1413	2175	1479	1923	3335	3523	3426	2914	1944	2344
TN	8335	10,895	9723	10,284	10,069	10,742	10,875	11,050	10,958	10,423	10,427	10,989
FP	2885	337	1509	948	1091	418	357	182	274	797	805	243
FN	2885	2807	2331	1569	2241	1797	409	221	318	826	1800	1400

Table 1.6 Two-classes dataset experiments with hyperparameters optimization.

Metric	VGG16	VGG19	Mobile Net	Mobile NetV2	Mobile NetV35small	Mobile NetV3Large	Efficient NetB0	Efficient NetB1	Efficient NetB2	Efficient NetB3	Efficient NetB4	Efficient NetB5
Loss	Poisson	Poisson	KLDivergence	KLDivergence	Poisson	Poisson	KLDivergence	Poisson	KLDivergence	KLDivergence	KLDivergence	Categorical Crossentropy
Batch size	16	40	36	8	20	28	48	8	20	36	4	12
Dropout	0.05	0.23	0.47	0.12	0.33	0.54	0.11	0.51	0.58	0.41	0.06	0.14
TF learn ratio	63	73	69	81	46	35	9	31	22	87	63	62
Optimizer	SGD	SGD	Nesterov SGD	SGD	AdaMax	AdaMax	AdaMax	AdaMax	Adam	AMSGrad	AdaMax	SGD
Scaling technique	Standardization	Standardization	Standardization	Standardization	Max-Abs	Max-Abs	Min-Max	Max-Abs	Max-Abs	Min-Max	Max-Abs	Min-Max
Apply augmentation	Yes	Yes	Yes	No	Yes	No	Yes	No	Yes	Yes	Yes	No
Rotation range	17	0	12	N/A	10	N/A	28	N/A	21	38	36	N/A
Width shift range	0.13	0.07	0.1	N/A	0.15	N/A	0.06	N/A	0.13	0.24	0.23	N/A
Height shift range	0.03	0.24	0.17	N/A	0.2	N/A	0.06	N/A	0.01	0.09	0.17	N/A
Shear range	0	0.14	0.13	N/A	0.07	N/A	0.22	N/A	0.09	0.05	0.04	N/A
Zoom range	0.18	0.08	0.01	N/A	0.13	N/A	0.25	N/A	0.05	0.09	0.01	N/A
Horizontal flip	TRUE	TRUE	TRUE	N/A	TRUE	N/A	FALSE	N/A	TRUE	FALSE	FALSE	N/A
Vertical flip	FALSE	TRUE	TRUE	N/A	TRUE	N/A	FALSE	N/A	TRUE	FALSE	TRUE	N/A
Brightness range	1.08–1.89	0.71–1.44	0.52–1.33	N/A	0.99–1.17	N/A	1.15–1.72	N/A	0.68–1.87	1.22–1.74	0.56–1.33	N/A
Loss	0.501	0.503	0.003	0.005	0.502	0.501	0.002	0.503	0.001	0.007	0.011	0.004
Accuracy	0.999	0.9987	0.9998	0.9995	0.9991	0.9999	0.9997	0.9978	0.9998	0.9997	0.998	0.9995
Precision	0.999	0.9987	0.9998	0.9995	0.9991	0.9999	0.9997	0.9978	0.9998	0.9997	0.998	0.9995
Recall	0.999	0.9987	0.9998	0.9995	0.9991	0.9999	0.9997	0.9978	0.9998	0.9997	0.998	0.9995
Specificity	0.999	0.9987	0.9998	0.9998	0.9991	0.9999	0.9997	0.9978	0.9998	0.9997	0.9998	0.9995
AUC	1	1	0.9998	0.9998	0.9998	0.9999	1	1	1	1	0.9998	0.9996
Sensitivity	0.999	0.9987	0.9998	0.9995	0.9991	0.9999	0.9997	0.9978	0.9998	0.9997	0.998	0.9995
IoU	0.9984	0.9969	0.9989	0.9973	0.9984	0.9994	0.9986	0.9971	0.999	0.9942	0.9926	0.9991
Dice	0.9988	0.9976	0.9992	0.998	0.9987	0.9996	0.999	0.9977	0.9993	0.9959	0.9946	0.9993
Cosine similarity	0.9993	0.9987	0.9997	0.9991	0.9988	0.9999	0.9997	0.9982	0.9997	0.9992	0.9979	0.9994
TP	11,173	11,146	11,158	11,178	11,170	11,171	11,181	11,159	11,178	11,157	11,166	11,178
TN	11,173	11,146	11,158	11,178	11,170	11,171	11,181	11,159	11,178	11,157	11,166	11,178
FP	11	14	2	6	10	1	3	25	2	3	22	6
FN	11	14	2	6	10	1	3	25	2	3	22	6

Table 1.7 Four-classes dataset experiments with hyperparameters optimization.

Metric	VGG16	VGG19	MobileNet	MobileNetV2	Mobile NetV3Small	Mobile NetV3Large	Efficient NetB0	Efficient NetB1	Efficient NetB2	Efficient NetB3	Efficient NetB4	EfficientNetB5
Loss	Poisson	Poisson	Poisson	Squared Hinge	Poisson	KLDivergence	KLDivergence	Poisson	Categorical Crossentropy	Categorical Crossentropy	Poisson	Categorical Crossentropy
Batch size	24	8	20	24	32	32	8	40	20	36	32	16
Dropout	0.6	0.08	0.01	0.55	0.21	0.5	0.55	0.5	0.14	0.39	0.36	0.39
TF learn ratio	83	69	67	71	6	33	99	99	91	73	60	85
Optimizer	SGD	Nesterov	AdaGrad	AdaMax	AdaGrad	AdaMax	AdaMax	SGD	Nesterov	NAdam	SGD	Nesterov
Scaling technique	Max-Abs	Standardization	Normalize	Normalize	Max-Abs	Min-Max	Max-Abs	Max-Abs	Min-Max	Min-Max	Min-Max	Min-Max
Apply augmentation	No	Yes	No	Yes	Yes	Yes	Yes	No	Yes	Yes	No	No
Rotation range	N/A	33	N/A	1	15	20	15	N/A	38	20	N/A	N/A
Width shift range	N/A	0.07	N/A	0.18	0.15	0.16	0.2	N/A	0.15	0.18	N/A	N/A
Height shift range	N/A	0.1	N/A	0.24	0.1	0.11	0.25	N/A	0.14	0.18	N/A	N/A
Shear range	N/A	0.19	N/A	0.17	0.09	0.09	0.18	N/A	0.12	0	N/A	N/A
Zoom range	N/A	0.06	N/A	0.23	0.16	0.23	0.12	N/A	0.03	0.18	N/A	N/A
Horizontal flip	N/A	TRUE	N/A	TRUE	TRUE	TRUE	FALSE	N/A	TRUE	TRUE	N/A	N/A
Vertical flip	N/A	FALSE	N/A	FALSE	TRUE	TRUE	FALSE	N/A	TRUE	FALSE	N/A	N/A
Brightness range	N/A	1.17–1.39	N/A	0.82–1.29	1.58–1.97	1.65–1.88	1.18–1.88	N/A	1.4–1.81	1.6–1.6	N/A	N/A
Loss	0.261	0.28	0.257	0.759	0.257	0.02	0.042	0.256	0.024	0.022	0.256	0.03
Accuracy	0.9941	0.9642	0.9955	0.992	0.992	0.9947	0.9933	0.9965	0.9973	0.9939	0.9952	0.9947
Precision	0.9941	0.97	0.9955	0.992	0.9919	0.9947	0.9941	0.9965	0.9973	0.9939	0.996	0.9949
Recall	0.9941	0.9559	0.9955	0.992	0.9912	0.9944	0.9933	0.9965	0.9973	0.9939	0.9952	0.9947
Specificity	0.998	0.9903	0.9985	0.9973	0.9973	0.9982	0.998	0.9988	0.9991	0.998	0.9987	0.9983
AUC	0.9976	0.9968	0.9988	0.9977	0.9994	0.9993	0.9983	0.9991	0.9992	0.9999	0.9997	0.999
Sensitivity	0.9941	0.9559	0.9955	0.992	0.9912	0.9944	0.9933	0.9965	0.9973	0.9939	0.9952	0.9947
IoU	0.9936	0.939	0.995	0.9915	0.9896	0.9947	0.9922	0.9971	0.9936	0.994	0.9909	0.9915
Dice	0.9945	0.9507	0.9958	0.9926	0.9916	0.9955	0.9933	0.9973	0.995	0.9948	0.9928	0.9929
Cosine similarity	0.9947	0.9662	0.9962	0.9925	0.9934	0.9958	0.9937	0.9966	0.9972	0.9945	0.9955	0.9943
TP	3722	3579	3723	3714	3711	3723	3719	3707	3730	3721	3726	3724
TN	11,210	11,123	11,203	11,202	11,202	11,212	11,210	11,147	11,210	11,209	11,217	11,213
FP	22	109	17	30	30	20	22	13	10	23	15	19
FN	22	165	17	30	33	21	25	13	10	23	18	20

pretrained model achieved the highest reported overall accuracy, attaining a remarkable 0.9973. The average accuracy across these experiments stands at 0.9919. Rounded average values for TP, TN, FP, and FN are 3,708, 11,196, 27, and 33, respectively. Augmentation was recommended in seven of the experiments, while the Poisson loss function received favor in six experiments. Furthermore, the stochastic gradient descent (SGD) Nesterov and AdaMax parameter optimizers garnered recommendations from three experiments, and the Min-Max scaling technique found favor in five experiments.

6. Conclusion, limitation, and future work

The field of automatic BT detection and segmentation remains a dynamic and evolving area of research, continually seeking improvements in results. This study explored several systems within the domain of related work. Leveraging CNNs, a SOTA approach for image classification and analysis, we developed a DL-based BT classification and segmentation system. Various experiments were meticulously conducted, and their results systematically recorded.

In the segmentation phase, we employed various U-Net models, each offering unique advantages. Among these models, the U-Net architecture exhibited superior performance across multiple metrics, excelling in loss, accuracy, specificity, IoU, and Dice coefficient values. Remarkably, the Attention U-Net model outperformed all others with respect to the Area Under the Curve (AUC) metric.

The U-Net architecture achieved notable scores, registering accuracy, specificity, AUC, IoU, and Dice coefficient values of 0.9973, 0.9993, 0.9935, 0.9978, and 0.9980, respectively, for the entire tumor region.

For the learning, classification, and optimization phase, we harnessed the power of the SpaSA, a metaheuristic optimizer, to fine-tune the hyperparameters of pretrained CNN models. These models encompassed a wide array of architectures, including VGG16, VGG19, MobileNet, MobileNetV2, MobileNetV3Small, MobileNetV3Large, EfficientNetB0, EfficientNetB1, EfficientNetB2, EfficientNetB3, EfficientNetB4, and EfficientNetB5.

Our dataset, meticulously curated from various public sources, was categorized into 2-classes and 4-classes datasets. To gauge the effectiveness of each pretrained CNN architecture, we considered an extensive set of performance metrics, including loss, accuracy, recall, specificity, AUC, sensitivity, IoU coefficient, Dice coefficient, cosine similarity, TP, TN, FP, and FN.

In the case of the 2-classes dataset, our experiments culminated in the MobileNetV3 Large pretrained model achieving an exceptional overall accuracy of 0.9999, with an average accuracy of 0.9992. The rounded average values for TP, TN, FP, and FN were 11,167, 11,167, 8, and 8, respectively.

Likewise, in the context of the 4-classes dataset, our experiments yielded an outstanding overall accuracy of 0.9973 through the deployment of the EfficientNetB2 pretrained model. The average accuracy for this dataset stood at 0.9919, with TP, TN, FP, and FN values approximating 3,708, 11,196, 27, and 33, respectively.

These results underscore the efficacy of our hybrid methodology in the realms of brain tumor detection, classification, and segmentation, outperforming previous studies. Nevertheless, our approach faces certain limitations, including computational time constraints during the classifier training phase, primarily due to the slow convergence of the boosting algorithm and the high dimensionality of features. Additionally, the study explored only 12 CNN architectures and 4 U-Net models, leaving room for further exploration and optimization.

In the future, we intend to expand our research by (1) assessing the performance of diverse algorithms, encompassing various machine and DL models, (2) enhancing the BT segmentation phase, (3) extending the applicability of our approach to other image types, such as T2-weighted and FLAIR images, (4) exploring alternative optimization techniques to identify the optimal metaheuristic optimizer, and (5) evaluating the system on additional available datasets. This multifaceted approach seeks to continually advance the field of BT recognition and segmentation.

References

[1] DeAngelis LM. Brain tumors. N Engl J Med 2001;344(2):114–23.

[2] Who Health Organization. Cancer today. 2021. https://gco.iarc.fr/today/online-analysis-map. [Accessed 1 October 2023].

[3] Ostrom QT, Patil N, Cioffi G, et al. CBTRUS statistical report: primary brain and other central nervous system tumors diagnosed in the United States in 2013–2017. Neuro Oncol 2020;22(Supplement_1):iv1–96.

[4] von Deimling A, editor. Gliomas, vol 171. Heidelberg: Springer; 2009.

[5] Stupp R, Tonn JC, Brada M, et al. High-grade malignant glioma: ESMO Clinical Practice Guidelines for diagnosis, treatment and follow-up. Ann Oncol 2010;21:v190–3.

[6] Davies E, Clarke C. Early symptoms of brain tumours. J Neurol Neurosurg Psychiatr 2004;75(8): 1205–6.

[7] Drevelegas A, Papanikolaou N. Imaging modalities in brain tumors. In: Imaging of brain tumors with histological correlations; 2011. p. 13–33.

[8] Menze BH, Jakab A, Bauer S, et al. The multimodal brain tumor image segmentation benchmark (BRATS). IEEE Trans Med Imaging 2014;34(10):1993–2024.

[9] Binaghi E, Omodei M, Pedoia V, Balbi S, Lattanzi D, Monti E. Automatic segmentation of MR brain tumor images using support vector machine in combination with graph cut. In: International conference on neural computation theory and applications, vol. 2. Scitepress; October 2014. p. 152–7.

[10] Sikka K, Sinha N, Singh PK, Mishra AK. A fully automated algorithm under modified FCM framework for improved brain MR image segmentation. Magn Reson Imag 2009;27(7):994–1004.

[11] Balaha HM, Shaban AO, El-Gendy EM, Saafan MM. A multi-variate heart disease optimization and recognition framework. Neural Comput Appl 2022;34(18):15907–44.

[12] El-Gendy EM, Saafan MM, Elksas MS, Saraya SF, Areed FF. New suggested model reference adaptive controller for the divided wall distillation column. Ind Eng Chem Res 2019;58(17):7247–64.

[13] Baldi P. Autoencoders, unsupervised learning, and deep architectures. In: Proceedings of ICML workshop on unsupervised and transfer learning. JMLR Workshop and Conference Proceedings; June 2012. p. 37–49.

[14] Ciregan D, Meier U, Schmidhuber J. Multi-column deep neural networks for image classification. In: 2012 IEEE conference on computer vision and pattern recognition. IEEE; June 2012. p. 3642—9.

[15] Schuster M, Paliwal KK. Bidirectional recurrent neural networks. IEEE Trans Signal Process 1997; 45(11):2673—81.

[16] Hinton GE. Deep belief networks. Scholarpedia 2009;4(5):5947.

[17] Bernal J, Kushibar K, Asfaw DS, Valverde S, Oliver A, Martí R, Lladó X. Deep convolutional neural networks for brain image analysis on magnetic resonance imaging: a review. Artif Intell Med 2019; 95:64—81.

[18] Bosch A, Munoz X, Oliver A, Marti J. Modeling and classifying breast tissue density in mammograms. In: 2006 IEEE computer society conference on computer vision and pattern recognition (CVPR'06), vol. 2. IEEE; June 2006. p. 1552—8.

[19] Avni U, Greenspan H, Konen E, Sharon M, Goldberger J. X-ray categorization and retrieval on the organ and pathology level, using patch-based visual words. IEEE Trans Med Imag 2010;30(3): 733—46.

[20] Yang W, Lu Z, Yu M, Huang M, Feng Q, Chen W. Content-based retrieval of focal liver lesions using bag-of-visual-words representations of single-and multiphase contrast-enhanced CT images. J Digit Imag 2012;25:708—19.

[21] Selvaraj H, Selvi ST, Selvathi D, Gewali L. Brain MRI slices classification using least squares support vector machine. Int J Intelligent Comput Med Sci Image Process 2007;1(1):21—33.

[22] John P. Brain tumor classification using wavelet and texture based neural network. Int J Sci Eng Res 2012;3(10):1—7.

[23] Ullah Z, Farooq MU, Lee SH, An D. A hybrid image enhancement based brain MRI images classification technique. Med Hypotheses 2020;143:109922.

[24] Cheng J, Yang W, Huang M, Huang W, Jiang J, Zhou Y, et al. Retrieval of brain tumors by adaptive spatial pooling and fisher vector representation. PLoS One 2016;11(6):e0157112.

[25] Papageorgiou EI, Spyridonos PP, Glotsos DT, Stylios CD, Ravazoula P, Nikiforidis GN, Groumpos PP. Brain tumor characterization using the soft computing technique of fuzzy cognitive maps. Appl Soft Comput 2008;8(1):820—8.

[26] Rajan PG, Sundar C. Brain tumor detection and segmentation by intensity adjustment. J Med Syst 2019;43:1—13.

[27] Smith SM, Jenkinson M, Johansen-Berg H, et al. Tract-based spatial statistics: voxelwise analysis of multi-subject diffusion data. Neuroimage 2006;31(4):1487—505.

[28] Deepak S, Ameer PM. Brain tumor classification using deep CNN features via transfer learning. Comput Biol Med 2019;111:103345.

[29] Çinar A, Yildirim M. Detection of tumors on brain MRI images using the hybrid convolutional neural network architecture. Med Hypotheses 2020;139:109684.

[30] Khawaldeh S, Pervaiz U, Rafiq A, Alkhawaldeh RS. Noninvasive grading of glioma tumor using magnetic resonance imaging with convolutional neural networks. Appl Sci 2017;8(1):27.

[31] Saxena P, Maheshwari A, Maheshwari S. Predictive modeling of brain tumor: a deep learning approach. In: Innovations in computational intelligence and computer vision: proceedings of ICICV 2020. Singapore: Springer Singapore; 2020. p. 275—85.

[32] Díaz-Pernas FJ, Martínez-Zarzuela M, Antón-Rodríguez M, González-Ortega D. A deep learning approach for brain tumor classification and segmentation using a multiscale convolutional neural network. Healthcare February 2021;9(2):153.

[33] Baghdadi NA, Malki A, Abdelaliem SF, Balaha HM, Badawy M, Elhosseini M. An automated diagnosis and classification of COVID-19 from chest CT images using a transfer learning-based convolutional neural network. Comput Biol Med 2022;144:105383.

[34] Liu T, Yuan Z, Wu L, Badami B. An optimal brain tumor detection by convolutional neural network and enhanced sparrow search algorithm. Proc IME H J Eng Med 2021;235(4):459−69.

[35] Liu T, Yuan Z, Wu L, Badami B. Optimal brain tumor diagnosis based on deep learning and balanced sparrow search algorithm. Int J Imag Syst Technol 2021;31(4):1921−35.

[36] Saouli R, Akil M, Kachouri R. Fully automatic brain tumor segmentation using end-to-end incremental deep neural networks in MRI images. Comput Methods Progr Biomed 2018;166:39−49.

[37] Havaei M, Davy A, Warde-Farley D, et al. Brain tumor segmentation with deep neural networks. Med Image Anal 2017;35:18−31.

[38] Gordillo N, Montseny E, Sobrevilla P. State of the art survey on MRI brain tumor segmentation. Magn Reson Imaging 2013;31(8):1426−38.

[39] White DR, Houston AS, Sampson WF, Wilkins GP. Intra-and interoperator variations in region-of-interest drawing and their effect on the measurement of glomerular filtration rates. Clin Nucl Med 1999;24(3):177−81.

[40] Foo JL. A survey of user interaction and automation in medical image segmentation methods. Iowa State University Human Computer Interaction Technical Report ISU-HCI-2006-02; 2006.

[41] Balaha HM, Hassan AES. A variate brain tumor segmentation, optimization, and recognition framework. Artif Intell Rev 2023;56(7):7403−56.

[42] Collins DL, Holmes CJ, Peters TM, Evans AC. Automatic 3-D model-based neuroanatomical segmentation. Hum Brain Mapp 1995;3(3):190−208.

[43] Cogswell M, Ahmed F, Girshick R, Zitnick L, Batra D. Reducing overfitting in deep networks by decorrelating representations. arXiv preprint arXiv:1511.06068 2015. https://doi.org/10.48550/arXiv.1511.06068.

[44] Balaha HM, Shaban AO, El-Gendy EM, Saafan MM. Prostate cancer grading framework based on deep transfer learning and Aquila optimizer. Neural Comput Appl 2024:1−26.

[45] Fukushima K. Neocognitron: a self-organizing neural network model for a mechanism of pattern recognition unaffected by shift in position. Biol Cybern 1980;36(4):193−202.

[46] Hubel DH, Wiesel TN. Receptive fields and functional architecture of monkey striate cortex. J Physiol 1968;195(1):215−43.

[47] Fu Y, Li C, Yu FR, Luan TH, Zhang Y. A survey of driving safety with sensing, vehicular communications, and artificial intelligence-based collision avoidance. IEEE Trans Intell Transport Syst 2021;23(7):6142−63.

[48] Balaha HM, El-Gendy EM, Saafan MM. CovH2SD: a COVID-19 detection approach based on Harris Hawks Optimization and stacked deep learning. Expert Syst Appl 2021;186:115805.

[49] Balaha HM, Saif M, Tamer A, Abdelhay EH. Hybrid deep learning and genetic algorithms approach (HMB-DLGAHA) for the early ultrasound diagnoses of breast cancer. Neural Comput Appl 2022;34(11):8671−95.

[50] Balaha HM, Ali HA, Badawy M. Automatic recognition of handwritten Arabic characters: a comprehensive review. Neural Comput Appl 2021;33:3011−34.

[51] LeCun Y, Bengio Y, Hinton G. Deep learning. Nature 2015;521(7553):436−44.

[52] Glorot X, Bordes A, Bengio Y. Deep sparse rectifier neural networks. In: Proceedings of the fourteenth international conference on artificial intelligence and statistics. JMLR Workshop and Conference Proceedings; June 2011. p. 315−23.

[53] Balaha MM, El-Kady S, Balaha HM, Salama M, Emad E, Hassan M, Saafan MM. A vision-based deep learning approach for independent-users Arabic sign language interpretation. Multimed Tool Appl 2023;82(5):6807−26.

[54] Balaha HM, Ali HA, Saraya M, Badawy M. A new Arabic handwritten character recognition deep learning system (AHCR-DLS). Neural Comput Appl 2021;33:6325−67.

[55] Ruder S. An overview of gradient descent optimization algorithms. arXiv preprint arXiv:1609.04747 2016. https://doi.org/10.48550/arXiv.1609.04747.

[56] Abd El-Khalek AA, Balaha HM, Alghamdi NS, Ghazal M, Khalil AT, Abo-Elsoud MEA, El-Baz A. A concentrated machine learning-based classification system for age-related macular degeneration (AMD) diagnosis using fundus images. Sci Rep 2024;14(1):2434.

[57] Badawy M, Balaha HM, Maklad AS, Almars AM, Elhosseini MA. Revolutionizing oral cancer detection: an approach using aquila and Gorilla algorithms optimized transfer learning-based CNNs. Biomimetics 2023;8(6):499.

[58] Dauphin Y, De Vries H, Bengio Y. Equilibrated adaptive learning rates for non-convex optimization. Adv Neural Inf Process Syst 2015;28.

[59] Xiang T, Wang J, Liao X. An improved particle swarm optimizer with momentum. In: 2007 IEEE congress on evolutionary computation. IEEE; September 2007. p. 3341−5.

[60] Balaha HM, El-Gendy EM, Saafan MM. A complete framework for accurate recognition and prognosis of COVID-19 patients based on deep transfer learning and feature classification approach. Artif Intell Rev 2022;55(6):5063−108.

[61] Bahgat WM, Balaha HM, AbdulAzeem Y, Badawy MM. An optimized transfer learning-based approach for automatic diagnosis of COVID-19 from chest x-ray images. PeerJ Comput Sci 2021; 7:e555.

[62] Balaha HM, Hassan AES. Skin cancer diagnosis based on deep transfer learning and sparrow search algorithm. Neural Comput Appl 2023;35(1):815−53.

[63] Baghdadi NA, Malki A, Balaha HM, Badawy M, Elhosseini M. A3c-tl-gto: Alzheimer automatic accurate classification using transfer learning and artificial gorilla troops optimizer. Sensors 2022; 22(11):4250.

[64] Abdulazeem Y, Balaha HM, Bahgat WM, Badawy M. Human action recognition based on transfer learning approach. IEEE Access 2021;9:82058−69.

[65] Balaha HM, Ayyad SM, Alksas A, et al. Early diagnosis of prostate cancer using parametric estimation of IVIM from DW-MRI. In: 2023 IEEE international conference on image processing (ICIP). IEEE; October 2023. p. 2910−4.

[66] Aljadani A, Alharthi B, Farsi MA, Balaha HM, Badawy M, Elhosseini MA. Mathematical modeling and analysis of credit scoring using the LIME explainer: a comprehensive approach. Mathematics 2023;11(19):4055.

[67] Lu J, Behbood V, Hao P, Zuo H, Xue S, Zhang G. Transfer learning using computational intelligence: a survey. Knowl Base Syst 2015;80:14−23.

[68] Balaha HM, Ali HA, Youssef EK, et al. Recognizing Arabic handwritten characters using deep learning and genetic algorithms. Multimed Tool Appl 2021;80:32473−509.

[69] Yousif NR, Balaha HM, Haikal AY, El-Gendy EM. A generic optimization and learning framework for Parkinson disease via speech and handwritten records. J Ambient Intell Hum Comput 2023;14(8): 10673−93.

[70] Baghdadi NA, Malki A, Balaha HM, AbdulAzeem Y, Badawy M, Elhosseini M. Classification of breast cancer using a manta-ray foraging optimized transfer learning framework. PeerJ Comput Sci 2022;8: e1054.

[71] Badawy M, Almars AM, Balaha HM, Shehata M, Qaraad M, Elhosseini M. A two-stage renal disease classification based on transfer learning with hyperparameters optimization. Front Med 2023;10: 1106717.

[72] Ma L, Cheng S, Shi Y. Enhancing learning efficiency of brain storm optimization via orthogonal learning design. IEEE Trans Syst Man Cybern 2020;51(11):6723−42.

[73] Ma L, Huang M, Yang S, Wang R, Wang X. An adaptive localized decision variable analysis approach to large-scale multiobjective and many-objective optimization. IEEE Trans Cybern 2021;52(7): 6684—96.

[74] Xue J, Shen B. A novel swarm intelligence optimization approach: sparrow search algorithm. Syst Sci Control Eng 2020;8(1):22—34.

[75] Azzam MT, Alksas A, Balaha HM, et al. A novel textural and morphological-based cad system for early and accurate diagnosis of vertebral tumors. In: 2023 IEEE 20th international symposium on biomedical imaging (ISBI). IEEE; April 2023. p. 1—4.

[76] Sharaby I, Alksas A, Nashat A, et al. An ai-based cap framework for Wilms' tumor preoperative chemotherapy susceptibility. In: 2023 IEEE 20th international symposium on biomedical imaging (ISBI). IEEE; April 2023. p. 1—4.

[77] Sharaby I, Alksas A, Nashat A, et al. Prediction of wilms' tumor susceptibility to preoperative chemotherapy using a novel computer-aided prediction system. Diagnostics 2023;13(3):486.

[78] Gerig G, Jomier M, Chakos M. Valmet: a new validation tool for assessing and improving 3D object segmentation. In: Medical image computing and computer-assisted intervention—MICCAI 2001: 4th international conference Utrecht, The Netherlands, October 14—17, 2001 proceedings 4. Springer Berlin Heidelberg; 2001. p. 516—23.

[79] Hanley JA, McNeil BJ. The meaning and use of the area under a receiver operating characteristic (ROC) curve. Radiology 1982;143(1):29—36.

[80] Bradley AP. The use of the area under the ROC curve in the evaluation of machine learning algorithms. Pattern Recogn 1997;30(7):1145—59.

[81] Powers DM. Evaluation: from precision, recall and F-measure to ROC, informedness, markedness and correlation. arXiv preprint arXiv:2010.16061 2020. https://doi.org/10.48550/arXiv.2010.16061.

[82] Balaha HM, Hassan AES. Comprehensive machine and deep learning analysis of sensor-based human activity recognition. Neural Comput Appl 2023;35(17):12793—831.

[83] Ronneberger O, Fischer P, Brox T. U-net: convolutional networks for biomedical image segmentation. In: Medical image computing and computer-assisted intervention—MICCAI 2015: 18th international conference, Munich, Germany, October 5—9, 2015, Proceedings, Part III 18. Springer International Publishing; 2015. p. 234—41.

[84] Zhou Z, Rahman Siddiquee MM, Tajbakhsh N, Liang J. Unet++: a nested u-net architecture for medical image segmentation. In: Deep learning in medical image analysis and multimodal learning for clinical decision support: 4th international workshop, DLMIA 2018, and 8th international workshop, ML-CDS 2018, held in conjunction with MICCAI 2018, Granada, Spain, September 20, 2018, proceedings 4. Springer International Publishing; 2018. p. 3—11.

[85] Oktay O, Schlemper J, Folgoc LL, et al. Attention u-net: learning where to look for the pancreas. arXiv preprint arXiv:1804.03999 2018. https://doi.org/10.48550/arXiv.1804.03999.

[86] Milletari F, Navab N, Ahmadi SA. V-net: fully convolutional neural networks for volumetric medical image segmentation. In: 2016 fourth international conference on 3D vision (3DV). IEEE; October 2016. p. 565—71.

[87] Balaha HM, Hassan AES, El-Gendy EM, ZainEldin H, Saafan MM. An aseptic approach towards skin lesion localization and grading using deep learning and harris hawks optimization. Multimed Tool Appl 2023;1—29.

[88] Balaha HM, Antar ER, Saafan MM, El-Gendy EM. A comprehensive framework towards segmenting and classifying breast cancer patients using deep learning and Aquila optimizer. J Ambient Intell Hum Comput 2023;14(6):7897—917.

[89] Balaha HM, Saafan MM. Automatic exam correction framework (aecf) for the mcqs, essays, and equations matching. IEEE Access 2021;9:32368—89.

2

The neural circuitry of PTSD—An RDOC approach

Anthony K. Allam, M. Benjamin Larkin, Ashwin Viswanathan,
Sameer A. Sheth and Garrett P. Banks

DEPARTMENT OF NEUROSURGERY, BAYLOR COLLEGE OF MEDICINE, HOUSTON, TX, UNITED STATES

1. Introduction

Posttraumatic stress disorder (PTSD) is a debilitating psychiatric illness that can emerge after direct trauma exposure, witnessing distressing events, or hearing about them. Hallmarks of PTSD include persistent traumatic memories, avoidance of trauma-related reminders, overwhelming negative feelings and thoughts, and prolonged heightened arousal that extends beyond a month [1]. While disruptive to daily life and emotional well-being, these symptoms can further lead to long-lasting mental and physical repercussions. In the US, PTSD affects approximately 6% of the general population [2–5]. Genetic, environmental, and psychological factors are also known to influence the onset of PTSD.

Two of the most common risk factors are the type of trauma experienced and the individual's gender. Childhood trauma and personal experiences with assault or violence are especially high-risk events [6,7]. For example, groups exposed to extreme traumas, like combat veterans, refugees, and assault victims, experience increased rates of PTSD (25%) compared to the general population. Moreover, PTSD resulting from combat often manifests with intensified intrusive and arousal symptoms compared to traumas of a civilian nature [8–10]. Research also indicates that women are twice as likely as men to develop PTSD [11,12]. This disparity is thought to occur due to differences in the types of trauma experienced and the role of gender-specific neurobiology [11].

In the US, PTSD poses a significant economic challenge, costing an estimated $232.2 billion each year for its management and treatment [13]. Yet, in the face of such immense expenditure, there is a lack of high-quality treatments tailored explicitly for PTSD [14–18]. The FDA has only approved two treatments, sertraline and paroxetine, which offer relief for specific PTSD symptoms rather than the entire disorder [15,16]. Currently, the most recognized therapeutic strategy is exposure-based cognitive behavioral therapy [19,20]. Despite these approaches, around 30%–60% of individuals fail to achieve remission [21]. Chronic PTSD correlates with diminished life satisfaction, increased depression, and a heightened risk for suicide [22]. Alarmingly, one study reported a 17% mortality rate over a span of 6 years, even with treatment [23]. The dearth of psychotropic drugs effective for PTSD underscores the urgent need for deeper insights into its neural pathways and overall neurobiology.

The current model of PTSD is frequently understood in the context of classical Pavlovian fear conditioning. In this paradigm, a neutral stimulus, such as a bell, is presented concurrently with an aversive unconditioned stimulus, such as an electrical shock [24]. With repeated exposure, the animal or human begins to link the neutral stimulus with the unconditioned stimulus, leading to a fear response toward what is now known as the conditioned stimulus [24]. The basolateral amygdala (BLA) processes information from unconditioned and conditioned stimuli, activating the amygdala's central nucleus (CE) [25–27]. Through efferent connections to downstream targets like the hypothalamus, periaqueductal gray, locus coeruleus, and other brainstem regions, the CE produces physiological (autonomic), behavioral (freezing), and generalized responses to fear [26,28]. Overcoming this ingrained fear—"extinction"—involves repeated exposure to the conditioned stimulus without the accompanying aversive event, allowing for new, nonfearful associations [29]. It is important to note that this process consists of the creation of new associations that suppress rather than erase the aversive memories [30–32] (Fig. 2.1).

However, this mechanism can fail either due to excessive encoding during or after a traumatic incident or an inability to diminish the associated fear following the event [29]. Studies on PTSD have corroborated that those with the disorder display heightened fear conditioning, a reduced capacity for extinction, and amplified physiological responses [33,34]. These findings underscore the relevance of the traditional Pavlovian fear conditioning paradigm in comprehending PTSD.

However, this model fails to capture the full spectrum of PTSD manifestations. Over the past decade, there has been a shift toward understanding psychiatric conditions through functional domains rather than symptomatic clusters [35,36]. Pioneered by the National Institute of Mental Health, the Research Domain Criteria (RDoC) offers a groundbreaking multidimensional framework, moving beyond the confines of conventional diagnoses [37]. The RDoC promotes a comprehensive understanding of psychiatric disorders by examining them as pathobiological entities interwoven with specific neural circuits that influence behavior, cognition, and emotions. Instead of focusing narrowly on the disorder's manifestations, this paradigm embraces a continuum from typical function to pathology [37,38]. Central to the RDoC are six core research domains:

(1) Negative valence systems (i.e., fear, anxiety, loss, and nonreward)
(2) Positive valence systems (i.e., reward learning and habituation)
(3) Cognitive systems (i.e., attention, memory, and cognitive control)
(4) Social process systems (i.e., attachment, social communication, and self-awareness)
(5) Arousal/modulatory systems (i.e., arousal and sleep-wake dynamics)
(6) Sensorimotor Systems (i.e., motor action, agency and ownership, and motor patterns)

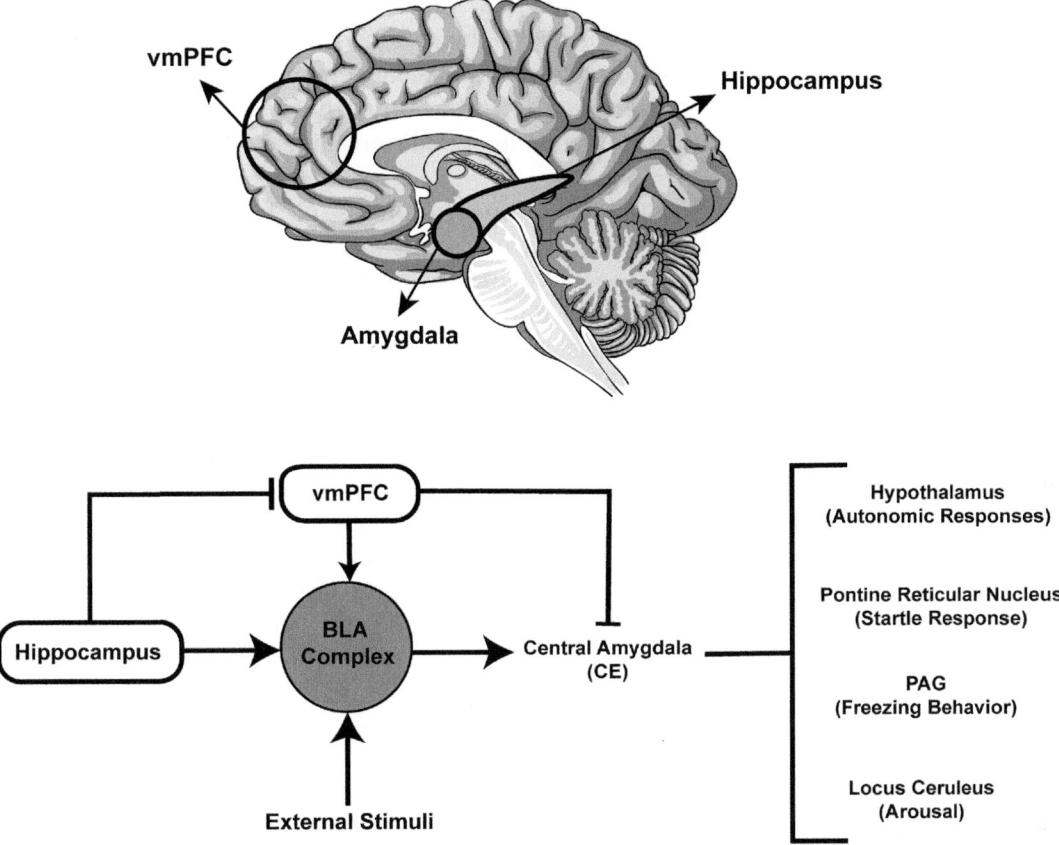

FIGURE 2.1 Diagram of the putative neurocircuitry and pathophysiology of PTSD. The hippocampus and vmPFC play essential roles in memory discrimination and fear conditioning. Their interaction with the BLA complex and central amygdala results in overgeneralized fear responses that are characteristic of PTSD.

In the context of the first five core research domains, this chapter will highlight the current understanding of neuronal circuit dysfunction in PTSD using neuroimaging and electrophysiological methods. Additionally, this chapter will explore both invasive and noninvasive neurotechnological interventions for PTSD treatment.

2. Pathophysiology of PTSD

2.1 Negative valence systems

PTSD is intricately connected to the negative valence system, as the predominant symptoms include intensified fear and anxiety, especially when avoiding trauma-related cues [1]. This aversion to conditioned cues can generalize to other neutral stimuli reminiscent of the trauma itself. In individuals with PTSD, anomalies in fear conditioning mean they may react more intensely to the previously neutral stimuli than to the original conditioned cues. While the natural fear-learning mechanism can become maladaptive through overgeneralization, it's worth noting that impaired fear extinction is more distinctively linked to PTSD than to conditions like generalized anxiety or depression.

In understanding the neurobiological underpinnings of fear conditioning and extinction, three principal brain structures emerge: the amygdala, the prefrontal cortex, and the hippocampus [39]. Within the amygdala, two nuclei stand out for their salience in the negative valence system: The CE and the BLA complex [26,27]. The BLA predominantly (approximately 80%) consists of glutamatergic spiny neurons under constant tonic inhibition by the remaining gamma-aminobutyric acidergic interneurons [27]. Contemporary optogenetic studies have discerned two discrete neuronal populations within the amygdala: the "fear on" cells, which facilitate activation of the CE, and the "fear off" cells, which are instrumental in CE inhibition [29]. Functionally, the BLA is an integrative hub of afferent projections from both cortical and subcortical regions. It integrates sensory information related to conditioned and unconditioned stimuli, thereby activating the CE [27,40]. Subsequently, the CE establishes efferent connections to several areas intrinsically implicated in fear-mediated responses. Notably, these include the periaqueductal gray (responsible for freezing behaviors), the hypothalamus (modulating heart rate and blood pressure), and the locus ceruleus (overseeing arousal, vigilance, and attention), among others [28]. Therefore, activation of the BLA is crucial to proper fear-learning responses, and dysregulation can lead to improper fear paradigms [27]. It is important to note that many of the neural circuits described above are modulated by orexinergic neurotransmission [41]. This notion originated from early mouse studies in which orexin knockout mice showed decreased freezing behavior in cued fear-conditioning tests [42]. Subsequent studies since then have solidified this view [42,43].

However, more so than fear learning, fear extinction is notably disrupted in individuals with PTSD. This is largely attributed to the inadequate top-down control from the ventromedial prefrontal cortex (vmPFC) to the amygdala [44–47]. The vmPFC assimilates a myriad of cues—sensory, emotional, environmental, and memory—and orchestrates the regulation of conditioned fear primarily through the activation of intercalated cells in the amygdala that inhibit the CE [40,48,49]. When functioning optimally, the vmPFC suppresses the CE output of the amygdala, thwarting the initiation of fear responses and facilitating the process of fear extinction [40,48]. However, for those with PTSD, a hypoactive vmPFC results in diminished inhibitory control, causing heightened activation of the CE [50,51]. This dysregulation subsequently impedes the proper extinction of trauma-associated stimuli. Aberrant orexigenic neurotransmission between the BLA, the CE, and the prefrontal cortex has been implicated in impaired fear extinction as well [41].

The hippocampus plays a pivotal component in fear conditioning and extinction mechanisms. Primarily recognized for its role in declarative memory and learning, the hippocampus, under standard extinction conditions, facilitates the activation of the vmPFC, inhibiting the amygdala's CE [40,48]. Furthermore, the hippocampus extends projections to the "fear off" neurons in the amygdala, reinforcing the inhibition of the CE [40,48]. Therefore, a dysregulated hippocampus results in diminished inhibition of fear responses, thereby impeding proper extinction learning. Notably, neuroimaging studies reveal reduced hippocampal volumes in PTSD-affected individuals [52,53]. However, it is unclear whether trauma leads to hippocampal shrinkage or if a smaller hippocampus predisposes an individual to PTSD. Insights from twin studies lean toward the notion that attenuated connectivity between the vmPFC and the hippocampus is an acquired characteristic, consequently diminishing the efficacy of the fear extinction process [54,55].

2.2 Positive valence systems

PTSD primarily impacts the negative valence system, but emerging research suggests that the dysregulation of the positive valence system, specifically the reward processing pathway, may play a crucial role in the manifestation of anhedonia and emotional numbing, characteristic symptoms of PTSD [56–59]. When exposed to negatively valenced stimuli, individuals with PTSD often find it challenging to modulate their emotional responses compared to those without the disorder [60,61]. Such a response suggests a broader dysfunction in the network responsible for recognizing and seeking rewarding stimuli and the subsequent experience of pleasure [56]. This network encompasses several interconnected brain regions: the ventral tegmental area (VTA), amygdala, orbitofrontal cortex, insula, anterior cingulate cortex, dorsomedial prefrontal cortex, striatum, and motor cortex [59,62]. Under normal physiological conditions, reward cues activate the meso-corticolimbic reward pathway, leading to approach behavior through dopaminergic activation of the ventral striatum and the motor cortices [57]. The meso-corticolimbic pathway consists of dopaminergic projections from the VTA that innervate limbic structures, including the BLA, medial prefrontal cortex (mPFC), and nucleus accumbens [57,63]. However, neuroimaging studies have identified hypoactivity in the striatum and PFC when individuals with PTSD are presented with rewarding stimuli [62]. Furthermore, decreased neural activity has been observed in the vmPFC, anterior insula, and inferior frontal gyrus when these individuals face challenges recognizing or labeling emotional states in PTSD-inducing scenarios [58,61,62]. Notably, the medial PFC is pivotal in reward processing and depression-like behavior [64].

2.3 Cognitive systems

Individuals with PTSD exhibit pronounced cognitive deficits encompassing the domains of attention, planning, declarative memory, and working memory [57,65–67]. Observations have indicated altered prefrontal cortex activity during tasks requiring sustained attention [62,68]. A notable decline in PFC activity when individuals with PTSD engage in intricate inhibition tasks suggests a potential impairment in modulating PFC functions under heightened cognitive demands [62,68]. Drawing insights from rodent studies, there's a significant correlation between hippocampal theta oscillations and both theta and single-unit PFC activity during spatial working memory evaluations [62]. Through the use of optogenetic manipulations, the ventral hippocampal-to-PFC pathway was implicated as a crucial network in spatial cue processing [61,62]. These findings collectively illustrate that anomalies in these neural networks could detrimentally influence memory processes, affecting normal fear learning and extinction paradigms.

PTSD is also associated with problems in executive function and excessive vigilance. Emerging research suggests that individuals with prior deficits in executive functioning experience a magnification of these deficits posttrauma, which could hinder recovery and elevate the risk of PTSD onset [69]. Additionally, PTSD patients demonstrate heightened vigilance toward potential environmental threats. This is evidenced by an attentional bias toward threat and a skewed memory inclination toward negative stimuli, often at the expense of other cognitive operations [57,67,70]. Neuroimaging studies have identified heightened activity in the amygdala, dorsal anterior cingulate cortex, and insula during tasks oriented toward negative attention in PTSD patients, indicating a hyperactivity of the salience network [57,66]. The ventromedial PFC has also been implicated, though the findings are mixed [57,66].

2.4 Systems for social processes

Patients with PTSD, especially those with complex manifestations, display marked alterations in foundational social processes, such as attachment formation, social communication, and self-awareness [1]. The neurobiological underpinnings of these alterations remain underexplored in the current literature. A prevailing hypothesis attributes these disturbances to a combination of factors: hyperreactivity of the amygdala, undermodulation of emotions by the vmPFC, and deactivation of the anterior cingulate cortex. Moreover, increased insular and decreased rostral ACC activity are observed in those with PTSD [71]. The insula is believed to be central to interoception and bodily awareness [50,72]. In contrast, the rostral ACC is pivotal in modulating attention, emotion, and arousal [50,72]. Together, these factors may amplify threat perceptions, making it difficult for patients with PTSD to trust others and form social attachments.

2.5 Arousal symptoms

Hyperarousal, characterized by intensified nervousness, sleep disturbances, nightmares, and an amplified startle response, is a hallmark symptom of PTSD [1]. Alterations in arousal and reactivity are often linked to the hypoactivity of the mPFC and hippocampus, alongside the hyperactivity of the amygdala and the bed nucleus of the stria terminalis [44]. Notably, individuals with PTSD frequently display heightened aggression in response to perceived threats [62]. The exact neural underpinnings of this aggression remain under investigation. Still, extant research implicates the amygdala, periaqueductal gray, and locus coeruleus—structures integral to threat detection and the orchestration of autonomic responses [45,73]. The LC stands out as a critical modulator of adrenergic signaling [73]. Elevated responsiveness within the noradrenergic pathway, specifically between the LC and the BLA, is hypothesized to play a pivotal role in the emergence of PTSD and analogous stress disorders [74,75].

Furthermore, a connection between aggression, impulsivity, and attenuated cortical modulation has been found, predominantly implicating PFC structures such as the ventral medial PFC and dorsal anterior cingulate cortex [46,76]. These structures play a role in threat discernment and response calibration. On the other hand, the orbitofrontal cortex and BLA are paramount in modulating and sustaining these responses [46,62,76]. An ancillary theory posits that PTSD's hallmark hyperarousal is rooted in an overactive sympathetic nervous system (SNS). A robust body of evidence, spanning animal models and clinical trials, underscores the role of SNS hormones, specifically adrenaline and noradrenaline, in augmenting memory consolidation [57,77]. This suggests that heightened SNS activity during trauma may exacerbate the embedding of traumatic memories, leading to their recurrent, intrusive manifestation [57,77]. Additionally, orexin, which is central to both fear conditioning and fear extinction processes, has been found to regulate sympathetic activity as well. This regulation amplifies the response of the SNS, resulting in the heightened arousal symptoms typically observed [41].

3. Neurotechnological strategies towards PTSD treatment

Posttraumatic stress disorder (PTSD) presents a neurobiological framework characterized by a network of intricate inhibitory and excitatory pathways involving pivotal brain regions such as the amygdala, prefrontal cortex, cingulate cortex, insula, and hippocampus [62]. The multifaceted nature of this network has historically rendered targeted therapeutic interventions challenging. Contemporary advancements in neuroimaging, paired with rigorous neuroscientific research, have provided a more detailed understanding of these circuits. Dysregulation in the amygdala, for example, can correlate with symptoms such as intrusion, avoidance, arousal, and reactivity [62]. In contrast, perturbations within the PFC might be linked to both intrusive symptoms and adverse cognitive and mood alterations [62]. With this refined comprehension of PTSD's neurobiological basis, the focus has shifted toward developing neurotechnological interventions. Leading this initiative are invasive and noninvasive techniques, such as neurofeedback, TMS, transcranial direct current stimulation (tDCS), DBS, and RNS.

3.1 Neurofeedback (nFb)

Functional magnetic resonance imaging (fMRI) and electroencephalography (EEG) neurofeedback are innovative, noninvasive techniques that allow individuals to gain insight into and potentially regulate their brain activity. For PTSD patients, understanding and modulating the neural circuits related to trauma could present a groundbreaking therapeutic avenue. Patients can visualize their brain activity by utilizing real-time feedback from fMRI or EEG and learn to alter specific neural patterns through training.

The underlying premise for using neurofeedback in PTSD treatments stems from observing altered neural circuitry in affected individuals. For instance, hyperactivity in the amygdala and hypoactivity in the prefrontal cortex are commonly reported in PTSD patients. Neurofeedback can target these irregularities, aiming to restore balance. Initial studies using fMRI neurofeedback have shown that PTSD patients can, with training, learn to modulate the activity of regions like the amygdala, potentially reducing symptom severity [78,79]. Similarly, EEG neurofeedback, which offers the advantage of temporal precision, has been used to train individuals to alter brainwave patterns associated with stress and hyperarousal.

Early results from both fMRI and EEG neurofeedback trials are promising but mixed. While many studies have successfully shown that patients are indeed able to regulate their brain activity, these findings rarely correlate back to clinical outcomes [78–87]. Such discrepancies between functional and clinical outcomes underscore the necessity for methodologically rigorous studies with larger sample sizes. Such research will help to refine protocols, establish long-term efficacy, and determine the optimal parameters for neurofeedback interventions in PTSD treatment.

3.2 Transcranial magnetic stimulation (TMS)

Transcranial magnetic stimulation (TMS) provides a noninvasive means of stimulating brain regions. In TMS, high-frequency stimulation (>1 Hz) activates the cortex, whereas low-frequency stimulation (≤ 1 Hz) dampens it [88]. Coils in TMS devices are designed to target specific superficial regions or broader, deep areas [89]. These coils generate a dynamic magnetic field, producing an electric field whose effects are determined by the energy's intensity, frequency, and temporal distribution [90].

Studies using EEG have identified that PTSD patients display reduced right hemisphere alpha power when exposed to trauma-related imagery, a finding further corroborated by SPECT studies [91]. Consequently, TMS research has focused mainly on the right dorsolateral prefrontal cortex (DLPFC) [91]. However, existing TMS studies show variable outcomes, with some revealing minimal clinical improvements, even in sham-controlled groups, highlighting the need for stringent controls and analysis [92–98].

Given TMS studies' diverse metrics and intervals, some positive findings might be incidental. Current evidence doesn't conclusively define TMS's efficacy or optimal parameters, underscoring the need for more rigorous research.

3.3 Transcranial direct current stimulation (tDCS)

tDCS is a noninvasive neuromodulatory technique that employs a constant, low current to modulate neuronal activity in brain regions. Its application for PTSD patients has garnered interest due to its relative safety profile and ease of administration. Unlike techniques such as DBS, tDCS does not require surgical implantation, making it more accessible and presenting fewer risks [99]. However, it's essential to determine the appropriate parameters, including the current's intensity and polarity, as these can influence the effects on neural excitability [99]. Anodal stimulation typically enhances cortical excitability, while cathodal stimulation tends to reduce it [99]. In the context of PTSD, targeted tDCS application aims to modulate hyperactive or hypoactive brain regions associated with trauma-related symptoms. Recent studies have started investigating the therapeutic potential of tDCS for PTSD, with some focusing on the DLPF, given its role in cognitive control and emotional regulation. A few early-stage trials hint at some benefits, such as diminished symptom severity and enhanced emotional regulation [100—104]. However, with only five studies to date reporting the use of tDCS for PTSD, there remains a need for more extensive and rigorously designed randomized controlled trials to definitively ascertain the efficacy of tDCS in treating PTSD.

3.4   Deep brain stimulation (DBS)/Responsive neurostimulation (RNS)

Deep brain stimulation (DBS) employs surgically embedded electrodes to administer continuous electrical stimulation to designated brain regions. Responsive neurostimulation (RNS) operates on a similar principle; however, it only delivers stimulation in an intermittent pattern. The RNS device has the capability to simultaneously stimulate and record brain activity from the same or different neural locations. Rather than providing continuous stimulation, the RNS device activates in response to predefined parameters identified in the recorded neural signals. This approach ensures that stimulation is ideally administered only when necessary. While both DBS and RNS are generally deemed safe, they are not without risks. There's an estimated 1%—3% per lead risk of hemorrhage and a 1%—9% risk of infection [105—108]. Compared to the previously discussed neuromodulatory strategies, DBS and RNS, although more invasive, facilitate the targeting of deeper brain structures and offer more precise energy delivery control. The outcomes of DBS and RNS are contingent upon the parameters set: the pulse width, the amplitude, and the frequency. High-frequency stimulation is the most common, and it is often perceived as replicating lesion effects. However, in recent years, even this assumption has faced scrutiny [109].

Currently, two clinical trials are evaluating the efficacy of DBS as a treatment modality for PTSD. One trial focuses on stimulating the subgenual cingulate, while the other targets the amygdala. Additionally, there is a single clinical trial exploring the use of RNS targeting the amygdala. Despite the encouraging outcomes observed in all the trials, a pressing need remains for more expansive, randomized studies to ensure robust and generalizable conclusions.

4. Conclusion

PTSD represents a multifaceted psychiatric pathology influenced by an intricate interplay of genetic, environmental, and psychological factors. The RDoC framework offers an integrative approach to PTSD, emphasizing a comprehensive assessment that extends beyond clinical symptoms to probe the foundational neural substrates, encompassing biological, cognitive, and socio-environmental dimensions. Such an analytical perspective is instrumental in formulating precision-targeted therapeutic interventions and identifying preventive strategies contingent upon discerning specific neural and genetic predispositions. Contemporary neurobiological research on PTSD underscores the involvement of various cortical and subcortical regions, including the amygdala, hippocampus, prefrontal cortex, and cingulate cortex. Drawing on this contemporary knowledge of PTSD's neural circuitry, emergent neurotechnological therapies, such as neurofeedback, TMS, tDCS, DBS, and RNS, are strategically developed to address the cardinal neurobiological mechanisms inherent to the disorder. While preliminary clinical trials indicate potential therapeutic efficacy, it is imperative to conduct further methodologically robust randomized clinical trials before endorsing these modalities as standard therapeutic interventions.

References

[1] Publishing AP. Diagnostic and statistical manual of mental disorders: dsm-5. American Psychiatric Publishing; 2013. https://doi.org/10.1176/appi.books.9780890425596.

[2] McLaughlin KA, Koenen KC, Friedman MJ, et al. Subthreshold posttraumatic stress disorder in the world Health organization world mental Health surveys. Biol Psychiatr 2015;77(4):375–84. https://doi.org/10.1016/j.biopsych.2014.03.028.

[3] Bromet E, Sonnega A, Kessler RC. Risk factors for DSM-III-R posttraumatic stress disorder: findings from the national comorbidity survey. Am J Epidemiol 1998;147(4):353–61. https://doi.org/10.1093/oxfordjournals.aje.a009457.

[4] Breslau N, Peterson EL, Poisson LM, Schultz LR, Lucia VC. Estimating post-traumatic stress disorder in the community: lifetime perspective and the impact of typical traumatic events. Psychol Med 2004;34(5):889–98. https://doi.org/10.1017/S0033291703001612.

[5] Breslau N, Kessler RC, Chilcoat HD, Schultz LR, Davis GC, Andreski P. Trauma and posttraumatic stress disorder in the community. Arch Gen Psychiatr 1998;55(7):626. https://doi.org/10.1001/archpsyc.55.7.626.

[6] Heim C, Nemeroff CB. The role of childhood trauma in the neurobiology of mood and anxiety disorders: preclinical and clinical studies. Biol Psychiatr 2001;49(12):1023–39. https://doi.org/10.1016/S0006-3223(01)01157-X.

[7] Kessler RC, Aguilar-Gaxiola S, Alonso J, et al. Trauma and PTSD in the WHO world mental Health surveys. Eur J Psychotraumatol 2017;8(Suppl. 5). https://doi.org/10.1080/20008198.2017.1353383.

[8] Laufer RS, Brett E, Gallops MS. Symptom patterns associated with posttraumatic stress disorder among Vietnam veterans exposed to war trauma. Am J Psychiatr 1985;142(11):1304–11. https://doi.org/10.1176/ajp.142.11.1304.

[9] Huckins LM, Chatzinakos C, Breen MS, et al. Analysis of genetically regulated gene expression identifies a prefrontal PTSD gene, SNRNP35, specific to military cohorts. Cell Rep 2020;31(9):107716. https://doi.org/10.1016/j.celrep.2020.107716.

[10] Brinker M, Westermeyer J, Thuras P, Canive J. Severity of combat-related posttraumatic stress disorder versus noncombat-related posttraumatic stress disorder. J Nerv Ment Dis 2007;195(8):655–61. https://doi.org/10.1097/NMD.0b013e31811f4076.

[11] Kornfield SL, Hantsoo L, Epperson CN. What does sex have to do with it? The role of sex as a biological variable in the development of posttraumatic stress disorder. Curr Psychiatr Rep 2018; 20(6):39. https://doi.org/10.1007/s11920-018-0907-x.

[12] Koenen KC, Ratanatharathorn A, Ng L, et al. Posttraumatic stress disorder in the world mental Health surveys. Psychol Med 2017;47(13):2260—74. https://doi.org/10.1017/S0033291717000708.

[13] Davis LL, Schein J, Cloutier M, et al. The economic burden of posttraumatic stress disorder in the United States from a societal perspective. J Clin Psychiatry 2022;83(3). https://doi.org/10.4088/JCP.21m14116.

[14] Raskind MA, Peterson K, Williams T, et al. A trial of prazosin for combat trauma PTSD with nightmares in active-duty soldiers returned from Iraq and Afghanistan. Am J Psychiatr 2013;170(9): 1003—10. https://doi.org/10.1176/appi.ajp.2013.12081133.

[15] Kelmendi B, Adams TG, Southwick S, Abdallah CG, Krystal JH. Posttraumatic stress disorder: an integrated overview of the neurobiological rationale for pharmacology. Clin Psychol Sci Pract 2017; 24(3):281—97. https://doi.org/10.1111/cpsp.12202.

[16] Krystal JH, Davis LL, Neylan TC, et al. It is time to address the crisis in the pharmacotherapy of posttraumatic stress disorder: a consensus statement of the PTSD psychopharmacology working group. Biol Psychiatr 2017;82(7):e51—9. https://doi.org/10.1016/j.biopsych.2017.03.007.

[17] Friedman MJ, Marmar CR, Baker DG, Sikes CR, Farfel GM. Randomized, double-blind comparison of sertraline and placebo for posttraumatic stress disorder in a department of veterans affairs setting. J Clin Psychiatry 2007;68(05):711—20. https://doi.org/10.4088/JCP.v68n0508.

[18] Hertzberg M, Feldman M, Beckham J, Kudler H, Davidson J. Lack of efficacy for fluoxetine in PTSD: a placebo controlled trial in combat veterans. Ann Clin Psychiatr 2000;12(2):101—5. https://doi.org/10.3109/10401230009147096.

[19] Resick PA, Nishith P, Weaver TL, Astin MC, Feuer CA. A comparison of cognitive-processing therapy with prolonged exposure and a waiting condition for the treatment of chronic posttraumatic stress disorder in female rape victims. J Consult Clin Psychol 2002;70(4):867—79. https://doi.org/10.1037/0022-006X.70.4.867.

[20] Monson CM, Schnurr PP, Resick PA, Friedman MJ, Young-Xu Y, Stevens SP. Cognitive processing therapy for veterans with military-related posttraumatic stress disorder. J Consult Clin Psychol 2006;74(5):898—907. https://doi.org/10.1037/0022-006X.74.5.898.

[21] Berger W, Mendlowicz MV, Marques-Portella C, et al. Pharmacologic alternatives to antidepressants in posttraumatic stress disorder: a systematic review. Prog Neuro-Psychopharmacol Biol Psychiatry 2009;33(2):169—80. https://doi.org/10.1016/j.pnpbp.2008.12.004.

[22] Panagioti M, Gooding PA, Tarrier N. A meta-analysis of the association between posttraumatic stress disorder and suicidality: the role of comorbid depression. Compr Psychiatr 2012;53(7): 915—30. https://doi.org/10.1016/j.comppsych.2012.02.009.

[23] Johnson DR, Fontana A, Lubin H, Corn B, Rosenheck R. Long-term course of treatment-seeking vietnam veterans with posttraumatic stress disorder: mortality, clinical condition, and life satisfaction. J Nerv Ment Dis 2004;192(1):35—41. https://doi.org/10.1097/01.nmd.0000105998.90425.6a.

[24] Lissek S, Powers AS, McClure EB, et al. Classical fear conditioning in the anxiety disorders: a meta-analysis. Behav Res Ther 2005;43(11):1391—424. https://doi.org/10.1016/j.brat.2004.10.007.

[25] LaBar KS, Gatenby JC, Gore JC, LeDoux JE, Phelps EA. Human amygdala activation during conditioned fear acquisition and extinction: a mixed-trial fMRI study. Neuron 1998;20(5):937—45. https://doi.org/10.1016/S0896-6273(00)80475-4.

[26] Davis M. The role of the amygdala in fear and anxiety. Annu Rev Neurosci 1992;15(1):353—75. https://doi.org/10.1146/annurev.ne.15.030192.002033.

[27] LeDoux J. The amygdala. Curr Biol 2007;17(20):R868—74. https://doi.org/10.1016/j.cub.2007.08.005.

[28] Gouveia F, Gidyk D, Giacobbe P, et al. Neuromodulation strategies in post-traumatic stress disorder: from preclinical models to clinical applications. Brain Sci 2019;9(2):45. https://doi.org/10.3390/brainsci9020045.

[29] Koek RJ, Schwartz HN, Scully S, et al. Treatment-refractory posttraumatic stress disorder (TRPTSD): a review and framework for the future. Prog Neuro-Psychopharmacol Biol Psychiatry 2016;70:170−218. https://doi.org/10.1016/j.pnpbp.2016.01.015.

[30] Milad MR, Rauch SL, Pitman RK, Quirk GJ. Fear extinction in rats: implications for human brain imaging and anxiety disorders. Biol Psychol 2006;73(1):61−71. https://doi.org/10.1016/j.biopsycho.2006.01.008.

[31] Myers KM, Davis M. Mechanisms of fear extinction. Mol Psychiatr 2007;12(2):120−50. https://doi.org/10.1038/sj.mp.4001939.

[32] Myers KM, Davis M. Behavioral and neural analysis of extinction. Neuron 2002;36(4):567−84. https://doi.org/10.1016/S0896-6273(02)01064-4.

[33] Jovanovic T, Ressler KJ. How the neurocircuitry and genetics of fear inhibition may inform our understanding of PTSD. Am J Psychiatr 2010;167(6):648−62. https://doi.org/10.1176/appi.ajp.2009.09071074.

[34] Jovanovic T, Kazama A, Bachevalier J, Davis M. Impaired safety signal learning may be a biomarker of PTSD. Neuropharmacology 2012;62(2):695−704. https://doi.org/10.1016/j.neuropharm.2011.02.023.

[35] Sharpe M, Walker J. Symptoms: a new approach. Psychiatry 2009;8(5):146−8. https://doi.org/10.1016/j.mppsy.2009.03.016.

[36] Schmidt U. A plea for symptom-based research in psychiatry. Eur J Psychotraumatol 2015;6(1). https://doi.org/10.3402/ejpt.v6.27660.

[37] Insel T, Cuthbert B, Garvey M, et al. Research domain Criteria (RDoC): toward a new classification framework for research on mental disorders. Am J Psychiatr 2010;167(7):748−51. https://doi.org/10.1176/appi.ajp.2010.09091379.

[38] Wildes JE, Marcus MD. Application of the research domain Criteria (RDoC) framework to eating disorders: emerging concepts and research. Curr Psychiatr Rep 2015;17(5):30. https://doi.org/10.1007/s11920-015-0572-2.

[39] Besnard A, Sahay A. Adult hippocampal neurogenesis, fear generalization, and stress. Neuropsychopharmacology 2016;41(1):24−44. https://doi.org/10.1038/npp.2015.167.

[40] Graham BM, Milad MR. The study of fear extinction: implications for anxiety disorders. Am J Psychiatr 2011;168(12):1255−65. https://doi.org/10.1176/appi.ajp.2011.11040557.

[41] Kaplan GB, Lakis GA, Zhoba H. Sleep-wake and arousal dysfunctions in post-traumatic stress disorder: role of orexin systems. Brain Res Bull 2022;186:106−22. https://doi.org/10.1016/j.brainresbull.2022.05.006.

[42] Soya S, Shoji H, Hasegawa E, et al. Orexin receptor-1 in the locus coeruleus plays an important role in cue-dependent fear memory consolidation. J Neurosci 2013;33(36):14549−57. https://doi.org/10.1523/JNEUROSCI.1130-13.2013.

[43] Soya S, Takahashi TM, McHugh TJ, et al. Orexin modulates behavioral fear expression through the locus coeruleus. Nat Commun 2017;8(1):1606. https://doi.org/10.1038/s41467-017-01782-z.

[44] Liberzon I, Sripada CS. The functional neuroanatomy of PTSD: a critical review. Prog Brain Res 2007;167:151−69. https://doi.org/10.1016/S0079-6123(07)67011-3.

[45] Blair RJR. Psychopathy, frustration, and reactive aggression: the role of ventromedial prefrontal cortex. Br J Psychol 2010;101(3):383−99. https://doi.org/10.1348/000712609X418480.

[46] Davidson RJ, Putnam KM, Larson CL. Dysfunction in the neural circuitry of emotion regulation–A possible prelude to violence. Science (1979) 2000;289(5479):591−4. https://doi.org/10.1126/science.289.5479.591.

[47] Lanius RA, Vermetten E, Loewenstein RJ, et al. Emotion modulation in PTSD: clinical and neurobiological evidence for a dissociative subtype. Am J Psychiatr 2010;167(6):640—7. https://doi.org/10.1176/appi.ajp.2009.09081168.

[48] Quirk GJ, Mueller D. Neural mechanisms of extinction learning and retrieval. Neuropsychopharmacology 2008;33(1):56—72. https://doi.org/10.1038/sj.npp.1301555.

[49] Wood JN, Grafman J. Human prefrontal cortex: processing and representational perspectives. Nat Rev Neurosci 2003;4(2):139—47. https://doi.org/10.1038/nrn1033.

[50] Britton JC, Phan KL, Taylor SF, Fig LM, Liberzon I. Corticolimbic blood flow in posttraumatic stress disorder during script-driven imagery. Biol Psychiatr 2005;57(8):832—40. https://doi.org/10.1016/j.biopsych.2004.12.025.

[51] Diekhof EK, Geier K, Falkai P, Gruber O. Fear is only as deep as the mind allows. Neuroimage 2011;58(1):275—85. https://doi.org/10.1016/j.neuroimage.2011.05.073.

[52] Bremner JD, Randall P, Scott TM, et al. MRI-based measurement of hippocampal volume in patients with combat-related posttraumatic stress disorder. Am J Psychiatr 1995;152(7):973—81. https://doi.org/10.1176/ajp.152.7.973.

[53] Logue MW, van Rooij SJH, Dennis EL, et al. Smaller hippocampal volume in posttraumatic stress disorder: a multisite ENIGMA-PGC study: subcortical volumetry results from posttraumatic stress disorder consortia. Biol Psychiatr 2018;83(3):244—53. https://doi.org/10.1016/j.biopsych.2017.09.006.

[54] Admon R, Milad MR, Hendler T. A causal model of post-traumatic stress disorder: disentangling predisposed from acquired neural abnormalities. Trends Cognit Sci 2013;17(7):337—47. https://doi.org/10.1016/j.tics.2013.05.005.

[55] Shin LM, Bush G, Milad MR, et al. Exaggerated activation of dorsal anterior cingulate cortex during cognitive interference: a monozygotic twin study of posttraumatic stress disorder. Am J Psychiatr 2011;168(9):979—85. https://doi.org/10.1176/appi.ajp.2011.09121812.

[56] Nawijn L, van Zuiden M, Frijling JL, Koch SBJ, Veltman DJ, Olff M. Reward functioning in PTSD: a systematic review exploring the mechanisms underlying anhedonia. Neurosci Biobehav Rev 2015;51:189—204. https://doi.org/10.1016/j.neubiorev.2015.01.019.

[57] Schmidt U, Vermetten E. Integrating NIMH research domain Criteria (RDoC) into PTSD research. Curr Top Behav Neurosci 2017:69—91. https://doi.org/10.1007/7854_2017_1.

[58] Litz BT. Emotional numbing in combat-related post-traumatic stress disorder: a critical review and reformulation. Clin Psychol Rev 1992;12(4):417—32. https://doi.org/10.1016/0272-7358(92)90125-R.

[59] Der-Avakian A, Markou A. The neurobiology of anhedonia and other reward-related deficits. Trends Neurosci 2012;35(1):68—77. https://doi.org/10.1016/j.tins.2011.11.005.

[60] New AS, Fan J, Murrough JW, et al. A functional magnetic resonance imaging study of deliberate emotion regulation in resilience and posttraumatic stress disorder. Biol Psychiatr 2009;66(7):656—64. https://doi.org/10.1016/j.biopsych.2009.05.020.

[61] Fenster RJ, Lebois LAM, Ressler KJ, Suh J. Brain circuit dysfunction in post-traumatic stress disorder: from mouse to man. Nat Rev Neurosci 2018;19(9):535—51. https://doi.org/10.1038/s41583-018-0039-7.

[62] Larkin MB, McGinnis JP, Snyder RI, et al. Neurostimulation for treatment-resistant posttraumatic stress disorder: an update on neurocircuitry and therapeutic targets. J Neurosurg 2021;134(6):1715—23. https://doi.org/10.3171/2020.4.JNS2061.

[63] Sharf R, Sarhan M, DiLeone RJ. Role of orexin/hypocretin in dependence and addiction. Brain Res 2010;1314:130—8. https://doi.org/10.1016/j.brainres.2009.08.028.

[64] Ferenczi EA, Zalocusky KA, Liston C, et al. Prefrontal cortical regulation of brainwide circuit dynamics and reward-related behavior. Science (1979) 2016;351(6268). https://doi.org/10.1126/science.aac9698.

[65] Schweizer S, Dalgleish T. The impact of affective contexts on working memory capacity in healthy populations and in individuals with PTSD. Emotion 2016;16(1):16–23. https://doi.org/10.1037/emo0000072.

[66] Hayes JP, VanElzakker MB, Shin LM. Emotion and cognition interactions in PTSD: a review of neurocognitive and neuroimaging studies. Front Integr Neurosci 2012;6. https://doi.org/10.3389/fnint.2012.00089.

[67] Block SR, Liberzon I. Attentional processes in posttraumatic stress disorder and the associated changes in neural functioning. Exp Neurol 2016;284:153–67. https://doi.org/10.1016/j.expneurol.2016.05.009.

[68] Falconer E, Bryant R, Felmingham KL, et al. The neural networks of inhibitory control in post-traumatic stress disorder. J Psychiatry Neurosci 2008;33(5):413–22.

[69] Aupperle RL, Melrose AJ, Stein MB, Paulus MP. Executive function and PTSD: disengaging from trauma. Neuropharmacology 2012;62(2):686–94. https://doi.org/10.1016/j.neuropharm.2011.02.008.

[70] Paunovic N, Lundh LG, Öst LG. Attentional and memory bias for emotional information in crime victims with acute posttraumatic stress disorder (PTSD). J Anxiety Disord 2002;16(6):675–92. https://doi.org/10.1016/S0887-6185(02)00136-6.

[71] Hopper JW, Frewen PA, van der Kolk BA, Lanius RA. Neural correlates of reexperiencing, avoidance, and dissociation in PTSD: symptom dimensions and emotion dysregulation in responses to script-driven trauma imagery. J Trauma Stress 2007;20(5):713–25. https://doi.org/10.1002/jts.20284.

[72] Craig AD. How do you feel? Interoception: the sense of the physiological condition of the body. Nat Rev Neurosci 2002;3(8):655–66. https://doi.org/10.1038/nrn894.

[73] Haden SC, Scarpa A. The noradrenergic system and its involvement in aggressive behaviors. Aggress Violent Behav 2007;12(1):1–15. https://doi.org/10.1016/j.avb.2006.01.012.

[74] Sara SJ, Bouret S. Orienting and reorienting: the locus coeruleus mediates cognition through arousal. Neuron 2012;76(1):130–41. https://doi.org/10.1016/j.neuron.2012.09.011.

[75] Bangasser DA, Wiersielis KR, Khantsis S. Sex differences in the locus coeruleus-norepinephrine system and its regulation by stress. Brain Res 2016;1641:177–88. https://doi.org/10.1016/j.brainres.2015.11.021.

[76] Dileo JF, Brewer WJ, Hopwood M, Anderson V, Creamer M. Olfactory identification dysfunction, aggression and impulsivity in war veterans with post-traumatic stress disorder. Psychol Med 2008;38(4):523–31. https://doi.org/10.1017/S0033291707001456.

[77] Zoladz PR, Diamond DM. Current status on behavioral and biological markers of PTSD: a search for clarity in a conflicting literature. Neurosci Biobehav Rev 2013;37(5):860–95. https://doi.org/10.1016/j.neubiorev.2013.03.024.

[78] Zhao Z, Duek O, Seidemann R, et al. Amygdala downregulation training using fMRI neurofeedback in post-traumatic stress disorder: a randomized, double-blind trial. Transl Psychiatr 2023;13(1):177. https://doi.org/10.1038/s41398-023-02467-6.

[79] Zotev V, Krueger F, Phillips R, et al. Self-regulation of amygdala activation using real-time fMRI neurofeedback. PLoS One 2011;6(9):e24522. https://doi.org/10.1371/journal.pone.0024522.

[80] Nicholson AA, Rabellino D, Densmore M, et al. The neurobiology of emotion regulation in post-traumatic stress disorder: amygdala downregulation via real-time fMRI neurofeedback. Hum Brain Mapp 2017;38(1):541–60. https://doi.org/10.1002/hbm.23402.

[81] Gerin MI, Fichtenholtz H, Roy A, et al. Real-time fMRI neurofeedback with war veterans with chronic PTSD: a feasibility study. Front Psychiatr 2016;7. https://doi.org/10.3389/fpsyt.2016.00111.

[82] Zotev V, Phillips R, Misaki M, et al. Real-time fMRI neurofeedback training of the amygdala activity with simultaneous EEG in veterans with combat-related PTSD. Neuroimage Clin 2018;19:106–21. https://doi.org/10.1016/j.nicl.2018.04.010.

[83] Peniston EG, Marrinan DA, Deming WA, Kulkosky PJ. EEG alpha-theta brainwave synchronization in Vietnam theater veterans with combat-related post-traumatic stress disorder and alcohol abuse. Adv Med Psychother 1993;6(7):37–50.

[84] van der Kolk BA, Hodgdon H, Gapen M, et al. A randomized controlled study of neurofeedback for chronic PTSD. PLoS One 2016;11(12):e0166752. https://doi.org/10.1371/journal.pone.0166752.

[85] Ros T, Frewen P, Théberge J, et al. Neurofeedback tunes scale-free dynamics in spontaneous brain activity. Cerebr Cortex 2016. https://doi.org/10.1093/cercor/bhw285. Published online September 12.

[86] Nicholson AA, Ros T, Frewen PA, et al. Alpha oscillation neurofeedback modulates amygdala complex connectivity and arousal in posttraumatic stress disorder. Neuroimage Clin 2016;12: 506–16. https://doi.org/10.1016/j.nicl.2016.07.006.

[87] Askovic M, Watters AJ, Aroche J, Harris AWF. Neurofeedback as an adjunct therapy for treatment of chronic posttraumatic stress disorder related to refugee trauma and torture experiences: two case studies. Australas Psychiatr 2017;25(4):358–63. https://doi.org/10.1177/1039856217715988.

[88] Maeda F, Keenan JP, Tormos JM, Topka H, Pascual-Leone A. Modulation of corticospinal excitability by repetitive transcranial magnetic stimulation. Clin Neurophysiol 2000;111(5):800–5. https://doi.org/10.1016/S1388-2457(99)00323-5.

[89] Rossi S, Hallett M, Rossini PM, Pascual-Leone A. Safety, ethical considerations, and application guidelines for the use of transcranial magnetic stimulation in clinical practice and research. Clin Neurophysiol 2009;120(12):2008–39. https://doi.org/10.1016/j.clinph.2009.08.016.

[90] Hemond CC, Fregni F. Transcranial magnetic stimulation in neurology: what we have learned from randomized controlled studies. Neuromodulation Technol Neural Interface 2007;10(4):333–44. https://doi.org/10.1111/j.1525-1403.2007.00120.x.

[91] Trevizol AP, Barros MD, Silva PO, Osuch E, Cordeiro Q, Shiozawa P. Transcranial magnetic stimulation for posttraumatic stress disorder: an updated systematic review and meta-analysis. Trends Psychiatr Psychother 2016;38(1):50–5. https://doi.org/10.1590/2237-6089-2015-0072.

[92] Ahmadizadeh MJ, Rezaei M. Unilateral right and bilateral dorsolateral prefrontal cortex transcranial magnetic stimulation in treatment post-traumatic stress disorder: a randomized controlled study. Brain Res Bull 2018;140:334–40. https://doi.org/10.1016/j.brainresbull.2018.06.001.

[93] Cohen H, Kaplan Z, Kotler M. Repetitive transcranial magnetic stimulation of the right dorsolateral prefrontal cortex in posttraumatic stress disorder: a double-blind, placebo-controlled study. Am J Psychiatr 2004;161(3):515–24. https://doi.org/10.1176/appi.ajp.161.3.515.

[94] Isserles M, Shalev AY, Roth Y. Effectiveness of deep transcranial magnetic stimulation combined with a brief exposure procedure in post-traumatic stress disorder—a pilot study. Brain Stimul 2013;6(3):377–83. https://doi.org/10.1016/j.brs.2012.07.008.

[95] Kozel FA, Trees K Van, Larson V. One hertz versus ten hertz repetitive TMS treatment of PTSD: a randomized clinical trial. Psychiatr Res 2019;273:153–62. https://doi.org/10.1016/j.psychres.2019.01.004.

[96] Osuch EA, Benson BE, Luckenbaugh DA. Repetitive TMS combined with exposure therapy for PTSD: a preliminary study. J Anxiety Disord 2009;23(1):54–9. https://doi.org/10.1016/j.janxdis.2008.03.015.

[97] Kozel FA. Clinical repetitive transcranial magnetic stimulation for posttraumatic stress disorder, generalized anxiety disorder, and bipolar disorder. Psychiatr Clin 2018;41(3):433–46. https://doi.org/10.1016/j.psc.2018.04.007.

[98]   Watts BV, Landon B, Groft A, Young-Xu Y. A sham controlled study of repetitive transcranial magnetic stimulation for posttraumatic stress disorder. Brain Stimul 2012;5(1):38−43. https://doi. org/10.1016/j.brs.2011.02.002.

[99]   Jog MV, Wang DJJ, Narr KL. A review of transcranial direct current stimulation (tDCS) for the individualized treatment of depressive symptoms. Precis Med Psychiatr 2019;17−18:17−22. https://doi.org/10.1016/j.pmip.2019.03.001.

[100] Saunders N, Downham R, Turman B, et al. Working memory training with tDCS improves behavioral and neurophysiological symptoms in pilot group with post-traumatic stress disorder (PTSD) and with poor working memory. Neurocase 2015;21(3):271−8. https://doi.org/10.1080/ 13554794.2014.890727.

[101] van't Wout M, Longo SM, Reddy MK, Philip NS, Bowker MT, Greenberg BD. Transcranial direct current stimulation may modulate extinction memory in posttraumatic stress disorder. Brain Behav 2017;7(5):e00681. https://doi.org/10.1002/brb3.681.

[102] van 't Wout-Frank M, Shea MT, Larson VC, Greenberg BD, Philip NS. Combined transcranial direct current stimulation with virtual reality exposure for posttraumatic stress disorder: feasibility and pilot results. Brain Stimul 2019;12(1):41−3. https://doi.org/10.1016/j.brs.2018.09.011.

[103] Ahmadizadeh MJ, Rezaei M, Fitzgerald PB. Transcranial direct current stimulation (tDCS) for post-traumatic stress disorder (PTSD): a randomized, double-blinded, controlled trial. Brain Res Bull 2019;153:273−8. https://doi.org/10.1016/j.brainresbull.2019.09.011.

[104] Hampstead BM, Mascaro N, Schlaefflin S, et al. Variable symptomatic and neurophysiologic response to HD-tDCS in a case series with posttraumatic stress disorder. Int J Psychophysiol 2020; 154:93−100. https://doi.org/10.1016/j.ijpsycho.2019.10.017.

[105] Sillay KA, Larson PS, Starr PA. Deep brain stimulator hardware-related infections: incidence and management in a large series. Neurosurgery 2008;62(2):360−7. https://doi.org/10.1227/01.neu. 0000316002.03765.33.

[106] Binder DK, Rau G, Starr PA. Hemorrhagic complications of microelectrode-guided deep brain stimulation. Stereotact Funct Neurosurg 2003;80(1−4):28−31. https://doi.org/10.1159/000075156.

[107] Weber PB, Kapur R, Gwinn RP, Zimmerman RS, Courtney TA, Morrell MJ. Infection and erosion rates in trials of a cranially implanted neurostimulator do not increase with subsequent neuro-stimulator placements. Stereotact Funct Neurosurg 2017;95(5):325−9. https://doi.org/10.1159/ 000479288.

[108] Razavi B, Rao VR, Lin C, et al. Real-world experience with direct brain-responsive neuro-stimulation for focal onset seizures. Epilepsia 2020;61(8):1749−57. https://doi.org/10.1111/epi. 16593.

[109] Benabid AL, Benazzous A, Pollak P. Mechanisms of deep brain stimulation. Mov Disord 2002; 17(S3):S73−4. https://doi.org/10.1002/mds.10145.

3

CNN-based artifact recognition from independent components of EEG signals

Matteo Polsinelli[1] and Giuseppe Placidi[2]

[1]*DEPARTMENT OF COMPUTER SCIENCE, UNIVERSITY OF SALERNO, SALERNO, ITALY;*
[2]*DEPARMENT OF LIFE, HEALTH & ENVIRONMENTAL SCIENCES, UNIVERSITY OF L'AQUILA, L'AQUILA, ITALY*

1. Introduction

EEG measures neuronal activity through electrodes placed on the scalp with excellent temporal resolution. Optimal temporal resolution and low invasiveness make EEG particularly suitable for real-time usage [1−3]. Extraneous signals, called artifacts, are produced by eye movements and blinking, muscular spasms, cardiac activity, and generic interferences (IF) [4,5]. Artifacts can obscure useful brain signals (UBS) since the skull and the scalp (including muscles) are between the brain and sensors.

Blinking and eye movements produce electrooculography (EOG) artifacts, mainly recorded by frontal sensors, which propagate across the scalp [6]. Three categories of EOG exist: Eye blinking (BEOG), vertical (VEOG), and horizontal (HEOG) eye movements. EOG often has a much higher amplitude than UBS at frequencies in the range 10−40 Hz, where also UBS is present.

Cardiac activity produces electrocardiography (ECG) artifacts [7]. ECG effects can be reduced by subtracting the signal of a peripheral sensor from those located on the scalp [5]. Residual ECG effects are lower than brain signals but are still present.

Cranial muscles produce electromyogram (EMG) artifacts. The main feature of EMG is the wide spectral distribution with maximum power in the range of 15−30 Hz, where also UBS insists [8].

Finally, artifacts are also due to generic discontinuities generated by impedance fluctuations or electric/electronic IF affecting single sensors with large fluctuations in amplitude [9].

Artifacts could be much more intense than UBS and propagate to large regions [6,9,10]: UBS could be completely obscured if selective preprocessing strategies are not

Signal Processing Strategies. https://doi.org/10.1016/B978-0-323-95437-2.00010-0

employed. As stated above, artifacts and UBS share frequencies, and preprocessing alternatives to frequency analysis are required to separate their respective contributions.

Different kinds of classical filtering techniques that have been designed and used for EEG artifact detection [4,11,12] can be grouped into the following categories: Regression methods; filtering algorithms; wavelet transform (WT), and empirical mode decomposition (EMD).

Regression methods assume that artifacts are measured through dedicated channels [12,13]. Measurements are necessary to estimate the propagation coefficients that need to be subtracted from brain signals. Though these strategies have good computational performance, they have two major drawbacks: one or more reference channels are needed, which is a severe limitation when reference channels are unavailable [14]; the reduction of artifacts also implies the removal of relevant UBS [9].

Filtering includes several approaches, the most widely used being the adaptive filter [15]. This method assumes that UBS are unrelated to artifacts and requires a dedicated channel to measure the artifacts to be subtracted from the signal. Filtering strategies suffer from the same limitations as regression methods [16].

Unlike regression and filtering, WT does not require reference signals. WT transforms the signals from the time domain to the time-frequency domain, and low-amplitude WT coefficients are zeroed before being inversely transformed. Their main limitation is that artifacts that overlap in frequency with UBS or that are too specific are not removed, or that UBS is also removed with them [4,12,17].

EMD [18] and multivariate EMD (MEMD) [19] are tools for signal decomposition into amplitude and frequency-modulated basis functions (intrinsic mode functions). EMD is specific for single-channel data, and MEMD is the extension of EMD to multichannel data. These methods are robust to noise and suitable for removing muscle artifacts [20] but they are ineffective for other kinds of artifacts (such as IF) and, being slow, are unsuitable for online applications [4].

An effective method to retrieve source components of EEG signals is independent component analysis (ICA) [10]. ICA, on a given temporal window of an n-channel EEG measurement, allows the calculation of at most n independent components (IC). A component is defined by an array of weights, each representing the contribution of a sensor to the component itself. Hence, EEG signals, measured in a given temporal window, can be viewed as a mixture of independent linear source components, some mainly due to artifacts and others to UBS [21]. Weights can be interpolated in 3D by using the spatial map of the sensors on the scalp and reprojected on a 2D topoplot [22] to allow a topographic component analysis (Fig. 3.1). In fact, components are artifact indicators just when they are reprojected on the scalp; shape and localization are fundamental information to classify component sources [23].

Usually, the visual inspection of topoplots by trained experts (humans) allows the production of source classification [22]. In some cases, other additional information such as power spectrum density (PSD) and autocorrelation integrate classification, though they imply a huge preprocessing overhead (visual inspection takes about 5–

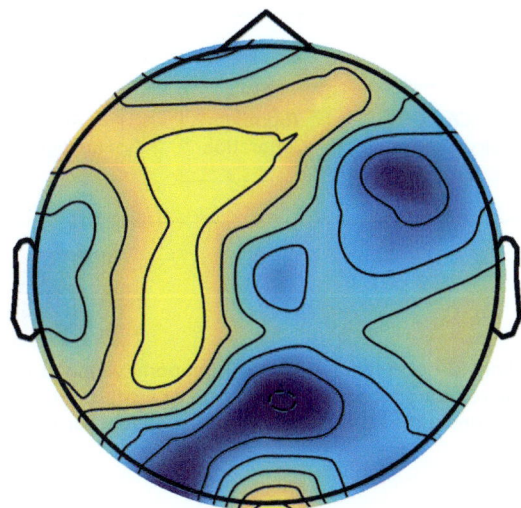

FIGURE 3.1 An example of an independent component reprojected in a scalp topography (the topoplot). Colors indicate the underlying brain activation: *Blue* represents low activation, and *yellow* corresponds to high activation.

7 seconds per topoplot; additional information could more than double this time). Visual inspection is considered the best practice for artifact inspection because it is effective, widely used for (off-line) EEG analysis, and therefore supported by most EEG-specific signal processing tools [24,25].

For this reason, despite the recent proposals for alternative preprocessing strategies, ICA and topoplot analysis are still the gold standard, thanks to their ability to deal with all kinds of artifacts [4,26,27].

In the last few years, automated ICA-based strategies have begun to appear. In what follows, we have considered a useful strategy first to present the general ICA-based artifact removal pipeline and then to address automatic strategies, mostly based on convolutional neural networks (CNN) approaches, which have recently been employed in EEG classification and artifact removal [28–32].

2. ICA-based artifact removal pipeline

Fig. 3.2 shows a sketch of the automatic preprocessing pipeline for artifact removal from EEG. In particular, the EEG signals in the consecutive, partially overlapping temporal windows (temporal windows overlap, as shown in Fig. 3.2, to avoid borders remaining unprocessed) are first passed to the ICA calculation module, then the resulting ICs are passed to the topoplot generation module to project them on the scalp, and, finally, the resulting topoplots are analyzed for artifact recognition (yellow box in Fig. 3.2). The last operation is carried out by a human expert or can be automated through computer-vision (CV)-based strategies. The resulting artifact IC is discarded (red box in Fig. 3.2),

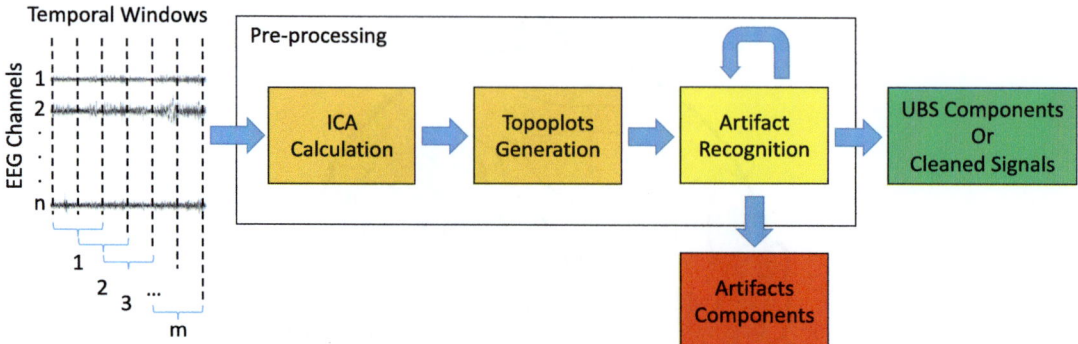

FIGURE 3.2 Preprocessing pipeline: ICA calculation related to temporal windows of EEG signals, enclosed by *curly brackets (left)*; generation of the resulting topoplots; recognition of artifacts from topoplot images. Artifacts are discarded, and the UBS, or the reconstructed cleaned signals, are used in a further analysis stage. *Adapted from Ref. [32].*

and the UBS components, or the related cleaned signals (green box in Fig. 3.2), are either used in this way or transformed back to signals before their following usage. In the following subsections, each block of the preprocessing pipeline is described and analyzed.

2.1   ICA calculation

Let's consider the situation where there are two people speaking simultaneously inside a room, and two microphones (positioned in different locations of the room) are recording the dialog.

The recorded signals (or observed signals), $x_1(t)$ and $x_2(t)$, are a weighted sum of the two voices (underlying source signals, or simply sources), $s_1(t)$ and $s_2(t)$.

In details:

$$x_1(t) = a_{11}s_1(t) + a_{12}s_2(t) \tag{3.1}$$

$$x_2(t) = a_{11}s_1(t) + a_{12}s_2(t) \tag{3.2}$$

where the coefficients a_{ij}, also called the mixing weights, are constants depending on the distance between the actors (microphones and persons). The coefficients a_{ij} are unknown, as are the underlying signals.

Generalizing to the case of multiple observed/underlying signals, Eqs. (3.1) and (3.2) can be written as

$$x = As \tag{3.3}$$

where A is a matrix containing all the mixing weights.

ICA allows us to estimate the mixing weights and, consequentially, solve the problem:

$$s = Wx \tag{3.4}$$

where $W = A^{-1}$. It is worth noticing that one of the conditions for the existence of the inverse of the matrix is that A is square. In these conditions, the number of recorded signals has to be equal to the number of sources. If the number of observed signals is greater than the number of sources, it is possible to reduce the dimensions of the problem before solving it through specific analytical strategies such as principal component analysis [33] (out of the scope of this chapter). In the EEG scenario, the observed signals are those recorded from the electrodes (recorded signals), and the sources (the "voices") are either generated by the brain (inside the brain) or are due to electrical activity external to the brain; these last generating artifacts.

The objective of ICA is to solve the above inverse problem by considering that the sources are nonGaussian and statistically independent of each other. Indeed, ICA imposes precisely these two characteristics to find a solution to the problem: it minimizes the mutual information of the sources while maximizing their nonGaussianity. In this sense, ICA is not effective in finding all the UBS because several UBS generated in the brain are not statistically independent. However, artifacts are all statistically independent, and ICA, by finding all the ICs, is very effective in considering, besides UBS owning this property, all the artifacts. For this reason, once ICA is calculated, artifacts have to be recognized and separated from UBS ICs. Once ICs related to artifacts are eliminated, cleaned signals undergo interpretation. Regarding the implemented algorithms for ICA calculation, the most commonly used are Infomax [34], Fastica [35], second-order blind identification [36] and JADE [37]. Using simulated, relatively low dimensional data sets for which all the assumptions of ICA are fulfilled, the algorithms return near-equivalent components [24]. Moreover, the physical significance of the results of 22 different ICA algorithms has been evaluated in Ref. [38] in terms of three parameters: Mutual information reduction (MIR), residual pairwise mutual information, and "dipolarity," that is, the number of returned components resembling an equivalent dipole.

The results show that all tested ICA algorithms return a similar decomposition. However, each ICA algorithm is more suitable for specific situations. For example, in Infomax, ICA gives stable decompositions with up to hundreds of channels and is particularly suited for sources with a superGaussian distribution, while JADE is fast and stable, especially for low numbers of channels [38]. An important aspect to consider is the minimum amount of data (minimum temporal window length) necessary to make ICA convergent to stable ICs. Typically, ICA requires more than kN^2 data sample points, where N is the number of weights in the unmixing matrix that ICA is trying to learn and k is a constant multiplier that depends on the number of channels.

The main issue with the above ICA algorithms is that they cannot be used for online applications (for example, BCI [39,40]). To this aim, it is fundamental to cite one of the most commonly used ICA algorithms for fast applications with acceptable convergence rates for relatively large numbers of channels: Online recursive ICA (ORICA) [41−43]. In particular, a recursive least squares (RLS) for incremental estimation of ICs from online data, derived as a fixed-point solution to the natural gradient Infomax learning rule, is proposed in Ref. [41]. In Ref. [42] an optimized implementation of ORICA with the online

RLS whitening filter is used. Moreover, a multiple measurement vector block-update rule is implemented to improve speed without sacrificing performance. Its stability and steady-state performance are analyzed in Ref. [44], where ORICA is compared with offline methods on real EEG data. Results show that ORICA decomposition is comparable with Infomax in terms of MIR and dipolarity. Moreover, the low computational complexity of ORICA enables its use for online processing.

A further extension of [41,42] is presented in Ref. [45], where an efficient computational pipeline for real-time, adaptive blind source separation (BSS) of EEG data using ORICA is presented. The pipeline was tested on three data sets: a simulated 64-channel stationary data set, a simulated 64-channel nonstationary data set, and a real 61-channel EEG data set. With the first data set, the effects of key parameters on convergence, the convergence speed, and the computational load of the algorithm were analyzed. With the second data set, it was demonstrated that the ORICA can adaptively track changes in the mixing matrix due to electrode displacements. With the last data set, it was demonstrated that the ORICA can decompose brain and artifact subspaces online, with comparable performance to the offline Infomax ICA.

It is important to know that ICA is not the only strategy to handle the BSS problem. In fact, there are several other methods, such as canonical correlation analysis (CCA) [46], morphological component analysis (MCA), and independent vector snalysis (IVA) [47].

CCA uses second-order statistics to calculate uncorrelated components. As previously stated, this is a weaker condition with respect to the statistical independence required for ICA. CCA is more effective than ICA in removing muscle artifacts via EEG signals [48]. CCA assumes a relatively low autocorrelation of muscle artifacts with respect to brain activity considered to be maximally autocorrelated. For this reason, CCA compares the current EEG signals to a delayed version of the past signals and preserves sources that maximize autocorrelation between the two data sets. This strategy has been shown to work well for the removal of muscle artifacts but must be combined with other strategies to cope with other sources of artifacts or with a few-channel EEG [49].

MCA decomposes EEG signals into components that have different morphological characteristics using a linear combination of waveforms collected in a dictionary. For this reason, it is only applicable to artifacts whose signature is defined by topology. MCA efficiently removes ocular artifacts and muscle artifacts.

IVA is an extension of ICA [50,51]. It includes in a single strategy both the need to assume that some artifacts have different autocorrelation properties than UBS (muscular activity) and that other artifacts are independent of UBS (ocular movements). IVA outperforms ICA and CCA in isolating muscle and ocular artifacts, especially in low-quality signals, but it does not handle all types of artifacts.

2.2   Topoplot generation

As discussed above, each IC calculated by ICA can potentially be an artifact, based on its location, that can be external or internal to the brain. External ICs are due to artifacts, and they have the classification defined in Section 1. Internal ICs are due to UBS and must be preserved.

Each IC is characterized by a topography (a set of inverse weights describing the projection of the independent source onto the electrode cap).

In fact, the possibility of representing the ICs calculated by using ICA in 2D scalp topographies allows them to be recognized and classified visually. Fig. 3.3 shows the characteristic topographies of the artifact classification defined above [9,22,52].

As it can be observed, BEOG concentrates on the frontal region (Fig. 3.3A) as well as VEOG (Fig. 3.3B), though VEOG spreads more than BEOG. Because of their similarity in shape and meaning, they can be grouped into a single class.

In the case of HEOG, two peaks of opposite signs are positioned around the nose (Fig. 3.3C). Similarly to HEOG, ECG (Fig. 3.3D) is composed of two peaks of opposite signs localized on the edges of the head, around the ears (ECG differs from HEOG only for the orientation). The similarity between HEOG and ECG suggests their inclusion in the same class (the recognition between them, outside the scope of this work, could be based on the orientation of the peaks).

EMG and IF consist of isolated peaks, the former usually found on the borders of the head near the neck and face where muscular activity is pronounced (Fig. 3.3E), while the latter is often located in the middle of the head (Fig. 3.3F). Due to their similar shape, EMG and IF can be included in the same class, although their nature is very different (EMG is due to muscle activation while IF is due to electrical disturbances). The distinction between a channel failure and a muscle artifact is difficult even for a human expert. Position, power, and frequency of occurrence increase the probability of one over the other, but for the purposes described therein, they must both be discarded.

As specified above, the role of IC topography is fundamental to the visual identification of artifacts. However, EEG measurement sessions often consist of several time windows (see Fig. 3.2), and the amount of IC topoplots could easily reach several thousand in a session. These numbers make offline visual inspection/recognition very hard and impossible when an online response is required. Automatic strategies are required to fill the gap with online applications, but the major numerical drawback, besides ICA calculation, is the generation of topoplots.

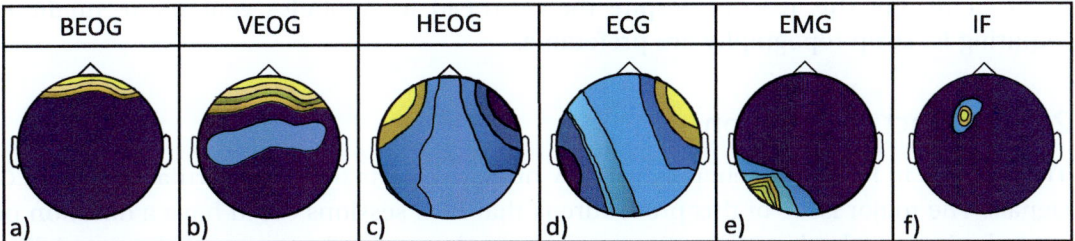

FIGURE 3.3 ICA components reprojected in the topoplots related to artifacts. BOEG (A) and VEOG (B) have similar shapes. The same occurs between HEOG (C) and ECG (D) and between EMG (E) and IF (F). EOG and ECG have well-defined locations on the head; EMG and IF are represented by isolated peaks. *Adapted from Ref. [32].*

Due to the overhead imposed by topoplot generation, several methods not using topoplots for the IC classification have been attempted.

For example, the framework proposed in Ref. [53] is capable of identifying BEOG, VEOG, and HEOG artifacts, but not the others. With this aim, the EEG signal is first converted into multivariate information using the combination of singular spectrum analysis (SSA) and ICA. Then, ICA is employed on the multivariate information, which separates the hidden sources as different ICs. After that, the SWT is applied to the artifact IC, which performs the thresholding to recognize artifacts.

In [54] a framework for BEOG artifact removal is proposed. It combines ICA with continuous WT, k-means clustering, SSA, and support vector machines (SVMs). The work used three evaluation data sets (one synthetic and two real). The SVM-based classifier has an accuracy of 99%; unfortunately, it is unable to recognize all artifact classes.

Another approach for BEOG artifact removal based on SVM is presented in Ref. [55]. In this work, SVM is used to identify ICA components and for the localization of the artifact windows in corrupted ICs. Finally, for the correction of the artifact windows, a denoising autoencoder is used. Similar results have been obtained in Ref. [56], in which wavelet multiresolution analysis (WMA) ICA and SVM are still used for BEOG artifact removal. The framework has been tested on recorded EEG data and publicly available data from EEGLAB.

Beyond SVM, fuzzy kernel SVM has also been used for EEG artifact removal in Ref. [57]. The method (which is wavelet ICA-based) is capable of recognizing BEOG, EMG, and IF artifacts, with a performance of about 80%.

Finally, in the work presented in Ref. [58], ICA components are characterized using symbolic aggregate approximations (SAX), which takes time series (ICA components) as input and produces an SAX word as output, a sequence of letters useful to characterize the time series. Then, a multiinstance classifier is used to find, for each component, the probability of being an artifact. The whole method has been implemented on an embedded system.

The above methods are, in general, faster than those using topoplots (topoplot generation is expensive), but they are unable to recognize all kinds of artifacts. Most of them, in fact, are very well suited for BEOG, VEOG, and HEOG artifacts but not for ECG, EMG, or IF. For this reason, despite the overhead for topoplot generation, methods requiring IC scalp topography are preferable.

2.3   Artifact recognition

The pipeline in Fig. 3.2 is generally used by human experts to remove artifacts from EEG signals. The major issue of this procedure is that EEG sessions could have a duration of several minutes (also hours), and manual removal, possible just as an off-line modality, could be very difficult (a very large number of topoplots have to be analyzed) and a source of errors (the possibility of making evaluation mistakes is high). For this reason, several tools have been developed to make this task easier for human experts. Despite that, automatic strategies are needed when fastness is required. In Subsection 2.3.1, two of the commonly used software programs for EEG signal processing are introduced: EEGLAB and FieldTrip. Moreover, the REST toolbox for online artifact removal is introduced in Subsection 2.3.2, when automatic strategies are described.

2.3.1   Software tools for human experts

EEGLAB is a Matlab toolbox for processing EEG signals and magnetoencephalography (MEG) signals. Its functions, including time/frequency analysis, ICA, artifact rejection, event-related statistics, standard averaging methods, etc., can be executed on Windows, Linux, and Mac OS X. It provides a graphical user interface (GUI), divided into interactive forms, to assist human experts during the analysis sessions. For example, it is possible to import the EEG signals and select the ICA algorithm to run directly in a specific form. The obtained components can be visualized through time courses or automatically represented in a topoplot. In this case, it is important to initialize the used channels in the 10−20 brain positioning system [59]. After that, a visualization form appears showing the topoplots. Besides topoplots, EEGLAB allows the inclusion of other information, such as the activity power spectrum and the time course of the component.

FieldTrip has almost the same functionalities as EEGLAB, though it uses a slightly different approach. The main differences between the two tools are the following: EEGLAB provides an interactive GUI for data exploration to guide users in writing custom analysis scripts; FieldTrip is designed for advanced users and does not provide any GUI. In this way, if the end-user has little experience in writing scripts, EEGLAB is the right choice, while users with considerable experience in script programming usually prefer FieldTrip to exploit its flexibility in providing a highly adaptive framework. Both tools are freely available, open-source software.

2.3.2   Automatic ICA-based strategies using topoplots

Recently, CNN has revolutionized object recognition [23,60,61], including EEG classification [2,31,62−64].

In particular, regarding artifact recognition, it is possible to mimic the visual recognition of human experts through CNN. With this aim, we report CNN-based automatic strategies that follow the pipeline in Fig. 3.2 to classify EEG artifacts by topoplots of ICs.

It is worth noticing that, besides topoplots, some automatic methods also use additional information. For example, in Ref. [29] normalized topographic images of the ICA mixing weights (ICMAPs) and spectrum of ICs are used as input for a framework based on two CNNs. The parallel structures consisted of convolutional, nonlinear, and pooling layers performing feature extraction and dimension reduction. The features extracted from both CNNs are conveyed to a fully connected layer and finally to a classification layer. The framework is capable of recognizing artifacts without explicitly defining their nature.

A similar approach is presented in Ref. [31]. The proposed framework is composed of three CNNs that classify three different types of inputs: a 2D topoplot represented with just one channel (the intensity); the median PSD; the autocorrelation function. The first CNN is 2-D, while the remaining are 1-D. The CNNs use weighted cross-entropy loss. The framework is generally referred to as ICLabel, and it is capable of recognizing all the defined artifacts above. ICLabel provides not only the classification of the components but also the estimated probabilities of being an artifact. The definition of a threshold allows the rejection of all components for which the framework has a probability higher than the threshold value.

An ensemble of deep neural networks (ensemble DNN) is proposed in Ref. [65]. It contains three parallel DNNs: (i) A convolutional neural network combined with bidirectional long short-term memory layers (CNNbiLSTM) to extract temporal features from the IC time series; (ii) a CNN to extract spectral features from the PSDs; and (iii) a CNN to extract spatial features from topographic maps. Two data sets were used, the EPILEPSIAE data set [66] and the BASE data set [67,68], for training the developed architecture and evaluating it, respectively. The obtained performance is about 90%.

There are also methods that use only topoplots for the classification of ICs; among them, we recall [69], in which a CNN for the recognition of artifacts, such as EMG, VEOG, and HEOG is presented. It consists of six convolutional layers and three fully connected layers. An attention module is added to the classifier to refine the spatial features of the classifier. The attention module is composed of three convolutional layers and a sigmoid activation function. The accuracy of the method reaches 96%.

A method that uses only topoplots but is capable of recognizing all kinds of artifacts is presented in Ref. [32]. The framework consists of three parallel CNNs: one CNN used to recognize BEOG and VEOG (B_V CNN); one CNN to recognize HEOG and ECG (H_E CNN); and one CNN to recognize EMG and IF (E_I CNN). The experimental data set is the DEAP data set [70], a public multicenetr database containing a collection of EEG signals of negative and positive emotional states. The reached performance is about 98%.

To the best of our knowledge, the above-described strategies, which use topoplots and sometimes also additional information, have poor performance compared to humans. All the presented methods use CNN for classification. However, ICA calculation and topoplot generation require high computational power and memory. It is worth mentioning that recently some optimization has been done regarding these methods. In particular, the tool REST [43,71] has been designed for real-time EEG signal processing. REST is coded in Matlab and uses ORICA for the calculation of the ICs. REST is a framework, like EEGLAB and FieldTrip, that uses a GUI like EEGLAB. The main window of REST can show the raw EEG signals or the estimated IC activation. It also shows topoplots and power spectra for the calculated ICs. All the calculated information is updated in real-time. In fact, REST implements EyeCatch [72], a real-time IC classifier that is capable of recognizing all the artifact-related ICs.

In the following section, we discuss the application of one of the above methods [32], which is one of the most effective online strategies and that, in principle, could become real-time if included in the REST framework by inheriting its optimization on ICA calculation and topoplot generation.

3. Report on a case of study method

In this section, we discuss in more detail the method presented in Ref. [32], for the following reasons:

1. It follows the ICA-based artifact removal pipeline described in Section 2;
2. it uses just IC topoplots without any additional information (PSD, autocorrelation, etc.);
3. it is capable of recognizing all the common artifacts;
4. it is based on CNN, one of the most promising tools in CV;
5. it is modular and has very good performance;
6. it is simple and suitable for online usage;
7. it completely fits the boundaries of this work.

3.1   The architecture

As previously stated, the method is composed of an ensemble of CNNs (Fig. 3.4), each one related to a specific class of the artifacts defined above. Each CNN is organized into three stages: An input stage and a classification stage, interleaved with a feature extraction stage. The input stage consists of only one input layer. The classification stage consists of a fully connected layer (of dimension 2), a softmax layer, and a classification layer. Input and classification stages are the same for all CNNs. The feature extraction stage is specific to each CNN and organized in "blocks." Each block contains a convolutional layer, a BatchNorm layer, a Relu layer, and a maxpool layer, except for the last block, where the MaxPool layer is absent (Fig. 3.4).

In the B_V CNN, the feature extraction stage consists of a first block, two inner blocks, and the last block. For each block, the convolutional layers use respectively 8, 16, 32, and 64 filters (kernel size 3×3), and the MaxPool layers have a size of 2×2, stride [4] and padding 0. H_E CNN contains a first block, three inner blocks, and the last block. In all blocks, the convolutional layers use respectively 8, 16, 32, 64, and 128 filters (kernel size 3×3), and the maxpool layers have the same size as B_V CNN. Finally, in the E_I CNN, the feature extraction stage consists of a first block, five inner blocks, and the last block. In all blocks, the convolutional layers use 8, 16, 32, 64, 128, 256, and 256 filters. In the latter case, the max-pool layers are still 2x2 in size but stride [2]. The choice of a small stride value is justified by the fact that IF and EMG artifacts are composed of a small, localized cluster of pixels.

Moreover, each class of artifacts has a different intrinsic difficulty: B_V patterns are well-defined and localized (they are easily identifiable); H_E patterns are less defined than B_V ones (consequentially, identification difficulty increases); and E_I patterns are

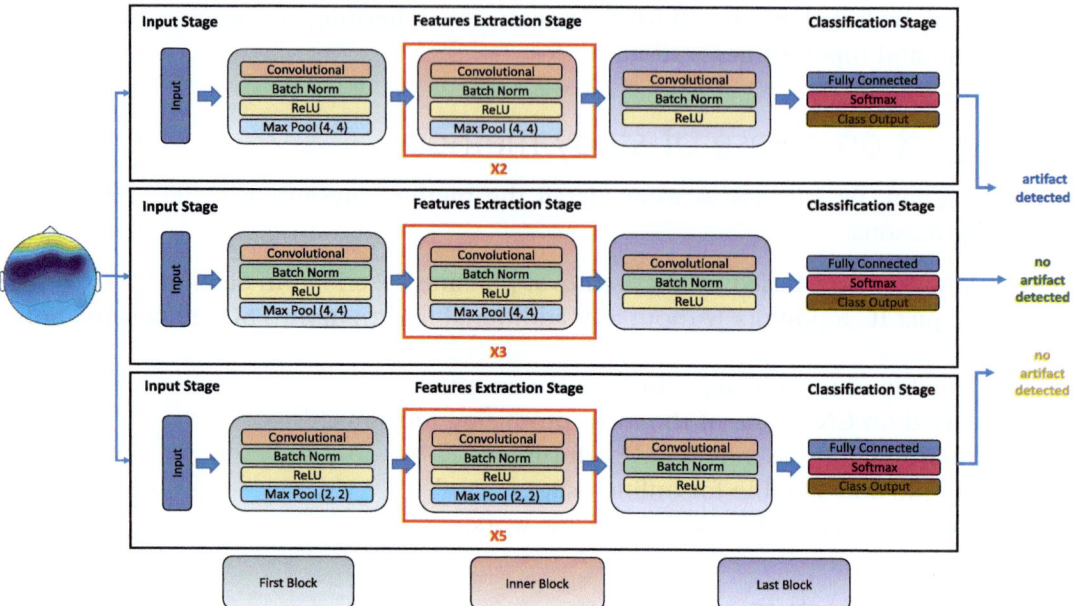

FIGURE 3.4 Framework architecture. The topoplot to be recognized is passed to three CNNs separately (the number of CNNs is the same as the artifact classes). The input and the classification stages have the same design in all CNNs. Regarding the feature extraction stage, B_V CNN contains 2 Inner blocks, H_E CNN 3 Inner blocks and E_I 5 Inner blocks. *Adapted from Ref. [32].*

neither well localized nor well defined (the most difficult to recognize). This justifies the incremental architectural complexity of the CNN.

Moreover, the number of inner blocks, filters in convolutional layers, and stride in MaxPool layers are optimized for each CNN. The whole architecture has 1.1x106 parameters, distributed as follows: 2.5×104 in B_V, 9.5×104 in H_E, and 9.5×105 in E_I. In terms of the number of parameters, the framework is lighter than SqueezeNet [73], one of the most competitive CNN architectures.

3.2   The experimental data set

The used EEG signal data set is the DEAP data set [70], a public multicenter database containing a collection of EEG signals of negative and positive emotional states recorded from 32 participants. The EEG signals were sampled at 512 Hz by 32 channels.

3.3   Data filtering and augmentation

Raw data underwent filtering with a notch filter [74] at 50 and 60 Hz to suppress powerline interference.

Data were divided into temporal 8-sec windows, overlapping by 4-sec each, each used to generate the ICs and the corresponding topoplots. With this strategy, the following requirements are fulfilled: Sufficient data is collected to ensure ICA convergence; signal acquisition is shortened to reduce acquisition time in fast applications; and window borders are covered in ICA calculation.

ICA was calculated on each window by obtaining a maximum of 32 components, and the topoplot corresponding to an IC was generated as a 134 × 136 RGB image (the best compromise between good spatial precision and low execution time) with a fixed position and orientation.

Data augmentation was avoided because: The rotations of the topoplot would change its meaning and, therefore, would be unjustified; scaling or translation would create redundancy because the interpretation of a topoplot always refers to the external silhouette of the head; brightness modifications would be wrong because the topoplot color map is fixed.

3.4 Data set distribution and training

Since the DEAP data set does not contain labeled topoplot, data from subjects one to eight was used for visual classification and manual labeling. This was made independently by five human experts who divided topoplots into the following four categories: BEOG ∪ VEOG (B_V), HEOG ∪ ECG (H_E), EMG ∪ IF (E_I), and UBS.

The ground truth (also called consensus) data set was obtained considering each topoplot belonging to the most voted class. In the case of a tie result (the case in which the most voted classes were 2 with 2 votes each), the resulting topoplot was discarded. The resulting labeled ground truth data set agreed at 95.7% with all human experts (its agreement with each of the experts oscillated between 97.2% and 99.1%).

All artifacts and UBS were randomly extracted from the labeled consensus by considering the reciprocal proportions of the artifact classes in reality but reducing the members of the UBS class with respect to the real situation. In fact, the common situation in EEG is that the number of UBS ICs is much greater than the number of artifact ICs, especially during laboratory acquisition sessions in which the subject is in a controlled environment. If not reduced, the unbalance could potentially cause training failure due to the saturation of UBS ICs with respect to those of the artifacts. Without a rebalance, each CNN could potentially classify all the topoplots as UBS.

Each CNN was trained separately, and for each CNN, the training set was organized by separating the topoplots into two classes: the first containing the specific artifacts that the CNN should recognize; the second containing all the others (other artifact kinds + UBS). This data organization was fundamental to training each CNN to recognize only its own artifacts with respect to UBS and other artifacts. Finally, the data set for each CNN was randomly split into 70% for training and 30% for validation, but the balance was maintained among classes. The composition of the training sets is illustrated in Table 3.1.

Regarding the training convergence, 100 epochs were sufficient, and the process took about 40 minutes using the Matlab environment on a PC with an Intel Core I7-6700, 32 GB of RAM, and Nvidia GeForce GTX 1080.

Table 3.1   Configuration of the data sets used to train the framework.

B_V CNN	H_E CNN	E_I CNN
1341 (B_V)	398 (H_V)	1592 (E_I)
5020 (H_E + E_I + UBS)	4823 (B_V + E_I + UBS)	6044(B_V + H_E + UBS)

Adapted from Ref. [32].

The process was repeated 10 times on randomly selected data sets, resulting in the following average accuracy: $99.4 \pm 0.4\%$ for B_V, $99.6 \pm 0.3\%$ for H_E, and $97.9 \pm 0.6\%$ for E_I.

3.5   Results

The proposed framework was tested on data from subjects 20–32 of the DEAP data set, corresponding to 340,890 images. The results, summarized in Table 3.2, show that among the topoplots considered artifacts, a very small percentage had double membership ("Double Detection"), that is, they were classified by two CNNs as their own at the same time. It is interesting to note that after a deep examination performed by human experts, most of them were actually found to be ambiguous and very difficult to be uniquely classified (additional information should be required). Moreover, the numbers are repeated in the "Double Detection" column: topoplots shared between classes j and k have been counted twice. Finally, no "Triple Detection" occurred, and, for

Table 3.2   Classification of 340,890 topoplots. The "Others" column contains other artifacts + UBS. The "Artifacts" column contains the number of artifacts recognized as their own by CNN, shown on the left. "Double detection" are artifacts considered to be their own by two CNN simultaneously and are reported twice. For instance, the topoplots belonging to B_V and, at the same time, to H_E (340) are the same ones that belong to H_E and, at the same time, to B_V. "Triple detection," never occurring, is not reported.

		Classification results			
		Others	Artifacts	Double detection	
CNN	B_V	310,720	30,120	H_E	E_I
				340	4480
	H_E	332,429	8411	B_V	E_I
				340	320
	E_I	296,669	44,171	B_V	H_E
				4480	320

Adapted from Ref. [32].

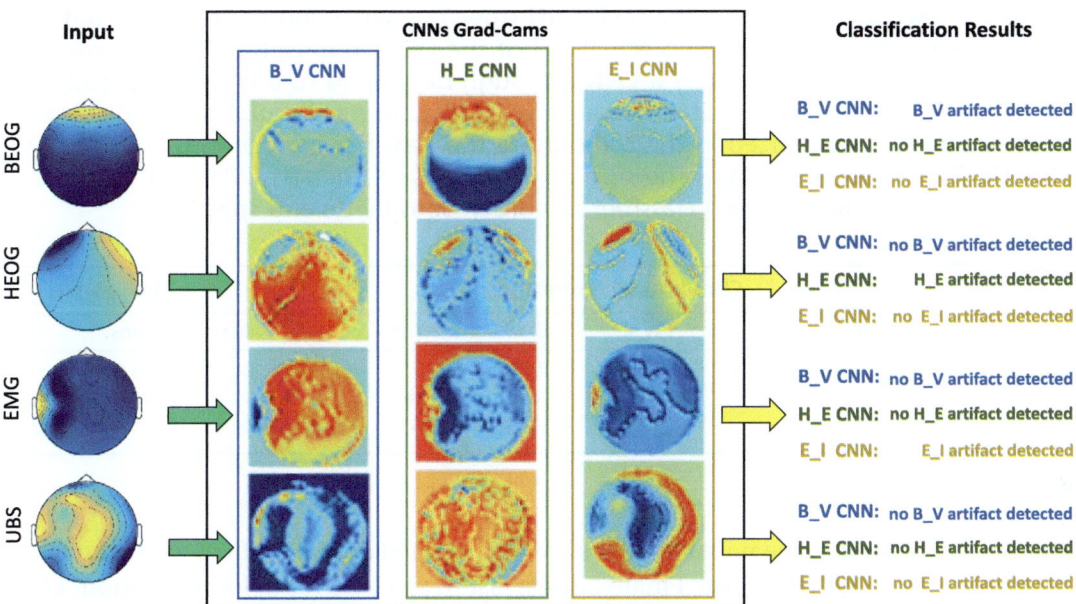

FIGURE 3.5 Grad-Cams: The topoplots belonging to the three classes of artifacts or to UBS are inputs of each CNN (*columns*); the same topoplot is the input of a different CNN (*rows*). The last two columns on the right show "output" and "classification results," respectively. *Adapted from Ref. [32].*

this reason, a "Triple Detection" column is not present in Table 3.2. It is worth noticing that the number of double detections was very low; this confirms that the computational overhead introduced by additional information is not justified.

A trustful artifact removal system should be 'transparent'. For this reason, gradient-weighted class activation mapping, or Grad-Cam (Fig. 3.5), was used to produce a coarse localization map showing which regions in the image are useful for prediction [75] and to demonstrate the correctness of the strategy. Grad-Cams are very important to explain why they predict what they predict. The convolutional layers naturally retain spatial information, which is lost in fully connected layers. The neurons in these layers look for semantic class-specific information in the image ('object parts'). As the name suggests, Grad-Cams use the gradient information flowing into the convolutional layer of the CNN to assign importance values to each neuron for a particular decision of interest.

Fig. 3.5 shows Grad-Cams in a table: Rows indicate the four different classes of topoplots (artifacts + UBS), and columns represent the respective Grad-Cams of each CNN. In the first row, a B_V artifact is the input of the three CNNs: the B_V CNN Grad-Cam shows that the activation is correctly localized in the frontal region of the head. In the second row, an H_E artifact is the input of the three CNNs: The H_E CNN Grad-Cam shows that the activation is correctly located on the lateral edges of the head. This confirms that the H_E CNN is not biased by the strongest positive activation (yellow) in the topoplot. In the third row, an E_I artifact is the input of the three CNNs: The E_I CNN

Table 3.3   Recognition errors for each CNN (rows) when acting on a labeled data set of topoplots generated from subjects 9–19 of the DEAP data set. False positives (FP) are those classified as proper by a CNN although belonging to another class (columns). False negatives (explicitly indicated with FN close to the number to distinguish them from FP) are reported in the cells where the classifier corresponds to the class. The "TOTAL" column contains the sum of the values on the left (FP + FN). The last three columns show, for each classifier, accuracy, sensitivity, and specificity, respectively.

		Classification errors (#)					PERF. (%)		
		UBS	B_V	H_E	E_I	Total	Acc.	Sens.	Spec.
CNN	B_V	346	20 (FN)	17	235	618	98.7	99.0	98.7
	H_E	114	16	9 (FN)	22	161	99.5	98.2	99.6
	E_I	162	134	138	73 (FN)	507	99.1	98.1	99.2

Adapted from Ref. [32].

Grad-Cam shows that the positive values are well located on the artifact, while the rest of the map shows negative values. Finally, in the last row, a UBS topoplot is the input of the three CNNs; none of the CNNs recognize it as its own.

To check in depth the behavior of the framework, another topoplot data set was generated from subjects 9–19 of the DEAP data set, and 1/10 of them (a total of 29,500), selected at random, was submitted to the five human experts for visual classification and labeling. The consensus contained 22,795 UBS, 2190 B_V, 526 H_E, and 3871 E_I, respectively. A set of 117 topoplots was discarded due to ties.

Then, automatic recognition was performed using the proposed framework; the resulting errors, compared to manual consensus, and the performance of each CNN are reported in Table 3.3.

For each classifier (rows), the number of topoplots wrongly classified as proper but belonging to the other classes, the false positives, is reported (columns). When the class matches the classifier, the corresponding cell contains the number of false negatives. For each classifier, the right part of Table 3.3 reports accuracy, sensitivity, and specificity.

Results show that the framework had very good performance (about 98%). Moreover, human experts disagreed on most topoplots misclassified by the framework, thus confirming that most of them exhibited ambiguous patterns.

In terms of computational performance, the proposed framework took about 0.3 seconds for ICS calculation (using the method presented in Ref. [76]), 0.9 seconds for topoplot generation, and 0.21 seconds for classification of the 32 ICs calculated for the problem at hand, thus corresponding to about.

1.4 seconds for the whole classification. Since the time necessary to collect new data is 4 seconds, it is completely feasible for online applications. As reported, most of the time is absorbed by the topoplot generation, but the good news is that it did not increase

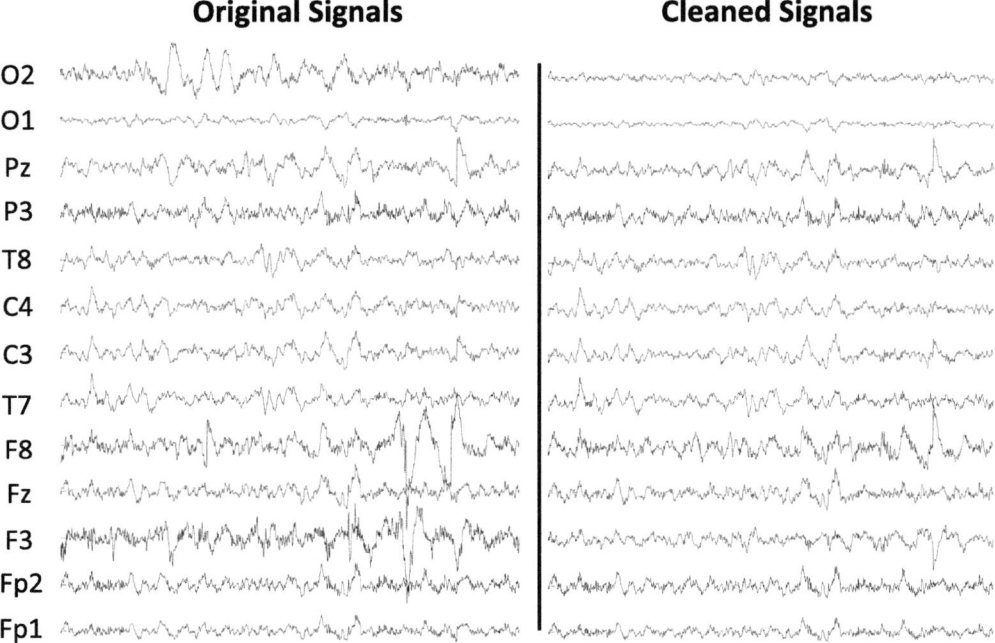

FIGURE 3.6 An example of artifact recognition and elimination is reported: Original signals (*left*); artifact-removed signals (*right*).

linearly with the number of images because some calculations, such as those required for channel positioning on the scalp, were executed just once.

The described strategy has been effectively applied in off-line experiments [77] and in online applications [39,40,78–80].

An example of EEG signals cleaned after artifact elimination by using the method above is reported in Fig. 3.6.

4. Conclusions

Artifact recognition/classification of IC topoplots from EEG signals is the most effective paradigm but suffers from high computational overhead and memory consumption. In this work, we tried to give a general overview of the process by discovering some fundamental aspects. First of all, we introduced the most effective pipeline used for EEG artifact recognition. We described the role of ICA and how it works, as well as the assumptions necessary to use ICA, and we gave some hints from the literature. Then we discussed the fundamental role of IC topography (topoplot) in discovering artifacts, the common artifact patterns in IC topoplots, how topoplots are used by human experts to reject artifacts from EEG signals, and the most commonly used software (EEGLAB and FieldTrip) to assist this procedure.

We highlighted the particular topological features of each artifact class.

We moved forward by introducing several CNN-based automatic methods for artifact removal from IC topoplots, miming human behavior. Since topoplot generation is time-consuming, we also introduced TEST, an optimized real-time ICA and topoplot-based tool.

Finally, as a case study, we described the strategy proposed in Ref. [32] because: it follows the ICA-based artifact EEG removal pipeline (in Fig. 3.2); it has performance comparable with human experts; it recognizes all kinds of actually known artifacts; it is modular (future discovered sources of artifacts could be treated with the addition of another CNN to the current architecture); it is usable for online applications.

References

[1] Suk H-I, Lee S-W. A novel Bayesian framework for discriminative feature extraction in brain-computer interfaces. IEEE Trans Pattern Anal Mach Intell 2012;35(2):286−99.

[2] Cecotti H, Graser A. Convolutional neural networks for p300 detection with application to brain-computer interfaces. IEEE Trans Pattern Anal Mach Intell 2010;33(3):433−45.

[3] Placidi G, Avola D, Petracca A, Sgallari F, Spezialetti M. Basis for the implementation of an EEG-based single-trial binary brain computer interface through the disgust produced by remembering unpleasant odors. Neurocomputing July 2015;160:308−18.

[4] Urigüen JA, Garcia-Zapirain B. EEG artifact removal—state-of-the-art and guide- lines. J Neural Eng June 2015;12:031001.

[5] Noureddin B, Lawrence PD, Birch GE. Online removal of eye movement and blink EEG artifacts using a high-speed eye tracker. IEEE (Inst Electr Electron Eng) Trans Biomed Eng 2011;59(8):2103−10.

[6] Joyce CA, Gorodnitsky IF, Kutas M. Automatic removal of eye movement and blink artifacts from EEG data using blind component separation. Psychophysiology March 2004;41:313−25.

[7] Lin P-F, Lo M-T, Tsao J, Chang Y-C, Lin C, Ho Y-L. Correlations between the signal complexity of cerebral and cardiac electrical activity: a multiscale entropy analysis. PLoS One 2014;9(2).

[8] Shibasaki H, Rothwell J. EMG-EEG correlation. the international federation of clinical neurophysiology. Electroencephalogr Clin Neurophysiol 1999;52:269−74. Supplement.

[9] Mognon A, Jovicich J, Bruzzone L, Buiatti M. ADJUST: an automatic EEG artifact detector based on the joint use of spatial and temporal features. Psychophysiology February 2011;48:229−40.

[10] Delorme A, Sejnowski T, Makeig S. Enhanced detection of artifacts in EEG data using higher-order statistics and independent component analysis. Neuroimage February 2007;34:1443−9.

[11] Islam MK, Rastegarnia A, Yang Z. Methods for artifact detection and removal from scalp EEG: a review. Neurophysiol Clin Clinic Neurophysiol November 2016;46:287−305.

[12] Mannan MMN, Kamran MA, Jeong MY. Identification and removal of physiolog- ical artifacts from electroencephalogram signals: a review. IEEE Access 2018;6:30630−52.

[13] van den Berg-Lenssen MMC, van Gisbergen JAM, Jervis BW. Comparison of two methods for correcting ocular artefacts in EEGs. Med Biol Eng Comput September 1994;32:501−11.

[14] Minguillon J, Lopez-Gordo MA, Pelayo F. Trends in EEG-BCI for daily-life: requirements for artifact removal. Biomed Signal Proc Control January 2017;31:407−18.

[15] Romero S, Mañanas MA, Barbanoj MJ. A comparative study of automatic techniques for ocular artifact reduction in spontaneous EEG signals based on clinical target vari- ables: a simulation case. Comput Biol Med March 2008;38:348−60.

[16] Sweeney KT, Ward TE, McLoone SF. Artifact removal in physiological signals—practices and pos-sibilities. IEEE Trans Inf Technol Biomed May 2012;16:488–500.

[17] Acharyya A, Jadhav PN, Bono V, Maharatna K, Naik GR. Low-complexity hard- ware design methodology for reliable and automated removal of ocular and muscular artifact from eeg. Comput Methods Progr Biomed 2018;158:123–33.

[18] Huang NE, Shen Z, Long S, Wu M, Shih H, Zheng Q, Tung C, Liu H. He empirical mode decom-position and Hilbert spectrum for nonlinear and nonstationary time series analysis. Proc R Soc A 1971;545:903–95. 1998.

[19] Rehman N, Mandic DP. Multivariate empirical mode decomposition. Proc R Soc A 2010;466(2117): 1291–302.

[20] Chen X, Liu A, Chiang J, Wang ZJ, McKeown MJ, Ward RK. Removing muscle artifacts from EEG data: multichannel or single-channel techniques? IEEE Sensor J 2015;16(7):1986–97.

[21] Jung T-P, Makeig S, Humphries C, Lee T-W, Mckeown M, Iragui V, Sejnowski T. Removing elec-troencephalographic artifacts by blind source separation. Psychophysiology 2000;37(2):163–78.

[22] Radüntz T, Scouten J, Hochmuth O, Meffert B. EEG artifact elimination by extrac- tion of ICA-component features using image processing algorithms. J Neurosci Methods March 2015;243: 84–93.

[23] Zhang D, Yao L, Chen K, Wang S, Chang X, Liu Y. Making sense of spatio-temporal preserving representations for EEG-based human intention recognition. In: IEEE transactions on cybernetics; 2019.

[24] Delorme A, Makeig S. EEGLAB: an open source toolbox for analysis of single-trial EEG dynamics including independent component analysis. J Neurosci Methods March 2004;134:9–21.

[25] Oostenveld R, Fries P, Maris E, Schoffelen J-M. FieldTrip: open source software for advanced analysis of MEG, EEG, and invasive electrophysiological data. Comput Intell Neurosci 2011;2011: 1–9.

[26] Urrestarazu E, Iriarte J, Alegre M, Valencia M, Viteri C, Artieda J. Independent component analysis removing artifacts in ictal recordings. Epilepsia September 2004;45:1071–8.

[27] Vigario R, Sarela J, Jousmiki V, Hamalainen M, Oja E. Independent component approach to the analysis of EEG and MEG recordings. IEEE (Inst Electr Electron Eng) Trans Biomed Eng May 2000; 47:589–93.

[28] Garg P, Davenport E, Murugesan G, Wagner B, Whitlow C, Maldjian J, Montillo A. Using con-volutional neural networks to automatically detect eye-blink artifacts in magne-toencephalography without resorting to electrooculography. In: International conference on medical image computing and computer-assisted intervention. Springer; 2017. p. 374–81.

[29] Croce P, Zappasodi F, Marzetti L, Merla A, Pizzella V, Chiarelli AM. Deep convolutional neural networks for feature-less automatic classification of independent components in multi-channel electrophysiological brain recordings. IEEE (Inst Electr Electron Eng) Trans Biomed Eng 2018;66(8): 2372–80.

[30] Acharya UR, Oh SL, Hagiwara Y, Tan JH, Adeli H. Deep convolutional neural network for the automated detection and diagnosis of seizure using EEG signals. Comput Biol Med 2018;100:270–8.

[31] Pion-Tonachini L, Kreutz-Delgado K, Makeig S. ICLabel: an automated electroencephalographic independent component classifier, dataset, and website. Neuroimage 2019;198:181–97.

[32] Placidi G, Cinque L, Polsinelli M. A fast and scalable framework for automated artifact recognition from EEG signals represented in scalp topographies of independent components. Comput Biol Med 2021;132:104347.

[33] Hyvärinen A, Oja E. Independent component analysis: algorithms and applications. Neural Network 2000;13(4–5):411–30.

[34] Bell AJ, Sejnowski TJ. An information-maximization approach to blind separation and blind deconvolution. Neural Comput 1995;7(6):1129–59.

[35] Hyvarinen A. Fast and robust fixed-point algorithms for independent component analysis. IEEE Trans Neural Network 1999;10(3):626–34.

[36] Belouchrani A, Abed-Meraim K, Cardoso J-F, Moulines E. A blind source separation technique using second-order statistics. IEEE Trans Signal Process 1997;45(2):434–44.

[37] Cardoso J-F, Souloumiac A. Blind beamforming for non-Gaussian signals. IEE Proceed F (Radar and Signal Process) 1993;140:362–70. IET.

[38] Delorme A, Palmer J, Onton J, Oostenveld R, Makeig S. Independent EEG sources are dipolar. PLoS One 2012;7(2):e30135.

[39] Giamberardino PD, Iacoviello D, Placidi G, Polsinelli M, Spezialetti M. A brain computer interface by EEG signals from self-induced emotions. In: European congress on com putational methods in applied sciences and engineering. Springer; 2017. p. 713–21.

[40] Placidi G, Polsinelli M, Spezialetti M, Cinque L, Di Giamberardino P, Iacoviello D. Self-induced emotions as alternative paradigm for driving brain–computer interfaces. In: Computer methods in biomechanics and biomedical engineering: imaging & visualization; 2018.

[41] Akhtar MT, Jung T-P, Makeig S, Cauwenberghs G. Recursive independent compo- nent analysis for online blind source separation. In: 2012 IEEE international symposium on circuits and systems (ISCAS); 2012. p. 2813–6. IEEE.

[42] Hsu S-H, Mullen T, Jung T-P, Cauwenberghs G. Online recursive independent component analysis for real-time source separation of high-density EEG. In: 2014 36th annual inter- national conference of the IEEE engineering in medicine and biology society. IEEE; 2014. p. 3845–8.

[43] Pion-Tonachini L, Hsu S-H, Makeig S, Jung T-P, Cauwenberghs G. Real-time EEG source-mapping toolbox (rest): online ica and source localization. In: 2015 37th annual international conference of the IEEE engineering in medicine and biology society (EMBC); 2015. p. 4114–7.

[44] Hsu S-H, Mullen T, Jung T-P, Cauwenberghs G. Validating online recursive inde- pendent component analysis on EEG data. In: 2015 7th international IEEE/EMBS conference on neural engineering (NER). IEEE; 2015. p. 918–21.

[45] Hsu S-H, Mullen TR, Jung T-P, Cauwenberghs G. Real-time adaptive EEG source separation using online recursive independent component analysis. IEEE Trans Neural Syst Rehabil Eng March 2016; 24:309–19.

[46] Hotelling H. Relations between two sets of variates. In: Breakthroughs in statistics. Springer; 1992. p. 162–90.

[47] Kim T, Eltoft T, Lee T-W. Independent vector analysis: an extension of ica to multivariate com- ponents. In: International conference on independent component analysis and signal separation. Springer; 2006. p. 165–72.

[48] De Clercq W, Vergult A, Vanrumste B, Van Paesschen W, Van Huffel S. Canonical correlation analysis applied to remove muscle artifacts from the electroencephalogram. IEEE (Inst Electr Electron Eng) Trans Biomed Eng 2006;53(12):2583–7.

[49] Chen X, Xu X, Liu A, McKeown MJ, Wang ZJ. The use of multivariate EMD and cca for denoising muscle artifacts from few-channel EEG recordings. IEEE Trans Instrum Meas 2017;67(2):359–70.

[50] Chen X, Peng H, Yu F, Wang K. Independent vector analysis applied to remove muscle artifacts in EEG data. IEEE Trans Instrum Meas 2017;66(7):1770–9.

[51] Chen X, Liu A, Chen Q, Liu Y, Zou L, McKeown MJ. Simultaneous ocular and muscle artifact removal from EEG data by exploiting diverse statistics. Comput Biol Med 2017;88:1–10.

[52] Radüntz T, Scouten J, Hochmuth O, Meffert B. Automated EEG artifact elimination by applying machine learning algorithms to ICA-based features. J Neural Eng August 2017;14:046004.

[53] Noorbasha SK, Sudha GF. Removal of EOG artifacts and separation of different cerebral activity components from single channel EEG—an efficient approach combining ssa— ica with wavelet thresholding for bci applications. Biomed Sig Proc Control 2021;63:102168.

[54] Maddirala AK, Veluvolu KC. Ica with cwt and k-means for eye-blink artifact removal from fewer channel EEG. In: IEEE transactions on neural systems and rehabilitation engineering; 2022.

[55] Phadikar S, Sinha N, Ghosh R. Automatic EEG eyeblink artefact identification and removal technique using independent component analysis in combination with support vector machines and denoising autoencoder. IET Sig Proc 2020;14(6):396—405.

[56] Sai CY, Mokhtar N, Arof H, Cumming P, Iwahashi M. Automated classification and removal of EEG artifacts with SVM and wavelet-ica. IEEE J Biomed Health Inform 2017;22(3):664—70.

[57] Yasoda K, Ponmagal R, Bhuvaneshwari K, Venkatachalam K. Automatic detection and classification of EEG artifacts using fuzzy kernel SVM and wavelet ica (wica). Soft Comput 2020;24(21):16011—9.

[58] Jafari A, Gandhi S, Konuru SH, Hairston WD, Oates T, Mohsenin T. An EEG artifact identification embedded system using ica and multi-instance learning. In: 2017 IEEE international symposium on circuits and systems (ISCAS). IEEE; 2017. p. 1—4.

[59] Jasper HH. The ten-twenty electrode system of the international federation. Electroencephalogr Clin Neurophysiol 1958;10:370—5.

[60] Al-Saffar AAM, Tao H, Talab MA. Review of deep convolution neural network in image classification. In: 2017 international conference on radar, antenna, microwave, electronics, and telecommunications (ICRAMET). IEEE; 2017. p. 26—31.

[61] Song L, Liu J, Qian B, Sun M, Yang K, Sun M, Abbas S. A deep multi-modal CNN for multi-instance multi-label image classification. IEEE Trans Image Process 2018;27(12):6025—38.

[62] Askari E, Setarehdan SK, Sheikhani A, Mohammadi MR, Teshnehlab M. Modeling the connections of brain regions in children with autism using cellular neural networks and electroencephalography analysis. Artif Intell Med 2018;89:40—50.

[63] Tang X, Wang T, Du Y, Dai Y. Motor imagery EEG recognition with KNN-based smooth autoencoder. Artif Intell Med 2019;101:101747.

[64] Gao X, Yan X, Gao P, Gao X, Zhang S. Automatic detection of epileptic seizure based on approximate entropy, recurrence quantification analysis and convolutional neural networks. Artif Intell Med 2020;102:101711.

[65] Lopes F, Leal A, Medeiros J, Pinto MF, Dourado A, Dümpelmann M, Teixeira C. Ensemble deep neural network for automatic classification of EEG independent components. IEEE Trans Neural Syst Rehabil Eng 2022;30:559—68.

[66] Ihle M, Feldwisch-Drentrup H, Teixeira CA, Witon A, Schelter B, Timmer J, Schulze-Bonhage A. EPILEPSIAE—a european epilepsy database. Comput Methods Progr Biomed 2012;106(3):127—38.

[67] Couceiro R, Barbosa R, Duráes J, Duarte G, Castelhano J, Duarte C, Teixeira C, Laranjeiro N, Medeiros J, Carvalho P, et al. Spotting problematic code lines using nonintrusive programmers' biofeedback. In: 2019 IEEE 30th international symposium on software reliability engineering (ISSRE). IEEE; 2019. p. 93—103.

[68] Medeiros J, Couceiro R, Duarte G, Duráes J, Castelhano J, Duarte C, Castelo-Branco M, Madeira H, de Carvalho P, Teixeira C. Can EEG be adopted as a neuroscience reference for assessing software programmers' cognitive load? Sensors 2021;21(7):2338.

[69] Lee SS, Lee K, Kang G. EEG artifact removal by Bayesian deep learning and ica. In: 2020 42nd annual international conference of the IEEE engineering in medicine and biology society (EMBC). IEEE; 2020. p. 932—5.

[70] Koelstra S, Muhl C, Soleymani M, Lee J-S, Yazdani A, Ebrahimi T, Pun T, Nijholt A, Patras I. DEAP: a database for emotion analysis; using physiological signals. IEEE Transac Affec Comp 2011;3(1):18—31.

[71] Pion-Tonachini L, Hsu S-H, Chang C-Y, Jung T-P, Makeig S. Online automatic artifact rejection using the real-time EEG source-mapping toolbox (rest). In: 2018 40th annual international conference of the IEEE engineering in medicine and biology society (EMBC). IEEE; 2018. p. 106−9.

[72] Bigdely-Shamlo N, Kreutz-Delgado K, Kothe C, Makeig S. EyeCatch: data-mining over half a million EEG independent components to construct a fully-automated eye-component detector. In: 2013 35th annual international conference of the IEEE engineering in medicine and biology society (EMBC). IEEE; 2013. p. 5845−8.

[73] Iandola FN, Han S, Moskewicz MW, Ashraf K, Dally WJ, Keutzer K. SqueezeNet: alexnet-level accuracy with 50x fewer parameters and¡ 0.5 mb model size. arXiv preprint arXiv:1602.07360; 2016.

[74] Leske S, Dalal SS. Reducing power line noise in EEG and meg data via spectrum interpolation. Neuroimage 2019;189:763−76.

[75] Selvaraju RR, Cogswell M, Das A, Vedantam R, Parikh D, Batra D. Grad-cam: visual explanations from deep networks via gradient-based localization. In: Proceedings of the IEEE international conference on computer vision; 2017. p. 618−26.

[76] Hsu S-H, Mullen TR, Jung T-P, Cauwenberghs G. Real-time adaptive EEG source separation using online recursive independent component analysis. IEEE Trans Neural Syst Rehabil Eng 2015;24(3):309−19.

[77] Invitto S, Grasso A, Lofrumento DD, Ciccarese V, Paladini A, Paladini P, Marulli R, De Pascalis V, Polsinelli M, Placidi G. Chemosensory event-related potentials and power spectrum could be a possible biomarker in 3m syndrome infants? Brain Sci 2020;10(4):201.

[78] Iacoviello D, Petracca A, Spezialetti M, Placidi G. A real-time classification algorithm for EEG-based bci driven by self-induced emotions. Comput Methods Progr Biomed 2015;122(3):293−303.

[79] Placidi G, Petracca A, Spezialetti M, Iacoviello D. A modular framework for EEG web based binary brain computer interfaces to recover communication abilities in impaired people. J Med Syst 2016; 40(1):1−14.

[80] Lozzi D, Mignosi F, Placidi G, Polsinelli M, Spezialetti M. A 4D LSTM network for emotion recognition from the cross-correlation of the power spectral density of EEG signals. In: The international workshop on affective computing and emotion recognition 2022 (ACER-EMORE2022); 2023.

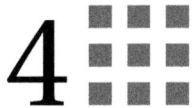

4

Deep multimodal representation learning for noninvasive neural speech decoding

Ciaran Cooney[1], Raffaella Folli[2] and Damien Coyle[1,3]

[1]INTELLIGENT SYSTEMS RESEARCH CENTRE, ULSTER UNIVERSITY, DERRY, UNITED KINGDOM; [2]SCHOOL OF COMMUNICATION AND MEDIA, ULSTER UNIVERSITY, BELFAST, UNITED KINGDOM; [3]THE BATH INSTITUTE FOR THE AUGMENTED HUMAN, UNIVERSITY OF BATH, BATH, UNITED KINGDOM

1. Introduction

A brain-computer interface (BCI) enabling communication by means of imagined speech would have potentially transformative implications for those suffering from neuropathologies such as amyotrophic lateral sclerosis, resulting in locked-in syndrome. However, there are many critical obstacles impeding the development of a functional BCI based on imagined speech. One of these critical obstacles is the discovery of algorithms with sufficient inferential capacity to facilitate the accurate decoding of language from the brain. Research to this point has often focused on heavily engineered features combined with a linear classifier [1,2], but is now trending more toward deep learning methods, which promise better use of big data, larger vocabularies, and multimodal representations of language in the brain.

Natural language processing and computer vision have delivered remarkable progress in recent years by incorporating novel deep learning training techniques and architecture design. Some of the most prominent developments include encoder-decoder architectures [3,4], attention [5], large-scale pretraining [6,7], contrastive training objectives [8], self-supervised learning [9] and multimodal deep learning [10,11]. Several of these techniques have already begun to be integrated into BCI research [12,13], and have recently provided the backbone for research into the intracranial decoding of language from the brain [14—20]. Makin et al. used an RNN encoder-decoder framework with temporal convolution on the input to map neural signals directly to word sequences [18]. At each time step, the encoded ECoG data is decoded as a word, demonstrating word error rates as low as 7%. Wilson et al. also used an RNN decoder to predict phonemes from neural data acquired as participants spoke English words [20]. Slow-varying feature drifts were added across time bins to help the RNN be more robust to nonstationarities, and the deep learning approach outperformed a linear decoder (33.9% vs. 29.4%; chance = 6%). Using an artificial neural network (ANN) for word classification, Moses et al. (2021) reported top-1 accuracy of 74.4% when decoding a vocabulary of 50 words from a patient with anarthria [19]. These studies have been able to utilize the relatively high signal-to-noise ratio (SNR) of intracranial recordings when applying deep learning to neural language decoding. However, intracranial electrodes require surgery as well as maintenance over extended time periods. Despite noisier signals, non-invasive recording techniques such as EEG and fNIRS offer potentially wearable recording setups to the entire population and easier recording of data at scale. The advantages of non-invasive recording techniques, along with the increasing power of deep learning methods, are leading to greater assimilation of these two areas in neural speech decoding [21–27].

Défossez et al. (2022) [22] apply modern deep learning methods with a data-driven approach to decode speech from M/EEG in an active listening task. Using the contrastive objective from CLIP [8] to guide latent representation learning through *incorrect* as well as *correct* samples and a pretraining scheme based on learned representations of speech, a model is trained to predict speech from thousands of possible segments. For MEG, the model achieves 72.5% top-10 accuracy over 1594 distinct segments and can be considered *zero-shot* in that it can predict audio segments that are not present in the training data. The adaptability of neural networks for decoding speech from non-invasive neural signals has also been demonstrated by the generalizability of these models across distinct datasets [25]. In another active listening task, Thornton et al. (2022) showed that deep neural networks performed significantly better than baseline linear models in reconstructing speech from the EEG. Both a convolutional neural network (CNN) inspired by EEGNet [12] with batch normalization, spatial dropout, and average pooling and a fully connected neural network significantly outperformed their linear analogs in the majority of subject-specific experiments. Pretraining, a common technique in deep learning methods for sharing information across tasks, has been shown to have potential for EEG-based decoding of language [21]. Pretraining on averaged EEG data has been shown to boost the ability of an encoder-based transformer to decode part-of-speech tags (word frequency, length, and class) from brain activity [21]. Statistical tests indicated that the best pretraining strategy was significantly better than single-trial training ($P < .01$).

In this chapter, we continue the development of deep learning approaches for non-invasive neural speech decoding by presenting a multimodal architecture trained to learn representations of speech corresponding to EEG and fNIRS. The design is based upon the collaboration of two sub-networks (subnets), which are fused to create a joint representation of the neural speech signal. In effect, each subnet is a data-specific CNN, which acts as a feature extractor for the different modalities. The approach is validated with EEG and fNIRS data simultaneously recorded as subjects produced overt and imagined speech. Given 800 ms segments from a 2-s trial period, the proposed approach outperformed comparable unimodal approaches for both overt and imagined speech.

Results were statistically significant for all except a unimodal network trained on overt speech EEG ($P = .098$).

2. Multimodal decoding of overt and imagined speech

2.1   Study population

The study population is composed of 19 native English speakers (9 male; mean age 26.63 ± 2.13). All had normal or corrected-to-normal vision and reported no history of neurological disorders. All subjects had been scheduled to complete two overt and two imagined speech sessions. Unfortunately, due to COVID-19 restrictions, experiments were not fully completed. Eight subjects completed all four planned sessions; five completed three; two completed two; and four completed one session. All subjects completed at least one overt speech session, and 15 completed at least one imagined speech session. Ulster University's Research Ethics Committee approved the study design, and all subjects signed informed consent forms.

2.2   Experimental protocol

The potential benefits of incorporating aspects of linguistics research and different techniques for presenting stimuli in speech decoding studies have been extensively highlighted in the literature [28–30]. Designing experimental procedures necessarily involves the interaction of several different factors, e.g., type of speech, neural recording apparatus, and stimulus presentation modality [30]. The experimental paradigm reported here is partly motivated by a desire to further our current understanding of how these factors impact neural speech decoding. The protocol uses three distinct modalities (text, audio, and image) to prompt subjects to begin speaking overt or imagined speech (Fig. 4.1A and B). Additionally, two linguistic categories are used: verbs versus modified noun phrases, in particular *action words* (verbs) and *combinations* (modified noun phrases) (Fig. 4.1C). Action words facilitate the investigation of the effects of embodiment on speech decoding, and combinations enable the study of the impact of syntactic modification.

Trials begin at -500 ms with the presentation of a fixation cross. At 0 s, one of three modalities is used to prompt subjects which word(s) to begin speaking. Text and image stimuli are presented on-screen for 1 second, followed by a blank gray screen for a further 1 second (Fig. 4.1A). Audio clips began playing at 0 seconds and were accompanied by a recognizable symbol on-screen for 1 second. The period from 0 to 2 s is the task execution period. A rest period is implemented post-task, with the duration randomized between 1.5 and 2.5 seconds. Participants were instructed to begin speaking immediately upon perceiving the word(s) associated with stimuli and to say a word or pair of words only once during each trial. Each combination of stimulus and word was presented to subjects 50 times each.

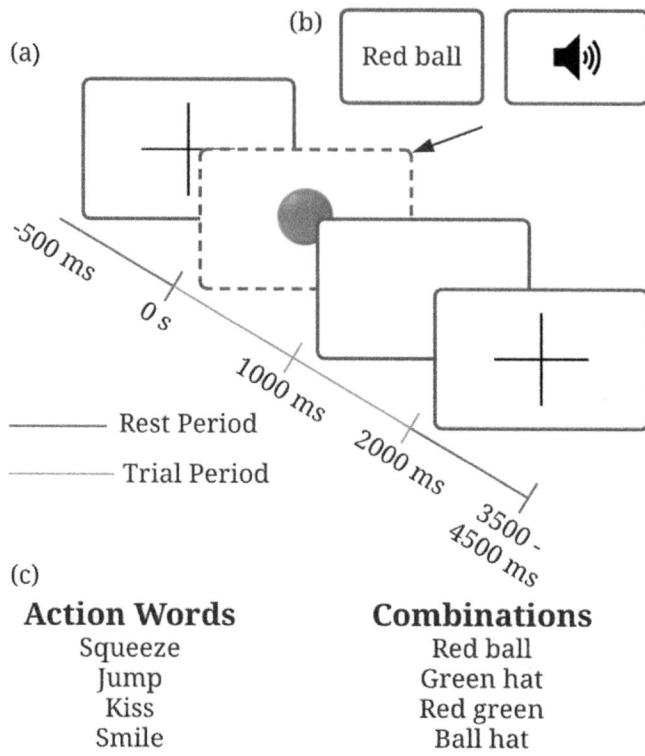

FIGURE 4.1 (A) Trial periods began at time −500 s, with a fixation cross presented for 500 ms. Stimuli were presented on-screen for 1 second, followed by a blank screen for 1 second. This 2-s period (*green*) was the trial period for experiments. (B) Three types of stimuli were used to present words: *Text, image,* and *audio.* (C) Words used for experiments are broadly categorized as action words and combinations. *Reprinted from Ref. [31] with permission from the authors.*

Sessions were split into 6 blocks with 2 runs each per block and 100 trials per run; therefore, there were 1200 trials per session.

2.3   Data acquisition and signal processing

EEG and fNIRS data were recorded using the g.Nautilus fNIRS-8 (g.tech medical engineering GmbH Austria), a fully integrated EEG and fNIRS recording device. Synchronous signal recording is achieved using the MATLAB-Simulink platform with bespoke Simulink blocks for EEG and fNIRS. A 64-channel EEG montage was configured using g.SCARABEO active wet electrodes. Electrodes were positioned according to the unified standard montage10-5 system to enable even distribution across scalp locations.

EEG data were sampled at 250 Hz with a 0.1 Hz high-pass filter used to remove slow drifts during recordings and a 48−52 Hz notch filter used to remove 50 Hz line noise. fNIRS data were recorded at 10 Hz and upsampled to 250 Hz during acquisition. Data were acquired using 8 LED-based transmitters, each of which emits light at wavelengths of 760 and 850 nm. Two receivers, each associated with 4 transmitter channels, produce 2×4 fNIRS channels. Each fNIRS channel recorded optical densities at both wavelengths. The g.Nautilus fNIRS-8 facilitates online conversion of optical densities into concentration changes of HbO and deoxyhemoglobin (HbR) using the Modified Beer–Lambert law [32,33]:

$$A(t;\lambda) = ln\frac{I_{in}(\lambda)}{I_{out}(t;\lambda)} = \alpha(\lambda) \times c(\lambda) \times d(\lambda) + \eta,$$ (4.1)

$$\begin{bmatrix} \Delta c_{HbO}(t) \\ \Delta c_{HbR}(t) \end{bmatrix} = \begin{bmatrix} \alpha_{HbO}(\lambda_1) & \alpha_{HbR}(\lambda_1) \\ \alpha_{HbO}(\lambda_2) & \alpha_{HbR}(\lambda_2) \end{bmatrix}^{-1} \times \begin{bmatrix} \Delta A(t;\lambda_1) \\ \Delta A(t;\lambda_2) \end{bmatrix} \frac{1}{l \times d(\lambda)}$$ (4.2)

where A is the optical density, t is time in seconds, λ_1 and λ_2 are the stated wavelengths, I_{in} is the incident intensity of light, I_{out} is the detected intensity of light, α is the extinction coefficient in $\mu M^{-1}cm^{-1}$, c is the absorber concentration in micromolars, l is the distance between source and detector optodes in centimeters, d is the differential path-length factor (6), and η is the loss of light due to scattering. The incident intensity of light is the initial intensity of light emitted from the g.Nautilus fNIRS-8 is a property of the device. Receiver optodes were positioned at C3 and C4, with each transmitter positioned 30 mm from the receivers.

EEG signal processing was undertaken with EEGLAB [34] in MATLAB 2017a (Mathworks, Natick, MA, USA). Channel rejection was applied following a visual inspection of the raw EEG. A Hamming windowed finite impulse response filter was used to bandpass EEG between 0.5 and 40 Hz, and all signals were rereferenced using common average referencing. Pre-trial EEGs (−500 ms–0 seconds) were averaged and used to apply baseline removal from the task-production period. Trials containing muscular artifacts were rejected by visual inspection. Independent component analysis was performed on the remaining preprocessed channels using the infomax algorithm to remove artifacts [35]. Components were visually inspected and removed if the frontal distribution of weights indicated ocular artifacts. Between one and three components were removed per session.

fNIRS data were processed in FieldTrip [36]. Due to poor fNIRS signal quality during setup (S5-Session 1, S6-Session 1 (Overt); S13-Session 1 (Imagined)) or signal dropout during experiments (S2-Session 2, S3-Session 1 (Overt); S2-Session 1 (Imagined)), several sessions were not used for further analysis. Channels with poor signal quality due to inadequate contact were eliminated from further analysis following visual inspection. Signals were bandpass filtered from 0.1 to 0.8 Hz to reduce artifacts from physiological signals. Data were epoched into periods of −500 ms–3.5 s (longer than EEG to account for slower fNIRS time courses) and baseline corrected. Trial rejection due to movement artifacts was applied through visual inspection.

The 2-s task execution period (Fig. 4.1A) was used for speech decoding. A temporal offset was applied to fNIRS for all classification tasks to account for the different time course in comparison with EEG. A recent study reported that peak correlation between

EEG and fNIRS signals occurred when the fNIRS lagged the EEG signal by approximately 1.7 seconds during a 3.5 seconds trial period [37]. Due to a relatively short task execution period (2 s), we applied a 1.5 seconds offset to fNIRS data, i.e., a 0–800 ms classification window corresponded to fNIRS data recorded 1.5–2.3 s postcue onset. As we applied trial-rejection to EEG and fNIRS independently, trial samples for multimodal classification were aligned by rejecting all independently rejected trials from both data types prior to training. Finally, the data were split into six different 4-class decoding tasks, facilitated by the experimental design. These were: *action-text*, *action-image*, *action-audio*, *combinations-text*, *combinations-image*, and *combinations-audio*.

2.4   Multimodal neural network

The multimodal neural network architecture presented here consists of two modality-specific subnets and an overarching network structure in which they are contained (Fig. 4.3). The subnets are two identical CNNs designed to learn feature representations from the two input data types. They are constructed with spatial and temporal convolutions, batch normalization, dropout, non-linear activation, and a fully connected layer output. The theoretical concepts underpinning our design are discussed in Sections 2.4.12.4.3 before the design itself is presented in Section 2.4.4.

2.4.1   *Convolution*

CNNs [38] are a specific type of deep learning architecture that specialize in processing data with a known grid-like topology [39]. For example, images, the most prominent data type associated with CNNs, can be considered a 2-D grid of pixels. Similarly, EEG may be thought of as a 2-D grid consisting of *channels* × *samples* in time. They are fundamentally different from standard ANNs in that CNNs learn local patterns within the data, whereas ANNs learn global patterns in the input feature space [40].

As the name indicates, CNNs substitute standard matrix multiplication for a mathematical operation known as *convolution* [39]. The reason for employing convolution rather than the *affine transformations* of a standard neural network is that it facilitates the exploitation of the intrinsic properties of a given data type. For example, EEG signals have spatial and temporal characteristics associated with channel locations and time samples. In a standard neural network, an input vector is multiplied by a matrix, producing an output (to which a bias is typically added) [41]. With this technique, all axes are considered equally, and therefore the transformation does not consider the implicit structure of the data, and thus, for certain data types, any topological information is lost. A discrete convolution is a *linear transformation* that preserves the inherent structure of the data. When applying convolution across two-dimensional input data (I), such as multichannel EEG or 2-D images, a two-dimensional kernel (K) is typically used, with the resulting formula:

$$S(i,j) = (K * I)(i,j) = \sum_m \sum_n I(i-m, j-n)K(m,n) \qquad (4.3)$$

where S is the 2-D output feature map. Convolution is performed in a CNN when a *kernel* slides across a feature map (this may be input data or the output feature map from a previous layer), computing the sum of the products of each element of the kernel and its corresponding input element to produce an output for each location [41]. The input to a convolutional layer is typically a multidimensional array (or tensor) of values, and the kernel is a multidimensional array of parameter values (or weights), which are adapted during training by a learning algorithm [39]. The basic architecture of a CNN consists of several stages. The first of these is the convolutional operation described in this section. Following this, the secondary and tertiary stages of the typical CNN procedure receive the linear activations returned from the convolution stage and apply them to a non-linear activation function and a pooling function [39]. Additional elements common to CNN structures are the regularization techniques of batch normalization [42] and dropout [43]. Each of these additional CNN components is described in the following sections:

2.4.2   Non-linear activation

Activation functions such as rectified linear units (ReLU) add non-linearity to a neural network so that it is not restricted to learning only linear transformations of the input data. The inclusion of non-linearity facilitates the training of deep neural networks by extending the possible representation space for features [40]. ReLU is a common activation function in various important CNN architectures, including AlexNet [44] and ResNet [45]. More recently, a modified version of ReLU, known as leaky ReLU, has become a more widely used approach. As ReLU sets the gradient for all inputs below zero to zero, the modification addresses the potential for neurons to be deactivated by instead assigning to the activation a very small non-negative value.

2.4.3   Batch normalization and dropout

Batch normalization [42] and dropout [43,46] are the two most common regularization algorithms applied to CNNs. Many of the most prominent CNN architectures have implemented these techniques [45,47,48], including those dedicated to EEG analysis and decoding [12,13].

Batch normalization is a method developed to combat a model's *internal covariant shift*, i.e., the fact that the distribution of a layer's input changes during training in response to changing parameters in previous layers [42]. The problem of internal covariant shift is addressed by normalizing the inputs of each layer during training for each mini-batch of data. This is achieved by computing an exponential moving average of the mean and variance of the mini-batch data during training [40]. One of the main effects of batch normalization is the improvement it gives to gradient propagation, an effect that also facilitates the training of deeper architectures.

Dropout is a technique explicitly designed to reduce overfitting (particularly when training with small datasets) and enhance a deep network's ability to generalize well to new data [43,46]. The basic premise of this approach is a simple one: to randomly drop

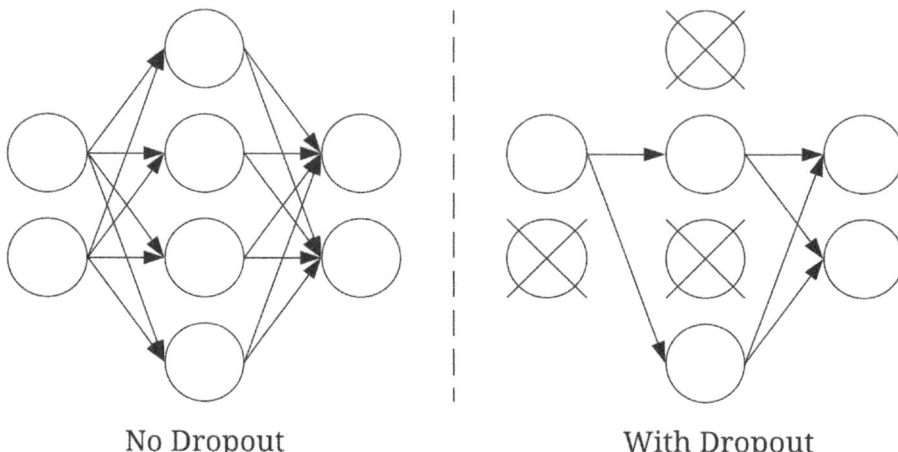

No Dropout   With Dropout

FIGURE 4.2 Visual depiction of dropout applied to neural networks. *Left* — All nodes are fully connected and therefore have the capacity to co-adapt through each pass of the network. *Right* — A random subset of nodes is zeroed for a single pass there for ensuing that they cannot be updated during that pass.

out a number of hidden units and their connections from a given layer during training [46] (Fig. 4.2). As a result of this dropping of a random sample of units, they do not contribute to downstream activations in the network or the updating of weights during backpropagation for a given pass of the network. This method prevents potential *co-adaptation* of features, an effect of features only being useful in the presence of other specific features, making the network much less beholden to the specific weights of individual neurons [43]. A *dropout rate* is a value between 0 and 1 selected by the model architect (or through hyperparameter (HP) optimization), which indicates the fraction of hidden units to be randomly dropped at each layer during a training epoch.

The application of dropout is depicted in Fig. 4.2. In Fig. 4.2A, the network uses all neurons and the connections between neurons to influence the final output of the network and consequently, its loss value. In Fig. 4.2B, when dropout is implemented, a fraction of neurons no longer has any influence on the network for a given pass, and thus only the weights of active neurons are updated during this epoch. As well as the excellent regularization properties that dropout exhibits, a secondary advantage of the method is that it is extremely computationally inexpensive [49].

2.4.4 Multimodal neural network implementation

Implementation of the CNN architecture for our two subnets begins with filters in the first layer being convolved with the input data along the time dimension (N filters = 40; filter size = 1×5). This is followed by spatial filtering, in which output weights from the first layer are filtered with weights for all possible pairs of electrodes. Batch normalization, an activation function, and dropout ($P = .1$) follow. To avoid diminishing spatial information in the data, no pooling operations were used. During HP optimization, an extension of this design with convolution, batch normalization, activation function, and

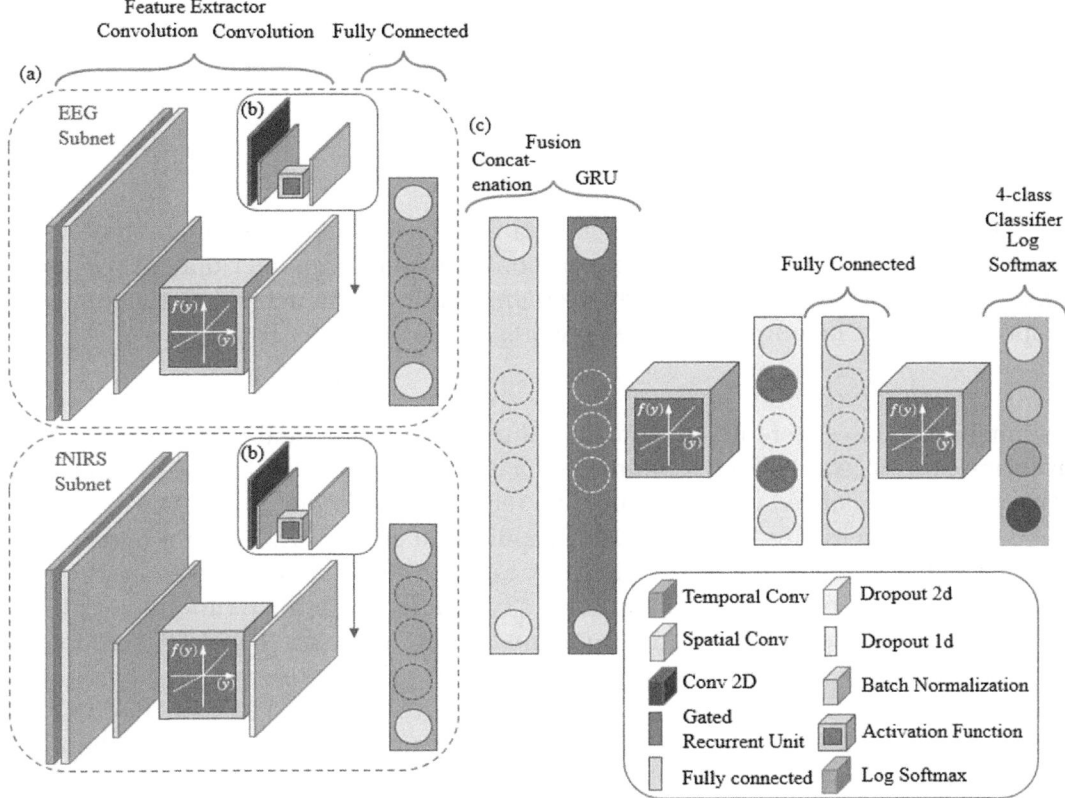

FIGURE 4.3 Bimodal network for training with EEG and fNIRS. (A) Two identical CNNs (EEG and fNIRS subnets) form a dual feature extractor. The CNNs' initial layers consist of combined temporal and spatial convolution. Batch normalization is then applied, followed by one of two possible activation functions (ELU or Leaky ReLU). This is followed by dropout ($P = .1$). The final layer of the subnets is an FC layer with 500 hidden units. (B) Parameters used to extend the depth of the CNN during HP optimization. This consists of 2D convolution, batch normalization, activation function, and dropout. (C) *Fusion* and classification layers. Subnet outputs are concatenated in *late fusion* and fed to a GRU. This is followed by the activation function, dropout, FC layer, and another activation function. The final layer is a log softmax classifier used here for 4-class classification. *Reprinted from Ref. [31] with permission from the authors.*

dropout layers (Fig. 4.3B) was evaluated. The final layer of each subnet is a fully connected layer with 500 hidden units.

The modality-specific outputs from the subnets are combined through *late fusion* [10]. This is where the model is trained to learn joint representations of the neural speech signal from two data modalities. The remainder of the model propagates through a gated-recurrent unit layer, activation function, dropout, a fully connected layer, and a final activation function. The multimodal design uses a log softmax classifier to classify overt and imagined speech segments:

$$p(l_k|f(X^j;\theta)) = \log\left(\frac{exp(f_k(X^j;\theta))}{\sum_{m=1}^{k} exp(f_k(X^j;\theta))}\right)$$ (4.4)

where X_j is the input, l_k is a class label, k is the number of classes, and θ are the parameters of the function.

2.4.5 Unimodal neural network

Baseline comparison is provided by unimodal EEG and fNIRS CNNs similar in design to the subnets. Temporal and spatial convolutions with the same filter dimensions as the subnets are followed by batch normalization, a non-linear activation function, and dropout ($P = .1$). The output of these networks is a log softmax classifier.

2.4.6 Training procedure

Training procedures were identical for multimodal and unimodal approaches. SMOTE oversampling [50] was applied to training data in cases where trials had been removed or classes were imbalanced due to incomplete recordings. Temporal and spatial convolution layers were initialized using the Xavier uniform method [51]. Uniform initialization was used in later layers due to its effectiveness with ReLU-based activation functions [52]. Training was optimized using Adam [53], a popular approach to gradient-based optimization of stochastic objective functions. Training was allowed to continue for up to 50 epochs, with an early stopping strategy (*patience* = 20) controlled by the validation loss. Training resumed with parameter values re-initialized to those that resulted in the best validation accuracy. Training was terminated when the validation loss dropped to the same value as the training loss achieved at the end of the first training phase [13]. Batch size was 32, the initial learning rate was 0.001, and the learning rate was decayed with multistep scheduling ($\gamma = 0.1$) applied at epochs 20 and 25. Categorical probability distributions were obtained by transforming the output from the final layer using the log softmax function. A negative log-likelihood loss was used.

2.4.7 Hyperparameter optimization

A nested cross-validation strategy (nCV) was used to optimize HPs during training. The approach uses outer and inner folds, with the data split into $k = 5$ for each fold. For a 200-sample classification task, the data were split into train/validation/test sets of 128/ 32/40. HPs are optimized during the inner training loop of the nCV procedure. Maximum inner-fold validation accuracy was used to select optimal HP values. Model performance was evaluated in the outer training loop with tuned HPs using validation accuracy.

The HPs selected for optimization were frequency band, classification window, number of network layers, and activation function. Frequency bands were optimized for both EEG and fNIRS: five bands for EEG (delta (0.5–4 Hz), theta (4–8 Hz), alpha (8–12 Hz), beta (12–28 Hz), and gamma (28–40 Hz)) and four for fNIRS (0.1–0.5 Hz, 0.2–0.6 Hz, 0.3–0.7 Hz, and 0.4–0.8 Hz). EEG bands were filtered during nCV using a fifth-order Butterworth filter and fNIRS using a second-order Butterworth filter. Three overlapping 800 ms classification windows (0–800 ms, 600–1400 ms, and 1200–2000 ms postcue-onset) were evaluated to determine optimal classification periods. Two non-linear activation functions were considered in the HP sweep: Exponential linear units (ELU) and leaky ReLU. ELU is defined as $f(x) = x$ for $x \geq 0$ and $f(x) = e^x - 1$ for $x < 0$ and leaky ReLU as $f(x) = x$ for $x \geq 0$ and $f(x) = \alpha x$ for $x < 0$, where α defines the extent to which the function "leaks," i.e., the slope of the function for $x < 0$. The depths of the networks were optimized by extending the original architectures (Fig. 4.3A) with additional layers (Fig. 4.3B). This consisted of an additional convolution layer, batch normalization, activation function, and dropout.

For multimodal training, HPs optimized with nCV were coupled across subnets, i.e., for each HP value, the entire multimodal network was instantiated with that value. For example, at no point was one subnet using ELU when the other was using Leaky ReLU as its activation function. EEG and fNIRS frequency bands were not coupled in this way, as each data type was associated with a single subnet.

3. Results

3.1 Multimodal learning outperforms unimodal

Multimodal decoding resulted in higher accuracies than either unimodal approach, with statistically significant improvement for all but overt speech EEG (Tables 4.1 and 4.2). For overt speech, the multimodal approach achieved the following scores for each task: AT = 49.61%, AI = 48.72%, AA = 45.02%, CT = 49.20%, CI = 46.76%, and CA = 38.52% (Table 4.1; Fig. 4.4), thus outperforming unimodal EEG decoding in all overt speech tasks. For imagined speech, the multimodal network also outperformed unimodal EEG in all tasks, with mean decoding accuracies of AT = 31.78%, AI = 38.37%, AA = 33.89%, CT = 32.21%, CI = 36.67%, and CA = 32.80%, respectively (Table 4.2; Fig. 4.4).

On average, multimodal decoding improved on unimodal EEG by 2.48% (overt) and 1.59% (imagined). This result suggests that combining modalities within a deep learning architecture can enhance neural speech decoding. A 2-way ANOVA *network* × *classification task* indicated differences between the two methods were significant for imagined speech ($F(1, 5) = 5.45$, $P = .0203$) while tending toward significance for overt speech

Table 4.1 Multimodal comparison with unimodal for overt speech.

Word-type	Action words			Combinations		
	Text	Image	Audio	Text	Image	Audio
Bimodal	49.61	48.72	45.02	49.20	46.76	38.52
EEG	46.04	46.66	41.55	46.90	45.08	36.72
fNIRS	32.46	32.46	33.66	31.73	31.91	33.49

Table 4.2   Multimodal comparison with unimodal for imagined speech.

Word-type	Action words			Combinations		
	Text	Image	Audio	Text	Image	Audio
Bimodal	31.78	38.37	33.89	32.21	36.67	32.80
EEG	30.25	36.21	32.11	31.55	34.56	31.60
fNIRS	30.62	28.61	29.72	31.32	28.95	28.64

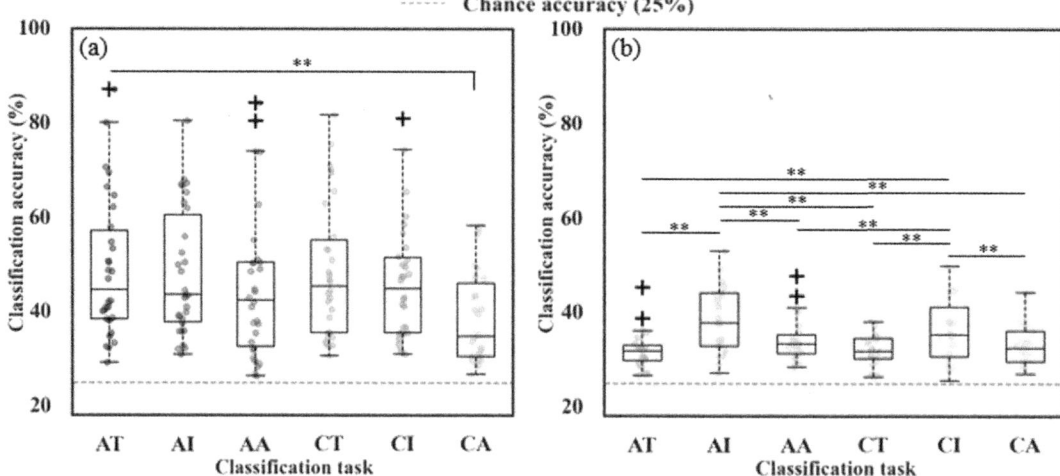

FIGURE 4.4 Classification results across all classification tasks for both overt and imagined speech. Each data point corresponds to classification accuracy for one of six conditions and for a single session (participants engaged in one or two sessions each). Boxplots visualize the distribution of results, indicating the median value (the point at which 50% of results are above and below), the interquartile range (*box heights*), and 1.5 times the interquartile range (whiskers extending beyond *box edges*). (A) Variability in performance across subjects and sessions for each classification task for overt speech. (B) Variability in performance across subjects and sessions for each classification task for imagined speech. **$P < .005$. *Reprinted from Ref. [31] with permission from the authors.*

($F(1, 5) = 2.75$, $P = .098$). 16 of the 21 imagined speech sessions resulted in improved classification using the multimodal approach. Despite a P-value $>.05$, 21 of the 28 overt speech sessions were also improved upon with multimodal decoding. Analysis of results suggested that negative transfer associated with several participants' fNIRS data reduced the absolute effectiveness of the multimodal network. For example, the overt speech scores for Subject 14 (Session 2) achieved a mean accuracy of 41.79% with EEG but dropped to 32.77% with hybrid decoding. In addition, there were instances of fNIRS data being classified at or below the chance level (25%), indicating that a small portion of the fNIRS data was not likely to benefit multimodal decoding. Reasons for this negative transfer are suggested by comparison with the fNIRS decoding below.

Comparison between the multimodal approach and unimodal fNIRS demonstrated that multimodal was significantly better for both overt and imagined speech (overt: $F(1, 5) = 131.13$, $P < .001$; imagined: $F(1, 5) = 69.11$, $P < .001$). The mean fNIRS decoding accuracy for overt speech was *AT*:

32.46%, *AI*: 32.46%, *AA*: 33.66%, *CT*: 31.73%, *CI*: 31.91%, *CA*: 33.49% (Table 4.1). fNIRS results for imagined speech were *AT*: 30.62%, *AI*: 28.61%, *AA*: 29.72%, *CT*: 31.32%, *CI*: 28.95%, and *CA*: 28.64% (Table 4.2). This result is likely due to the overall effectiveness of the fNIRS signal due to experimental constraints, rather than a property of the neural network itself. With a 2s task-production period, the fNIRS signal does not exhibit the typical time course associated with longer trial periods. Instead, the HbO signal only begins its expected rise associated with task production at approximately 2–2.5 seconds post cues. As stated in Section 2.3, task-related fNIRS is usually expressed over longer periods. However, the relatively short task period required to investigate the different stimuli and word groups means that there are potential performance gains to be made from a longer fNIRS period.

3.2   Decoding performance of the multimodal approach

Scattered boxplots visualizing the accuracies obtained from the multimodal neural network for overt and imagined speech are shown in Fig. 4.4. Two results are clear from these plots: (1) The multimodal network classifies overt and imagined speech with accuracy substantially greater than the chance level while exhibiting significant variance between classification tasks. (2) There is a clear performance gap between the two speech types, with overt speech resulting in significantly better decoding accuracy ($F(1, 5) = 3.06$, $P < .05$; 2-way ANOVA). The mean decoding accuracy across all tasks was 46.31% for overt speech and 34.29% for imagined speech, resulting in a 12.02% difference. Overt speech achieved a maximum decoding performance of 87.18% for *AT,* and imagined achieved a best score of 53% for *AI*.

3.3   Stimulus effects greater than word-type

Results exhibited variation in decoding performance depending on the type of stimuli used to prompt tasks. Additionally, these trends were not consistent across speech types. A 2-way ANOVA *stimulus* × *word-type* indicated that the main effects of different stimuli were significant ($F(2,162) = 4.59$, $P < .05$) but that the effects of different word types were not ($F(1,162) = 1.87$, $P = .174$). *Post hoc* tests attributed significance to the inferior scores obtained from audio trials (AA, CA) ($P < .05$), with differences between text and images negligible ($P = .80$).

A 2-way ANOVA *stimulus* × *word-type* indicated that the main effects of stimuli were highly significant for imagined speech ($F(2,120) = 12.27$, $P = 1.42 \times 10^{-5}$), although the main effect of words was not ($F(1,120) = 1.22$, $P = .272$). *Post hoc* tests revealed that superior accuracies obtained from trials using image stimuli were significant with respect to both text ($P = 1.44 \times 10^{-5}$) and audio ($P < .005$) trials. Comparison of text v audio revealed no significance ($P = .312$).

4. Discussion

In this chapter, we present a multimodal deep neural network for decoding speech from brain signals. At present, multimodal deep learning is not a common approach to neural speech decoding or other BCI applications. The primary reason for this is the simple fact that, in comparison with unimodal approaches, the quantity of available data is much lower and the relative difficulty in acquiring such data is much higher. Nevertheless, multimodal BCI research is growing [54–60], and it is therefore important that decoding approaches modeled on deep learning techniques are developed. Our approach uses two convolutional subnets that are trained to extract modality-specific features before late fusion is used to force the network to learn a joint representation of the neural speech signal. The network then learns to classify combined EEG and fNIRS signals as words or phrases spoken in overt or imagined speech.

Our multimodal approach proved more successful than unimodal baselines at decoding speech from brain activity. This result was statistically significant for three of the four tasks. This is an important initial validation for this type of network design and indicates that there is potential for multimodal approaches to speech decoding. In particular, the approach taken was conceived as a method for decoding non-invasively acquired neural signals and vindicates our contention that novel deep learning methods need not be confined to signals acquired with implanted electrodes.

However, performance improvements were relatively minor, and further research and development are required to fully optimize the decoding potential of the network. One such approach may be the development of a distinct subnet for fNIRS decoding. Our subnets were based on the *shallow* CNN designed specifically for raw EEG signals and were likely not ideal for decoding fNIRS [13]. Related to this is the time course of the fNIRS signals and the fact that we were unable to utilize the entire task-dependent response in our decoding pipeline (Fig. 4.5). Longer trial periods associated with overt and imagined speech would enable a better objective measure of how effective combining EEG and fNIRS can be for decoding.

Other design decisions, such as the omission of pooling layers and the positioning of fusion within the network, are also subject to revision should experiments indicate that these are aspects that can improve decoding performance. The training objective also warrants further consideration following developments in the utilization of contrastive learning objectives for deep learning tasks. Training schemes such as CLIP have demonstrated significant performance improvements when applying negative sampling with a contrastive objective [8], and this has recently been applied to neural speech decoding with impressive results [22]. Large-scale pretraining and downstream fine-tuning have demonstrated tremendous results in other deep learning domains. These approaches use generic training objectives to optimize models using large-scale datasets before they are fine-tuned for a variety of related tasks. This facilitates the transfer of knowledge from one task to many others, while also facilitating task-specific training. Few studies to date have evaluated pretraining as a potential enhancement to current

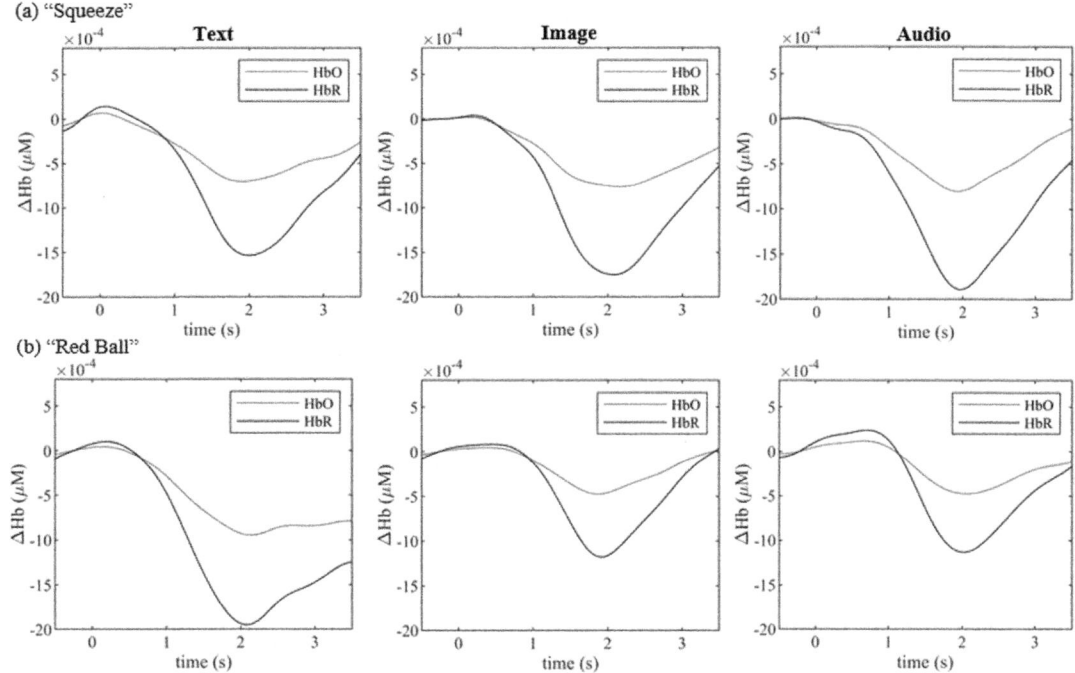

FIGURE 4.5 HbO and HbR signal time courses for the period −0.5 to 3.5 seconds about cue onset taken from subject one for the words "squeeze" and *"red ball"* for all stimulus methods. fNIRS data is not fully utilized here as the timing constraints of the experiments meant that the complete rise and fall of a typical HbO signal associated with task production was not possible. *Reprinted from Ref. [31] with permission from the authors.*

deep learning methods for BCI beyond using cross-subject training paradigms [26,61−63]. Reasons for this include the low SNR associated with the various data acquisition protocols, difficulty obtaining high volumes of relevant data, and high variation in task-related brain activity between BCI users. However, large-scale pre-training may be one way to unlock the potential benefits of deep learning for speech decoding. Défossez et al. (2022) introduced a "subject layer" consisting of a 1 x 1 convolution as a way to enforce subject-specific learning while also leveraging knowledge from pretraining on over 150 subjects [22].

Beyond the multimodal architecture itself, it would be pertinent to investigate the degree to which models trained on overt speech can transfer knowledge to the imagined speech task. If decoders developed for overt speech exhibit strong transferability to imagined speech decoding, the possibility of sequence-to-sequence models for imagined speech decoding would likely offer significant potential. These models are common in overt speech decoding tasks or passive listening tasks that facilitate time-locking and labeling of discrete segments within a sequence [16,18]. This is something that is extremely difficult to do to a high standard with imagined speech, but would allow for a more naturalistic communication style.

The field of deep learning is rapidly evolving, offering opportunities to transfer new techniques to other domains. It is our expectation that deep learning techniques will continue to form a major part of research into neural speech decoding and that the work presented in this chapter is a significant contribution to this research.

5. Conclusion

In conclusion, we present a multimodal neural network architecture designed to enable neural speech decoding from non-invasively acquired signals. Motivated by advances in deep learning methods and impressive enhancements in speech decoding using invasive recording techniques, our network demonstrates the feasibility of applying sophisticated decoding approaches to noisy, non-invasive signals. Features of the architecture include data-specific subnets, which include spatial and temporal convolutions designed for decoding brain activity, and feature fusion, which enables the learning of joint representations of neural speech signals. Our design is validated using multimodal EEG and fNIRS data recorded from subjects as they performed overt and imagined speech tasks. Results indicated that the multimodal approach significantly outperformed unimodal baselines in three out of four tasks, despite evidence of negative transfer for some subjects' data. This work demonstrates the potential for deep learning techniques to be successfully applied to non-invasive neural speech decoding.

References

[1] Nguyen CH, Karavas G, Artemiadis P. Inferring imagined speech using EEG signals: a new approach using Riemannian Manifold features. J Neural Eng 2017;15:016002. https://doi.org/10.1088/1741-2552/aa8235.

[2] Cooney C, Folli R, Coyle D. Mel frequency cepstral coefficients enhance imagined speech decoding accuracy from EEG. In: 29th Irish signals syst. Conf. ISSC 2018, institute of electrical and electronics engineers inc; 2018. https://doi.org/10.1109/ISSC.2018.8585291.

[3] Sutskever I, Vinyals O, Le QV. Sequence to sequence learning with neural networks. Adv Neural Inf Process Syst 2014;4:3104−12.

[4] Cho K, van Merrienboer B, Gülçehre Ç, Bahdanau D, Bougares F, Schwenk H, et al. Learning phrase representations using RNN encoder-decoder for statistical machine translation. EMNLP; 2014.

[5] Vaswani A, Shazeer N, Parmar N, Uszkoreit J, Jones L, Gomez AN, et al. Attention is all you need. Adv Neural Inf Process Syst 2017:5999−6009. 2017-Decem.

[6] Devlin J, Chang MW, Lee K, Toutanova K. BERT: pre-training of deep bidirectional transformers for language understanding. In: Naacl hlt 2019 - 2019 conf north Am chapter assoc comput linguist hum lang technol - proc conf, vol. 1; 2019. p. 4171−86.

[7] Brown TB, Mann B, Ryder N, Subbiah M, Kaplan J, Dhariwal P, et al. Language models are few-shot learners. In: Advances in neural information processing systems; 2020. 2020-Decem.

[8] Radford A, Kim JW, Hallacy C, Ramesh A, Goh G, Agarwal S, et al. Learning transferable visual models from natural language supervision. Int Conf Mach Learn 2021:8748−63.

[9] Lan Z, Chen M, Goodman S, Gimpel K, Sharma P, Soricut R. ALBERT: a lite bert for self-supervised learning of language representations. In: International conference on learning representations; 2019.

[10] Eitel A, Springenberg JT, Spinello L, Riedmiller M, Burgard W. Multimodal deep learning for robust RGB-D object recognition. In: IEEE international conference on intelligent robots and systems; 2015. p. 681–7. https://doi.org/10.1109/IROS.2015.7353446. 2015-Decem.

[11] Xu Y, Li M, Cui L, Huang S, Wei F, Zhou M. Layoutlm: pre-training of text and layout for document image understanding. In: Proc. 26th ACM SIGKDD int. Conf. Knowl. Discov. \and data min; 2020. p. 1192–200.

[12] Lawhern VJ, Solon AJ, Waytowich NR, Gordon SM, Hung CP, Lance BJ. EEGNet : a compact convolutional network for EEG - based brain - computer interfaces. J Neural Eng 2018;15:056013. https://doi.org/10.1088/1741-2552/aace8c.

[13] Schirrmeister RT, Springenberg JT, Fiederer LDJ, Glasstetter M, Eggensperger K, Tangermann M, et al. Deep learning with convolutional neural networks for EEG decoding and visualization. Hum Brain Mapp 2017;38:5391–420. https://doi.org/10.1002/hbm.23730.

[14] Akbari H, Khalighinejad B, Herrero JL, Mehta AD, Mesgarani N. Towards reconstructing intelligible speech from the human auditory cortex. Sci Rep 2019;9:1–12. https://doi.org/10.1038/s41598-018-37359-z.

[15] Angrick M, Herff C, Mugler E, Tate MC, Slutzky MW, Krusienski DJ, et al. Speech synthesis from ECoG using densely connected 3D convolutional neural networks. J Neural Eng 2019;16:36019.

[16] Anumanchipalli GK, Chartier J, Chang EF. Speech synthesis from neural decoding of spoken sentences. Nature 2019;568:493–8. https://doi.org/10.1038/s41586-019-1119-1.

[17] Moses DA, Leonard MK, Makin JG, Chang EF. Real-time decoding of question-and-answer speech dialogue using human cortical activity. Nat Commun 2019;10:3096. https://doi.org/10.1038/s41467-019-10994-4.

[18] Makin JG, Moses DA, Chang EF. Machine translation of cortical activity to text with an encoder–decoder framework. Nat Neurosci 2020;23:575–82. https://doi.org/10.1038/s41593-020-0608-8.

[19] Moses DA, Metzger SL, Liu JR, Anumanchipalli GK, Makin JG, Sun PF, et al. Neuroprosthesis for decoding speech in a paralyzed person with anarthria. N Engl J Med 2021;385:217–27. https://doi.org/10.1056/nejmoa2027540.

[20] Wilson GH, Stavisky SD, Willett FR, Avansino DT, Kelemen JN, Hochberg LR, et al. Decoding spoken English from intracortical electrode arrays in dorsal precentral gyrus. J Neural Eng 2020;17. https://doi.org/10.1088/1741-2552/abbfef.

[21] Murphy A, Bohnet B, McDonald R, Noppeney U. Decoding part-of-speech from human EEG signals, vol. 1; 2022. p. 2201–10. https://doi.org/10.18653/v1/2022.acl-long.156.

[22] Défossez A, Caucheteux C, Rapin J, Kabeli O, King J-R. Decoding speech from non-invasive brain recordings. 2022. p. 1–15.

[23] Kiroy VN, Bakhtin OM, Krivko EM, Lazurenko DM, Aslanyan EV, Shaposhnikov DG, et al. Spoken and inner speech-related EEG connectivity in different spatial direction. Biomed Signal Proc Control 2022;71:103224. https://doi.org/10.1016/j.bspc.2021.103224.

[24] Li F, Chao W, Li Y, Fu B, Ji Y, Wu H, et al. Decoding imagined speech from EEG signals using hybrid-scale spatial-temporal dilated convolution network. J Neural Eng 2021;18. https://doi.org/10.1088/1741-2552/ac13c0.

[25] Thornton M, Mandic D, Reichenbach T. Robust decoding of the speech envelope from EEG recordings through deep neural networks. J Neural Eng 2022;19:46007.

[26] Cooney C, Folli R, Coyle D. Optimizing layers improves CNN generalization and transfer learning for imagined speech decoding from EEG. Conf Proc IEEE Int Conf Syst Man Cybern 2019;2019: 1311−6. https://doi.org/10.1109/SMC.2019.8914246. October, IEEE.

[27] Cooney C, Korik A, Folli R, Coyle D. Evaluation of hyperparameter optimization in machine and deep learning methods for decoding imagined speech EEG. Sensors 2020;20:4629. https://doi.org/ 10.3390/s20164629.

[28] Iljina O, Derix J, Schirrmeister RT, Schulze-Bonhage A, Auer P, Aertsen A, et al. Neurolinguistic and machine-learning perspectives on direct speech BCIs for restoration of naturalistic communication. Brain-Comput Interfac 2017;4:186−99. https://doi.org/10.1080/2326263X.2017.1330611.

[29] Cooney C, Folli R, Coyle D. Neurolinguistics research advancing development of a direct-speech brain-computer interface. iScience 2018;8:103−25. https://doi.org/10.1016/j.isci.

[30] Cooney C, Folli R, Coyle D. Opportunities, pitfalls and trade-offs in designing protocols for measuring the neural correlates of speech. Neurosci Biobehav Rev 2022;140:104783. https://doi. org/10.1016/j.neubiorev.2022.104783.

[31] Cooney C, Folli R, Coyle DH. A bimodal deep learning architecture for EEG-fNIRS decoding of overt and imagined speech. IEEE (Inst Electr Electron Eng) Trans Biomed Eng 2021;69:1983−94. https:// doi.org/10.1109/TBME.2021.3132861.

[32] Baker WB, Parthasarathy AB, Busch DR, Mesquita RC, Greenberg JH, Yodh AG. Modified Beer-Lambert law for blood flow. Biomed Opt Express 2014;5:4053. https://doi.org/10.1364/boe.5. 004053.

[33] Bhatt M, Ayyalasomayajula KR, Yalavarthy PK. Generalized Beer−Lambert model for near-infrared light propagation in thick biological tissues. J Biomed Opt 2016;21:076012. https://doi.org/10.1117/ 1.jbo.21.7.076012.

[34] Delorme A, Makeig S. EEGLAB: an open source toolbox for analysis of single-trial EEG dynamics including independent component analysis. J Neurosci Methods 2004;134:9−21. https://doi.org/10. 1016/j.jneumeth.2003.10.009.

[35] Lee TW, Girolami M, Sejnowski TJ. Independent component analysis using an extended infomax algorithm for mixed subgaussian and supergaussian sources. Neural Comput 1999;11:417−41. https://doi.org/10.1162/089976699300016719.

[36] Oostenveld R, Fries P, Maris E, Schoffelen JM. FieldTrip: open source software for advanced analysis of MEG, EEG, and invasive electrophysiological data. Comput Intell Neurosci 2011;2011. https:// doi.org/10.1155/2011/156869.

[37] Ge S, Wang P, Liu H, Lin P, Gao J, Wang R, et al. Neural activity and decoding of action observation using combined EEG and fNIRS measurement. Front Hum Neurosci 2019;13:1−15. https://doi.org/ 10.3389/fnhum.2019.00357.

[38] Lecun Y, Bottou L, Bengio Y, Haffner P. Gradient-based learning applied to document recognition. Proc IEEE 1998;86:2278−324.

[39] Goodfellow I, Bengio Y, Courville A. Deep learning, vol. 1. MIT press Cambridge; 2016.

[40] Chollet F. Deep learning with Python. Manning Publications; 2018.

[41] Dumoulin V, Visin F. A guide to convolution arithmetic for deep learning. In: ArXiv; 2018. https:// doi.org/10.1051/0004-6361/201527329. Prepr ArXiv160307285 2016:1−28.

[42] Ioffe S, Szegedy C. Batch normalization: accelerating deep network training by reducing internal covariate shift. 2015. ArXiv Prepr ArXiv150203167.

[43] Hinton GE, Srivastava N, Krizhevsky A, Sutskever I, Salakhutdinov RR. Improving neural networks by preventing co-adaptation of feature detectors. 2012. p. 1−18. ArXiv Prepr ArXiv12070580.

[44] Krizhevsky A, Sutskever I, Hinton GE. ImageNet classification with deep convolutional neural networks. Adv Neural Inf Process Syst 2012:1097−105.

[45] He K, Zhang X, Ren S, Sun J. Deep residual learning for image recognition. Proc IEEE Int Conf Comput Vis 2016:770–8.

[46] Srivastava N, Hinton G, Krizhevsky A, Sutskever I, Salakhutdinov R. Dropout : a simple way to prevent neural networks from overfitting. J Mach Learn Res 2014;15:1929–58.

[47] Szegedy C, Vanhoucke V, Shlens J. Rethinking the inception architecture for computer vision. Proc IEEE Int Conf Comput Vis 2016:2818–26.

[48] Chollet F. Xception: deep learning with depthwise separable convolutions. Proc IEEE Int Conf Comput Vis 2017:1251–8.

[49] Goodfellow, et al. Deep learning. 2016. https://doi.org/10.1007/s13218-012-0198-z.

[50] Chawla NV, Bowyer KW, Hall LO, Kegelmeyer WP. SMOTE: synthetic minority over-sampling technique. J Artif Intell Res 2002;16:321–57. https://doi.org/10.1002/eap.2043.

[51] Glorot X, Bengio Y. Understanding the difficulty of training deep feedforward neural networks. J Mach Learn Res 2010;9:249–56.

[52] He K, Zhang X, Shaoqing R, Sun J. Delving deep into rectifiers: surpassing human-level performance on ImageNet classification. Proc IEEE Internl Conf Comput Vis 2015:1026–34. https://doi.org/10.1109/ICCV.2015.123.

[53] Kingma DP, Ba J. Adam: a method for stochastic optimization. ArXiv Prepr ArXiv14126980, https://doi.org/10.1063/1.4902458; 2014.

[54] Buccino AP, Keles HO, Omurtag A. Hybrid EEG-fNIRS asynchronous brain-computer interface for multiple motor tasks. PLoS One 2016;11:1–16. https://doi.org/10.1371/journal.pone.0146610.

[55] Khan MJ, Hong KS. Hybrid EEG-FNIRS-based eight-command decoding for BCI: application to quadcopter control. Front Neurorob 2017;11. https://doi.org/10.3389/fnbot.2017.00006.

[56] Chiarelli AM, Croce P, Merla A, Zappasodi F. Deep learning for hybrid EEG-fNIRS brain-computer interface: application to motor imagery classification. J Neural Eng 2018;15:36028. https://doi.org/10.1088/1741-2552/aaaf82.

[57] Ahn S, Jun SC. Multi-modal integration of EEG-fNIRS for brain-computer interfaces – current limitations and future directions. Front Hum Neurosci 2017;11:1–6. https://doi.org/10.3389/fnhum.2017.00503.

[58] Kwon J, Shin J, Im CH. Toward a compact hybrid brain-computer interface (BCI): performance evaluation of multi-class hybrid EEG-fNIRS BCIs with limited number of channels. PLoS One 2020;15:1–14. https://doi.org/10.1371/journal.pone.0230491.

[59] Sereshkeh AR, Yousefi R, Wong AT, Rudzicz F, Chau T, Wong AT, et al. Development of a ternary hybrid fNIRS-EEG brain – computer interface based on imagined speech. Brain-Comput Interfac 2019;00:1–13. https://doi.org/10.1080/2326263X.2019.1698928.

[60] Hong K-S, Khan MJ, Hong MJ. Feature extraction and classification methods for hybrid fNIRS-EEG brain-computer interfaces. Front Hum Neurosci 2018;12:1–25. https://doi.org/10.3389/fnhum.2018.00246.

[61] Behncke J, Schirrmeister RT, Volker M, Hammer J, Marusic P, Schulze-Bonhage A, et al. Cross-paradigm pretraining of convolutional networks improves intracranial EEG decoding. In: 2018 IEEE International Conference on systems, man, and Cybernetics; 2018. p. 1046–53.

[62] Cimtay Y, Ekmekcioglu E. Investigating the use of pretrained convolutional neural network on cross-subject and cross-dataset EEG emotion recognition. Sensors 2020;20:2034.

[63] Sadiq MT, Aziz MZ, Almogren A, Yousaf A, Siuly S, Rehman AU. Exploiting pretrained CNN models for the development of an EEG-based robust BCI framework. Comput Biol Med 2022;143:105242.

5

Neural signals processing using deep learning for diagnosis of cognitive disorders

Hamid Jahani[1] and Ali Asghar Safaei[1,2]

[1]DEPARTMENT OF DATA SCIENCE, FACULTY OF INTERDISCIPLINARY SCIENCE AND TECHNOLOGY, TARBIAT MODARES UNIVERSITY, TEHRAN, IRAN; [2]DEPARTMENT OF MEDICAL INFORMATICS, FACULTY OF MEDICAL SCIENCES, TARBIAT MODARES UNIVERSITY, TEHRAN, IRAN

1. Introduction

Neurological disabilities include a wide range of disorders, such as epilepsy, learning disability, neuromuscular disorders, attention deficit hyperactivity disorder (ADHD), and so many other disabilities, just to name a few. The connection between neural signals and neurological disabilities is a complex and intriguing area of study [1]. These disabilities are characterized by impairments in various neural functions, including motor control, sensory processing, and cognitive abilities [2]. Neural signals, which are electrical impulses transmitted between neurons, play a crucial role in facilitating communication within the nervous system. Disruptions or abnormalities in the transmission of these signals can lead to the manifestation of neurological disabilities [1].

Research in this field has shed light on the underlying mechanisms that contribute to neurological disabilities. For instance, studies have highlighted the role of altered neural signaling in conditions like Parkinson's disease, where a depletion of dopamine-producing cells disrupts the normal functioning of the basal ganglia, resulting in motor symptoms such as tremors and rigidity [3]. Similarly, in epilepsy, abnormal neural activity can trigger recurrent seizures due to disruptions in the balance between excitatory and inhibitory signals in the brain [2]. Understanding the connection between neural signals and neurological disabilities is crucial for the development of effective treatments and interventions. Advances in neuroscience, including techniques like neuroimaging and electrophysiology, have provided valuable insights into the intricate workings of the brain and how neural signals are affected in different disorders [1]. With respect to all of these subjects, processing neural signals can help to understand and diagnose disabilities.

Signal Processing Strategies. https://doi.org/10.1016/B978-0-323-95437-2.00005-7

Deep learning, a subfield of artificial intelligence, has emerged as a powerful tool for processing and analyzing neural signals in the context of neurological disabilities and understanding brain functions. Deep learning algorithms are designed to automatically learn and extract relevant features from complex datasets, such as electroencephalography (EEG) signals or neuroimaging data, without the need for explicit programming [4]. This capability makes deep learning particularly well-suited for capturing intricate patterns and relationships within neural signals, facilitating a deeper understanding of neurological disorders.

One significant application of deep learning in processing neural signals is in the field of brain-computer interfaces (BCIs). BCIs enable individuals with severe motor disabilities to communicate with or control external devices directly through their neural signals. Deep learning models can be trained to decode neural activity recorded from invasive or noninvasive sensors, allowing users to perform actions by leveraging their intentions alone [5]. By extracting meaningful information from neural signals, deep learning algorithms can enhance the accuracy and efficiency of BCIs, offering new possibilities for individuals with neurological disabilities to regain independence and improve their quality of life. Moreover, deep learning has shown promise in the diagnosis and classification of neurological disorders based on neural signals. This chapter will focus mainly on feature extraction techniques and classification problems. Researchers have developed deep learning models that can analyze EEG or neuroimaging data to distinguish between different conditions, such as epilepsy subtypes or neurodegenerative diseases [4]. These models leverage the hierarchical representations learned through multiple layers of neural networks to capture subtle patterns that may be indicative of specific neurological abnormalities. By providing automated and accurate diagnostic support, deep learning-based approaches hold the potential to improve early detection and personalized treatment strategies for individuals with neurological disabilities. While the application of deep learning to process neural signals in neurological disabilities is still an active area of research, it offers exciting opportunities for advancing our understanding of these conditions and developing innovative therapeutic interventions. By leveraging the power of deep learning algorithms, researchers can unravel the complex relationships within neural signals and pave the way for more precise and effective treatments.

2. Deep learning-aided medical diagnosis systems

Deep learning-aided medical diagnosis systems have emerged as a promising approach for improving the accuracy and efficiency of medical diagnostics. Deep learning, a subfield of artificial intelligence, leverages neural networks with multiple layers to automatically learn and extract complex patterns and features from medical data. These systems have shown remarkable potential in various medical domains, including radiology, pathology, and cardiology, among others. By analyzing large datasets and learning

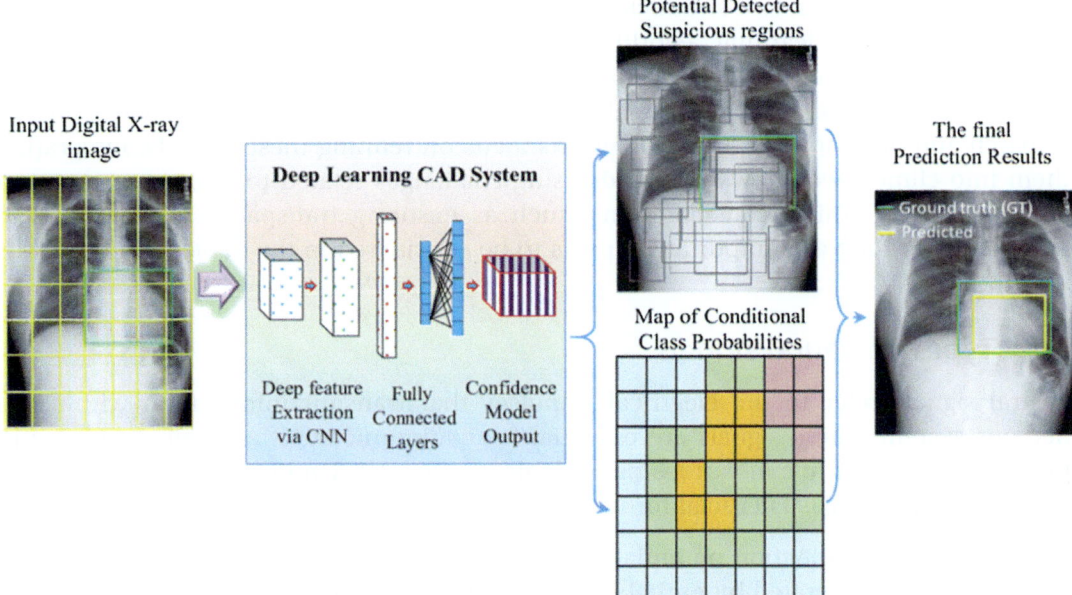

FIGURE 5.1 Segmentation of X-ray images [8].

from expert annotations, deep learning models can assist healthcare professionals in making more accurate diagnoses.

One notable application of deep learning in medical diagnosis is in radiology, specifically in the interpretation of medical images such as X-rays, CT scans, and MRI scans. Deep learning models can be trained to detect and classify abnormalities, such as tumors, fractures, or other pathologies, in medical images [6]. For example, deep learning algorithms have demonstrated high accuracy in detecting breast cancer from mammograms [7]. These systems have the potential to augment radiologists' expertise by providing reliable and efficient support in image interpretation, enabling earlier detection, and improving patient outcomes (Fig. 5.1).

Deep learning also holds promise in the field of pathology. Pathologists can benefit from deep learning models that aid in the analysis of histopathology slides. These models can automatically identify and classify different types of cells, tissues, or diseases, helping pathologists diagnose conditions like cancer [9]. By leveraging large-scale annotated datasets, deep learning models have achieved performance levels comparable to or even surpassing those of human experts in some cases, demonstrating their potential as valuable tools in pathology diagnostics.

In cardiology, deep learning models have shown promise in the detection and diagnosis of various cardiac conditions. For instance, these models can analyze electrocardiogram (ECG) data to identify patterns associated with arrhythmias, heart failure, or other cardiac abnormalities [10]. Deep learning models can also predict patient outcomes and assist in risk stratification for cardiovascular diseases [11]. By leveraging the power of deep learning algorithms, cardiologists can potentially enhance their diagnostic capabilities and improve patient care.

The application of deep learning-aided medical diagnosis systems is an active area of research and development. Ongoing efforts focus on refining these models, integrating them into clinical workflows, and addressing challenges related to interpretability and generalizability. Ethical considerations, such as ensuring transparency, fairness, and patient privacy, are also important aspects to be considered in the deployment of these systems.

2.1   Types of neural signals

Neural signals refer to the electrical impulses that transmit information within the nervous system. These signals are essential for communication between neurons and play a fundamental role in various physiological processes. Understanding the different types of neural signals is crucial for unraveling the complexities of the nervous system.

1. Action potentials: Action potentials, also known as nerve impulses, are brief electrical signals generated by excitable cells, primarily neurons. They result from the rapid depolarization and repolarization of the cell membrane. Action potentials allow for the rapid and long-distance transmission of signals throughout the nervous system. They follow the all-or-none principle, meaning they occur at full strength once the threshold for activation is reached [12].

As you can see in Fig. 5.2 after reaching the threshold, depolarization will happen and then repolarization.

2. Synaptic potentials: Synaptic potentials are electrical signals that occur at the synapses, the junctions between neurons. They can be either excitatory or inhibitory, depending on their effect on the postsynaptic neuron. Excitatory postsynaptic potentials depolarize the postsynaptic neuron, making it more likely to generate an action potential. In contrast, inhibitory postsynaptic potentials hyperpolarize the postsynaptic neuron, reducing the likelihood of an action potential [13].
3. Local field potentials: Local field potentials (LFPs) are low-frequency electrical signals that reflect the synchronized activity of a population of neurons in a specific brain region. LFPs are typically recorded using electrodes placed in or near the brain. They provide insights into the collective activity and information processing within neural circuits, particularly in relation to cognitive processes, such as attention and memory [14].
4. Electroencephalogram (EEG): EEG measures the electrical activity of the brain through electrodes placed on the scalp. It records the summation of postsynaptic potentials generated by thousands of neurons. EEG signals are characterized by different frequency bands, such as delta, theta, alpha, beta, and gamma waves. These frequency bands are associated with different brain states and cognitive

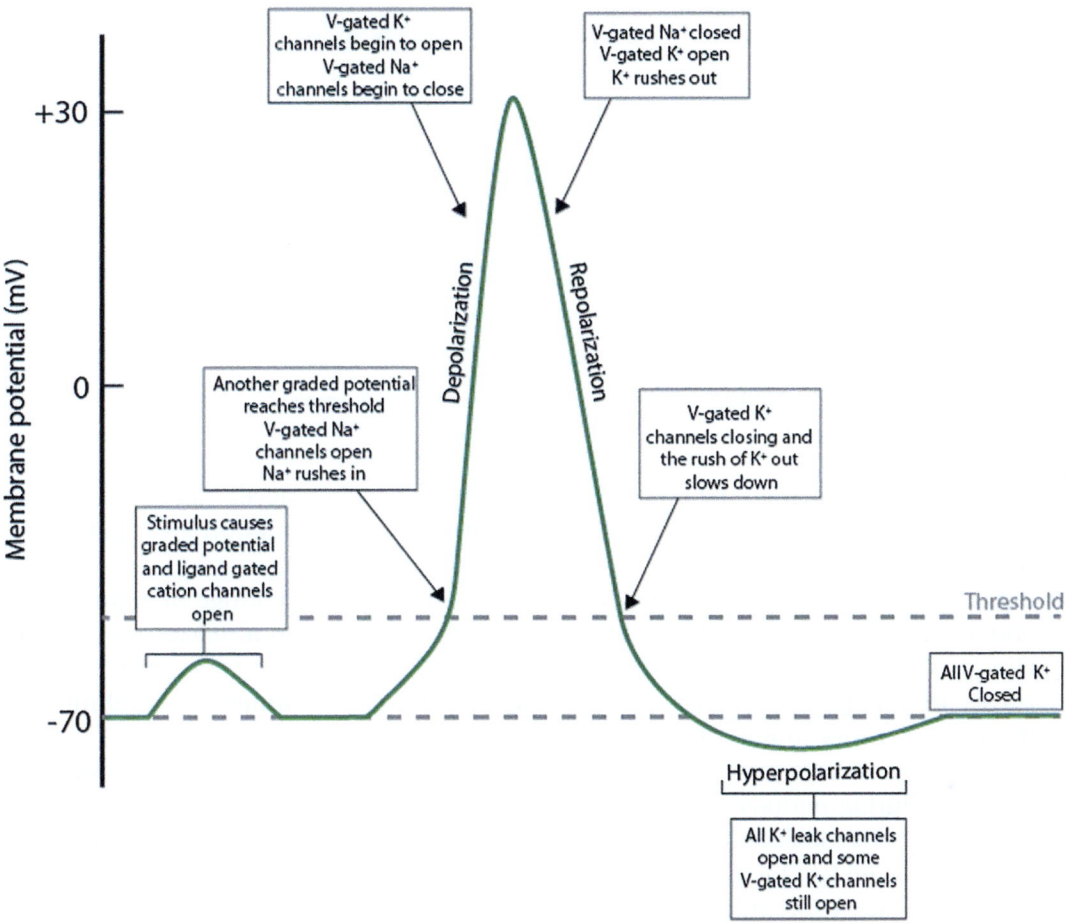

FIGURE 5.2 Visualization of action potentials.

processes, providing valuable information about brain function and dysfunction [15] (Fig. 5.3).

Understanding the characteristics of these bands is crucial for interpreting EEG data and gaining insights into brain function.

- Delta (δ) waves: Delta waves are the slowest EEG oscillations, ranging from 0.5 to 4 Hz. They are typically observed during deep sleep stages and are associated with restorative processes such as physical recovery and memory consolidation [16]. Excessive delta activity during wakefulness may indicate brain dysfunction or pathology.
- Theta (θ) waves: Theta waves have a frequency range of 4–8 Hz and are often observed during drowsiness, daydreaming, and light sleep. They are also present in

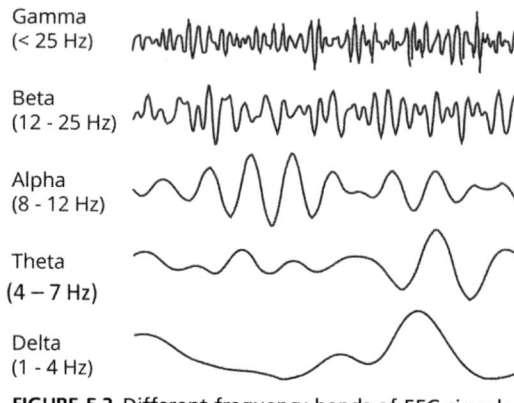

FIGURE 5.3 Different frequency bands of EEG signals.

certain cognitive states, such as during meditation or when accessing deeply stored memories [17]. Theta waves are associated with memory formation and spatial navigation processes.

- Alpha (α) waves: Alpha waves are prominent in the frequency range of eight–13 Hz and are typically observed when an individual is awake but in a relaxed state with closed eyes. They are most prominent in the posterior regions of the brain. Alpha waves are thought to reflect a state of mental relaxation, readiness, or inhibition of irrelevant sensory input [18].
- Beta (β) waves: Beta waves have a frequency range of 13–30 Hz and are commonly observed during periods of focused attention, active thinking, or sensory stimulation. They are associated with cognitive and motor processing and are often observed in the frontal and central regions of the brain [19].
- Gamma (γ) waves: Gamma waves have the highest frequency range among EEG bands, typically above 30 Hz. They are associated with complex cognitive processes such as attention, memory encoding, and sensory perception [20]. Gamma oscillations are thought to reflect the coordination of neural activity across different brain regions.

It is important to note that the specific frequency ranges for each band may vary slightly across studies and individuals. Additionally, EEG signals often exhibit a combination of multiple frequency bands simultaneously, reflecting the dynamic and complex nature of brain activity.

Understanding the different frequency bands of EEG signals provides valuable insights into brain states, cognitive processes, and neurological disorders. Analyzing these bands in conjunction with task performance or clinical assessments can help unravel the underlying mechanisms of brain function and dysfunction.

5. Electrooculogram (EOG): EOG measures the electrical signals generated by eye movements and is used to study eye-related activities, such as rapid eye movements during sleep and tracking eye movements during visual tasks. EOG signals are recorded using electrodes placed around the eyes and are useful for studying oculomotor control and eye movement disorders. EOG signals are obtained by placing electrodes near the eyes to capture the electrical potential differences resulting from the movement of the eye muscles

2.2   Paradigms for neural signals processing and analytics

Paradigms for neural signal processing and analytics refer to the different approaches and methodologies used to analyze and extract meaningful information from neural signals. These paradigms encompass a variety of techniques that are employed to study brain function, develop neurotechnologies, and advance our understanding of the nervous system. There are some key paradigms in neural signal processing and analytics:

1. Time-domain analysis: This paradigm involves analyzing neural signals in the time domain to extract temporal characteristics and patterns. It includes techniques such as event-related potentials (ERPs), which focus on identifying and analyzing specific electrical responses to stimuli or events [21]. Time-domain analysis is particularly useful for studying cognitive processes, sensory perception, and evoked responses.
2.Frequency-domain analysis: Frequency-domain analysis examines neural signals in the frequency spectrum to understand the distribution of power across different frequency bands. Techniques such as Fourier transform and wavelet analysis are commonly used to analyze neural oscillations and assess spectral features. This paradigm provides insights into brain rhythms, functional connectivity, and cognitive states associated with different frequency bands.
3. Spectral analysis: Spectral analysis involves decomposing neural signals into their frequency components to identify specific spectral features and their dynamics over time. Power spectral density (PSD) estimation, coherence analysis, and cross-spectral analysis are techniques employed to examine the power, coherence, and phase relationships between different frequency components. Spectral analysis enables the investigation of functional connectivity and synchronization among brain regions.
4. Multivariate pattern analysis (MVPA): MVPA focuses on decoding and extracting information from neural signals using machine learning algorithms and pattern recognition techniques. By training models on labeled data, MVPA allows for the classification, prediction, and decoding of cognitive states, mental processes, or stimulus categories based on neural activity patterns. MVPA is widely used in BCI, neuroimaging studies, and cognitive neuroscience research.
5. Connectivity analysis: Connectivity analysis aims to understand the functional and structural connections between different brain regions based on neural signals. It involves techniques such as coherence analysis, cross-correlation, graph theory, and network analysis to assess the strength and patterns of connectivity.

Connectivity analysis provides insights into brain network organization, information flow, and integration across different brain regions.

6. Deep learning: Deep learning is a subset of machine learning that utilizes artificial neural networks (ANNs) with multiple layers to automatically learn and extract hierarchical representations from neural signals. Deep learning models, such as convolutional neural networks (CNNs) and recurrent neural networks (RNNs), have shown great success in tasks such as image classification, speech recognition, and natural language processing. In the context of neural signal processing, deep learning techniques are employed for tasks such as EEG-based emotion recognition, motor imagery classification, and BCI.

These paradigms are not mutually exclusive, and often multiple approaches are combined to gain a comprehensive understanding of neural signals. They provide powerful tools for investigating brain function, developing diagnostic tools, and advancing neurotechnologies for clinical and research applications.

2.3 Feature extraction techniques

To extract meaningful information from neural signals, it is essential to employ effective feature extraction techniques. These techniques aim to identify relevant features from the raw signals, enabling subsequent analysis and interpretation. There are some methods to do that, as follows:

1. Time-Domain Features: Time-domain analysis focuses on extracting features directly from the temporal characteristics of neural signals. Some commonly used time-domain features include [22]:
    a. Mean: Represents the average value of the signal over a specific time interval.
    b. Variance: Indicates the dispersion of the signal values.
    c. Skewness and Kurtosis: Describe the asymmetry and peakedness of the signal's probability distribution, respectively.
    d. Autocorrelation: Measures the similarity between a signal and its delayed versions.
    e. Zero-crossing rate: It is not a statistical feature; it is a simple feature. It is the number of times that the signal crosses the x-axis.
2. Frequency-domain features: Frequency-domain analysis involves transforming the neural signals from the time domain to the frequency domain using techniques such as the Fourier transform. Extracting features from the frequency domain provides insights into the signal's spectral characteristics. Common frequency-domain features include.
    a. PSD: Represents the power distribution across different frequency bands. It can be calculated with several parametric and nonparametric methods. Nonparametric methods are used more often and include methods like the Fourier transform.
    b. Spectral entropy: Quantifies the signal's complexity or irregularity.

  c. Peak frequency: Identifies the frequency with the highest power in a given signal.

   Statistical features like mean, median, variance, standard deviation, skewness, kurtosis, and similar are also used in the frequency domain. Relative powers of certain frequency bands are the most commonly used frequency-domain features in all fields of analysis of EEG signals.

3. Time-Frequency Features: Neural signals often exhibit time-varying frequency content. Time-frequency analysis techniques, such as the short-time Fourier transform and continuous wavelet transform, enable the extraction of features that capture both temporal and spectral information. Some common time-frequency features include:
   a. Spectrogram: Displays the time-varying power spectrum of a signal.
   b. Wavelet coefficients: Represents the energy distribution across different time-frequency scales.
   c. Cross-wavelet transform: Enables the comparison of multiple neural signals to identify synchronized activity.
4. Nonlinear features: Neural signals often exhibit nonlinear behavior, and capturing these nonlinear features can enhance signal processing outcomes. Nonlinear feature extraction methods are include:

a. Recurrence plot analysis: Captures the presence of recurrent patterns in a signal.
b. Fractal dimension: Quantifies the signal's self-similarity or complexity.
c. Phase space reconstruction: Recreates the system's underlying dynamics from a single-channel signal.

Feature extraction is a critical step in neural signal processing. By extracting relevant features from neural signals, researchers can uncover valuable information about brain activity and facilitate various applications in neuroscience and medical fields. The discussed techniques provide a foundation for further exploration and customization to suit specific signal processing requirements.

2.4   Deep learning essentials

Deep learning has emerged as a powerful subset of machine learning, revolutionizing various domains such as computer vision, natural language processing, and speech recognition. In this section, we will discuss its essentials briefly. Deep learning can be the subject of a complete book. Deep learning is an approach to AI. Specifically, it is a type of machine learning, a technique that allows computer systems to improve with experience and data [23] (Fig. 5.4).

2.4.1   Artificial neural networks

ANNs serve as the backbone of deep learning. ANNs are composed of interconnected nodes, known as neurons, organized in layers. Each neuron performs a weighted sum of its inputs, applies an activation function to introduce nonlinearity, and passes the output

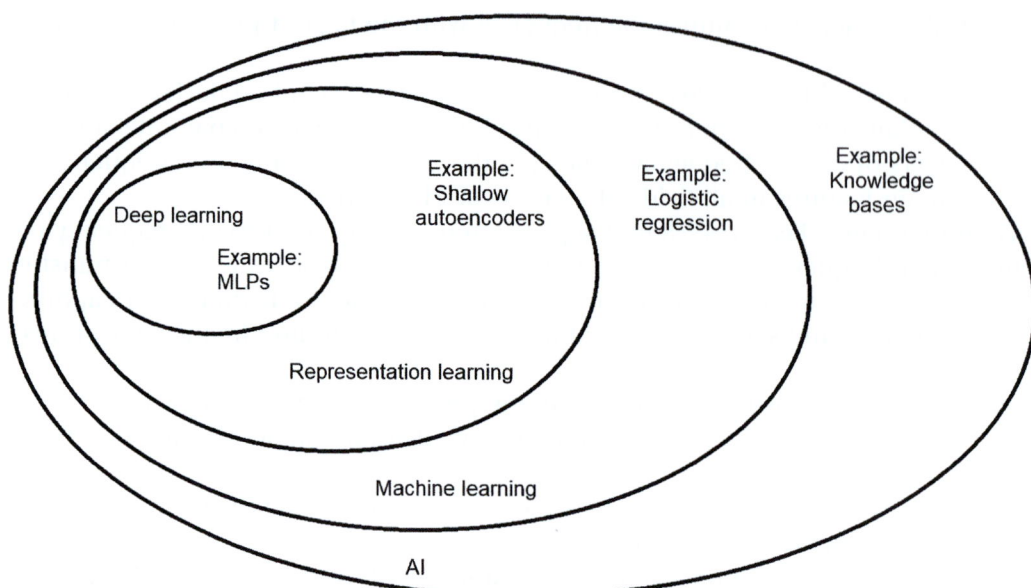

FIGURE 5.4 Relationship between deep learning, representation learning, machine learning, and AI

to the next layer. Deep learning networks often consist of multiple hidden layers, enabling the extraction of complex representations from input data. The goal of a feedforward network is to approximate some function f*. For example $y = f^*(\chi)$ can be a classifier that maps χ to a category y.

As it's obvious in Fig. 5.5. An ANN consists of three part:

1. **Input layer:** The input layer is the first layer of the neural network. It receives the input data and passes it forward to the subsequent layers. Each neuron in the input layer represents a feature or attribute of the input data. The number of neurons in the input layer is determined by the dimensionality of the input data.

FIGURE 5.5 Structure of an artificial neural network.

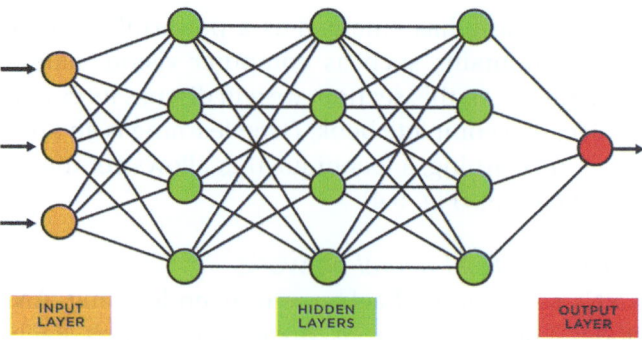

2. **Hidden layers:** Hidden layers are the intermediate layers between the input and output layers in an ANN. They are called "hidden" because their activations are not directly observable from the input or output. Hidden layers enable the network to learn complex representations and extract higher-level features from the input data. The number of hidden layers and the number of neurons in each hidden layer can vary based on the architecture and complexity of the problem being addressed.

3. **Output layer:** The output layer is the final layer of the neural network. It produces the network's predictions or outputs based on the information processed through the preceding layers. The number of neurons in the output layer depends on the nature of the problem being solved. For example, in a binary classification task, there would typically be one neuron representing the probability of one class, while in a multiclass classification task, there would be multiple neurons representing the probabilities of each class.

2.4.1.1 Activation functions

Activation functions introduce nonlinearities to neural networks, allowing them to model complex relationships. Commonly used activation functions include the sigmoid function, the hyperbolic tangent (tanh) function, and the rectified linear unit (ReLU). Each activation function possesses unique characteristics and is suited for different scenarios [23].

The activation function g is typically chosen to be a function that is applied element-wise, with $h_i = g(x^T W_{:,i} + c_i)$.

As an example, ReLU is defined as $g(z) = \max\{0, z\}$. We can now specify our complete network as $f(\mathbf{x}; \mathbf{W}, \mathbf{c}, \mathbf{w}, b) = \mathbf{w}^\top \max\{0, \mathbf{W}^\top \mathbf{x} + \mathbf{c}\} + b$.

2.4.1.2 Loss functions

Loss functions quantify the discrepancy between predicted and actual values during the training process. They guide the learning algorithm by measuring the error and driving the network toward better predictions. Popular loss functions include mean squared error (MSE) for regression tasks and categorical cross-entropy for classification tasks.

Some common loss functions include:

- MSE: MSE calculates the average squared difference between the predicted and true values. It is often used in regression problems.
- Binary cross-entropy: Binary cross-entropy measures the dissimilarity between two probability distributions. It is commonly used in binary classification tasks.
- Categorical cross-entropy: Categorical cross-entropy is used in multiclass classification problems. It quantifies the dissimilarity between the predicted class probabilities and the true class labels.

- Sparse categorical cross-entropy: Similar to categorical cross-entropy, this loss function is used in multiclass classification, but when the true labels are provided as integers instead of one-hot encoded vectors.
- Kullback-Leibler divergence: KL divergence measures the difference between two probability distributions. It is often used in generative models or when training a model to approximate a specific distribution.

2.4.1.3 Backpropagation

Backpropagation is a fundamental algorithm used to train deep neural networks. It leverages the chain rule of calculus to efficiently compute the gradients of the network's parameters with respect to the loss function. These gradients are then used to update the weights and biases of the network through gradient descent, optimizing the network's performance. Speaking about backpropagation is beyond the scope of this chapter. For more reading, take a look at [23].

2.4.2 Deep learning architectures

2.4.2.1 Convolutional neural networks (CNNs)

CNNs have revolutionized computer vision tasks by leveraging convolutional layers to extract spatial features hierarchically. They have achieved breakthroughs in image classification, object detection, and semantic segmentation. Notable architectures, such as AlexNet, VGGNet, and ResNet, have propelled the field forward.

CNN has been designed to solve computer vision problems. Fig. 5.6 shows a CNN model that can detect COVID-19 by processing CT scan images. CNNs use a special function called the convolutional function to extract features from images.

2.4.2.2 Recurrent neural networks

RNNs excel in sequential data analysis by maintaining internal memory, allowing them to capture contextual dependencies. They have found success in tasks such as language modeling, speech recognition, and machine translation. Long short-term memory

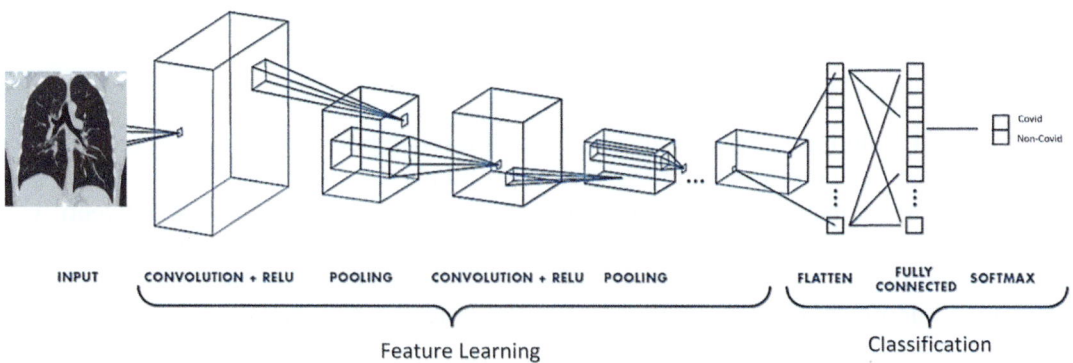

FIGURE 5.6 Diagnosis of COVID-19 from CT scan images by CNNs.

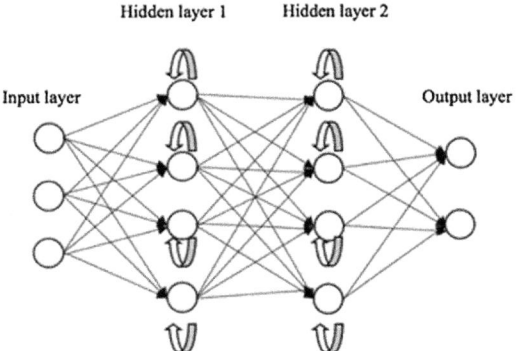

Hidden layer 1   Hidden layer 2

Input layer   Output layer

FIGURE 5.7 General form of RNNs.

(LSTM) and gated recurrent units are popular variants of RNNs that address the vanishing gradient problem (Fig. 5.7).

2.4.2.3 Transformer networks

Transformer networks have revolutionized natural language processing tasks by introducing the attention mechanism. By attending to relevant parts of the input sequence, transformers achieve state-of-the-art results in machine translation, text generation, and sentiment analysis. The Transformer model, proposed in the seminal paper "Attention is All You Need," has become a cornerstone in this domain [24] (Fig. 5.8).

These are just a few examples of how transformers can be applied to neural signal processing tasks. EEG classification, BCI, neuroimaging data analysis, sleep stage classification, Neural Signal Denoising, and reconstruction [24].

2.4.3 Training deep neural networks

2.4.3.1 Optimization algorithms

Stochastic gradient descent (SGD) and its variants, such as Adam and RMSprop, are widely used optimization algorithms for training deep neural networks. These algorithms iteratively update the network's parameters to minimize the loss function by considering small subsets of data (mini-batches) [23]. SGD is an extension of the gradient descent algorithm. The cost function used by a machine learning algorithm often decomposes as a sum over training examples of some per-example loss function. For example, the negative conditional log-likelihood of the training data can be written as

$$J(\boldsymbol{\theta}) = \mathbb{E}_{\mathbf{x},y \sim \hat{p}_{data}} L(\mathbf{x}, y, \boldsymbol{\theta}) = \frac{1}{m} \sum_{i=1}^{m} L\left(\mathbf{x}^{(i)}, y^{(i)}, \boldsymbol{\theta}\right)$$

Where $L(\mathbf{x}, y, \boldsymbol{\theta}) = -\log p(y|x; \boldsymbol{\theta})$.

FIGURE 5.8 The transformer model
architecture [24].

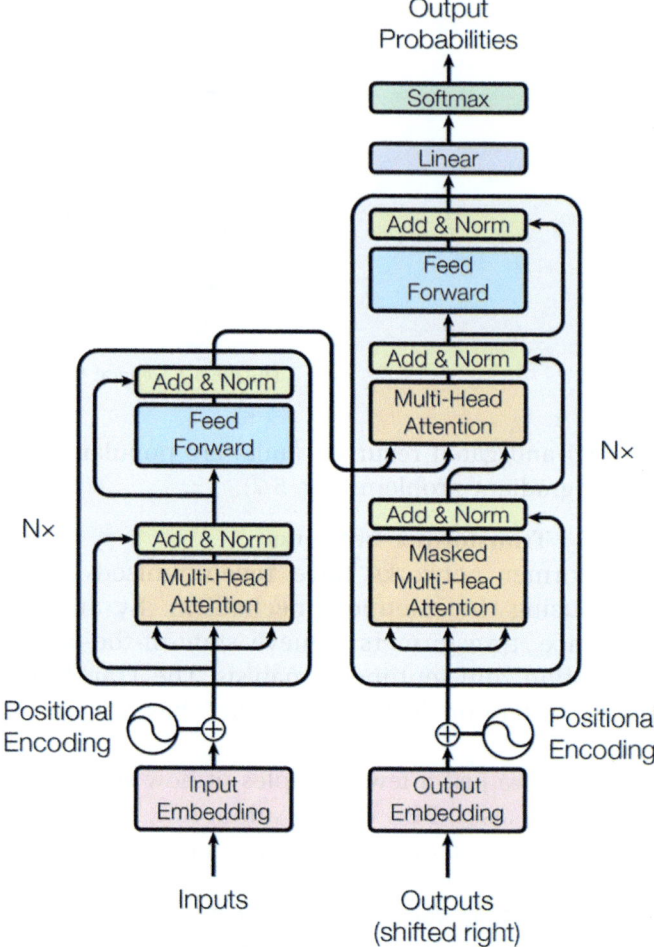

For these additive cost functions, gradient descent requires computing

$$\nabla_\theta J(\theta) = \frac{1}{m} \sum_{i=1}^{m} \nabla_\theta L\left(\mathbf{x}^{(i)}, \mathbf{y}^{(i)}, \theta\right)$$

We can sample a minibatch of examples by size m$'$. The estimate of the gradient is formed as

$$\mathbf{g} = \frac{1}{m'} \nabla_\theta \sum_{i=1}^{m'} L\left(\mathbf{x}^{(i)}, \mathbf{y}^{(i)}, \theta\right)$$

The SGD algorithm then follows the estimated gradient downhill:

$$\theta \leftarrow \theta - \varepsilon\theta$$

where ε is the learning rate.

2.4.3.2 Regularization techniques

To prevent overfitting, regularization techniques are employed. Dropout randomly deactivates neurons during training, while L1 and L2 regularization introduce penalties on the network's weights. Batch normalization normalizes the inputs of each layer, improving the stability and speed of convergence.

2.4.4 Deep learning applications

This section consists of a few applications of deep learning.

2.4.4.1 Computer vision

Deep learning has transformed computer vision tasks, including object detection, image segmentation, and facial recognition. State-of-the-art models, such as Faster R—CNN, Mask R—CNN, and ArcFace, have achieved remarkable accuracy and efficiency.

2.4.4.2 Natural language processing

In natural language processing, deep learning has revolutionized machine translation, sentiment analysis, and text generation. Transformers, such as the GPT series and BERT, have set new benchmarks in language understanding and generation.

2.4.4.3 Speech recognition

Deep learning has made significant strides in speech recognition, enabling applications such as speech-to-text conversion and speaker identification. Models like Listen, Attend and Spell (LAS) and connectionist temporal classification have achieved impressive performance.

2.4.4.4 Signal processing

Deep learning revolutionizes signal processing by extracting meaningful information from complex data. Applications include time series analysis, biomedical signal analysis, and radar/sonar signal processing, enabling accurate predictions and enhanced insights from diverse signal sources.

2.5 DL models for neural signal analytics

Deep models leverage the power of ANNs to decode and interpret neural signals, enabling novel applications such as motor control, prosthetic devices, and neuro-rehabilitation. In this section, some of the deep learning models for neural signal processing will be shown briefly.

2.5.1 Deep convolutional neural networks

In Section 2.4.2.1, CNNs have been introduced. Deep CNNs have shown great promise in analyzing EEG signals, which measure electrical activity in the brain. These models can capture spatial and temporal patterns in EEG data, leading to advancements in tasks

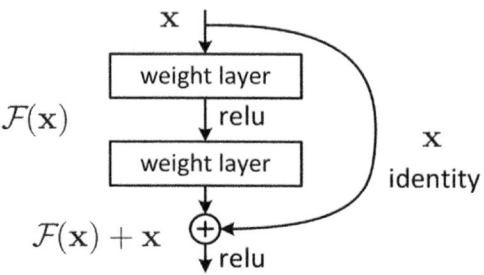

FIGURE 5.9 A residual block [25].

such as mental state classification, seizure detection, and emotion recognition. There are too many model structures for CNNs, one of the best of them is ResNet.

2.5.1.1   ResNet

In recent years, significant advances have been made in the field of machine vision, especially with the introduction of deep CNNs. We have achieved very good results in problems such as image classification and image recognition. Over the years, researchers have sought to create deeper neural networks (add more layers) to solve complex problems, but the point is that adding more layers to the neural network makes them more difficult to train, and the accuracy of the network begins to decline. To solve this problem, researchers introduced the ResNet network. Before the introduction of this network, the use of very deep neural networks was problematic. As the number of layers increases, the network faces the vanishing gradient problem. The ResNet network was able to solve this problem to a large extent by providing a solution; for this reason, this network can have up to 152 layers. Skip connection was the solution that ResNet provided to solve the vanishing gradient problem. As can be seen from Fig. 5.9, the difference between this network and conventional networks is that it has a short circuit that passes through one or more layers and does not take them into account; it takes a shortcut and connects a layer to a farther layer.

By adding these blocks to a standard conventional network, a ResNet network is formed.

As a result, we can introduce ResNet-50 in Fig. 5.10 (in this example for the COVID-19 classification).

There are methods from 2.3 Feature Extraction techniques that can be used to extract relevant images from an EEG signal to serve into ResNet.

2.5.2   Recurrent neural networks for EEG and ECoG analysis

RNNs have proven effective in analyzing both EEG and electrocorticography (ECoG) signals, which provide higher temporal resolution than EEG. RNNs, especially variants like LSTM, enable the modeling of sequential dependencies in neural signals, aiding in tasks such as motor imagery decoding, speech synthesis, and brain wave prediction.

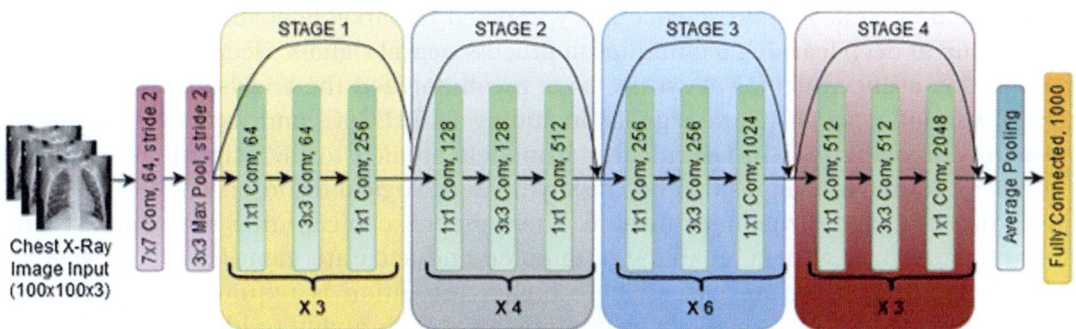

FIGURE 5.10 Structure of ResNet-50 for the classification of COVID-19 from chest X-ray images.

2.5.3   Generative adversarial networks for brain image synthesis

Generative adversarial networks (GANs) have emerged as powerful tools for brain image synthesis, enabling the generation of realistic and diverse brain images. GANs have been used to create synthetic EEG signals, brain MRI scans, and functional connectivity maps. These synthesized data can be valuable for data augmentation, domain adaptation, and addressing limitations in data availability.

2.5.4   Deep reinforcement learning for neurofeedback and brain-controlled interfaces

Deep reinforcement learning (DRL) techniques have been applied to neurofeedback and brain-controlled interfaces, where users can learn to modulate their brain activity to control external devices or interact with virtual environments. DRL algorithms combined with EEG or ECoG signals have shown promise in tasks such as neurorehabilitation, brain-computer gaming, and assistive technology.

2.6   Clinical application

With the progress of medical technology, the biomedical field ushered in the era of big data, based on which, driven by artificial intelligence technology, computational medicine has emerged. People need to extract the effective information contained in these big biomedical data to promote the development of precision medicine. Traditionally, machine learning methods are used to dig out biomedical data to find features, which generally rely on feature engineering and domain knowledge of experts, requiring tremendous time and human resources. Different from traditional approaches, deep learning, as a cutting-edge machine learning branch, can automatically learn complex and robust features from raw data without the need for feature engineering. The applications of deep learning in medical imaging, electronic health records, genomics, and drug development are studied, with the suggestion that deep learning has an obvious advantage in making full use of biomedical data and improving medical health. One of the most important parts of medical processing is clinical neuroscience [26].

The field of clinical neuroscience has witnessed remarkable advancements with the application of deep learning techniques to process neural signals. Deep learning models, powered by artificial neural networks, have revolutionized the analysis and interpretation of various types of neural signals, including EEG, ECoG, and functional magnetic resonance imaging (fMRI). Deep learning models applied to EEG signals have shown promising results in diagnosing and classifying neurological disorders such as epilepsy, sleep disorders, and Alzheimer's disease. These models can capture subtle patterns and abnormalities in EEG data, enabling automated and accurate identification of specific neurological conditions. In cases where higher resolution is required, deep learning models applied to ECoG signals have demonstrated their ability to identify and localize seizure activity, detect motor intentions, and provide insights into cognitive processes. These models leverage the temporal and spatial information encoded in ECoG data, facilitating precise diagnosis and treatment planning for patients with neurological disorders. Deep learning models have been successfully employed in decoding motor intentions from neural signals, allowing individuals with motor impairments to control external devices using their brain activity. This breakthrough in BCIs holds tremendous potential for neurorehabilitation, enabling patients to regain motor function and improve their quality of life. By leveraging deep learning techniques, BCIs can convert neural signals into synthesized speech, offering a means of communication for individuals with severe speech impairments. These models extract linguistic features from neural signals, enabling the real-time generation of spoken words or sentences. Deep learning models applied to longitudinal neural signal data, such as repeated EEG recordings, can provide valuable insights into disease progression and help clinicians predict the trajectory of neurological disorders. These models integrate temporal dependencies in the data to forecast disease outcomes, aiding in treatment planning and personalized patient care. Deep learning models have also been utilized to assess the response to specific treatments or interventions in neurological disorders. By analyzing changes in neural signals over time, these models can provide objective measures of treatment efficacy, allowing clinicians to tailor therapies and optimize patient outcomes.

3. Case study: Diagnosis of ADHD

ADHD is a behavioral condition observed in children that may persist into adulthood. The primary challenges faced by children with this disorder include hyperactivity and difficulties in maintaining focus. Research indicates that approximately 5% of children are affected by ADHD [27]. In China, Hong Kong, and Taiwan, the estimated prevalence rates are 6.5%, 6.4%, and 4.2%, respectively [28]. Children with ADHD commonly struggle with tasks requiring concentration and relaxation. Symptoms typically emerge during the preschool years, but significant difficulties arise during schooling. The key issue experienced by children with this disorder lies in attention deficits and difficulties in meeting expected behavioral standards. Here are some of the challenges faced by these children [29]. Consequently, early diagnosis of ADHD is crucial for securing a better future for affected children.

3.1   Introduction to ADHD

People with ADHD have three common symptoms:

- **Inattention:**

Frequently demonstrates a lack of attention to detail or makes careless errors in schoolwork, work, or other activities (e.g., overlooking or missing important information, resulting in inaccuracies). Often struggles to maintain focus and sustain attention during tasks or play activities (e.g., difficulty staying engaged during lectures, conversations, or lengthy reading). Frequently appears inattentive when directly spoken to (e.g., mind seems elsewhere, even without any apparent distractions). Often fails to complete instructions and has difficulty finishing school assignments, chores, or work duties (e.g., starts tasks but quickly becomes distracted and easily sidetracked). Frequently experiences challenges in organizing tasks and activities (e.g., struggles with managing sequential tasks, maintaining order with materials and belongings, produces messy and disorganized work, has poor time management, and fails to meet deadlines). Often avoids or shows reluctance toward tasks that demand sustained mental effort (e.g., schoolwork or homework; for older adolescents and adults, tasks such as report preparation, form completion, or reviewing lengthy documents). Frequently misplaces items necessary for tasks or activities (e.g., school supplies, pencils, books, tools, wallets, keys, paperwork, eyeglasses, or mobile phones). Is often easily distracted by unrelated stimuli (for older adolescents and adults, this may include unrelated thoughts).

- **Hyperactivity:**

Regularly engages in fidgeting with hands or feet, or displays restlessness by squirming in their seat. Frequently, they leave their seat in situations where remaining seated is expected, such as in the classroom, office, or other workplace. Often demonstrates excessive running or climbing in inappropriate settings (note: In adolescents or adults, this may manifest as subjective feelings of restlessness). Frequently struggles to engage in leisure activities or play quietly. Often exhibits a constant need for movement and appears as if compelled by a motor, finding it difficult or uncomfortable to remain still for extended periods (e.g., experiencing restlessness in restaurants or meetings) and may be perceived by others as being restless or challenging to keep up with. Frequently engages in excessive talking.

- **Impulsivity:**

Consistently interrupts or blurts out answers before questions have been fully stated (e.g., finishing other people's sentences, lacking patience to wait for their turn in a conversation). Regularly experiences difficulty waiting their turn (e.g., while waiting in line). Frequently interrupts or intrudes on others (e.g., interjecting in conversations, games, or activities; using other people's belongings without permission; for adolescents and adults, intruding on or taking over others' tasks or activities).

Early identification of this disorder plays a crucial role in ensuring successful treatment outcomes. The standard approach for treating individuals with ADHD involves a combination of medication and psychotherapy. The diagnostic and statistical manual of mental disorders is widely used as the primary method for diagnosing ADHD. However, this method heavily relies on the honesty and input of teachers and parents. A study revealed that many physicians admitted a lack of sufficient knowledge to diagnose ADHD accurately. Among the 400 surveyed doctors, 44% found the diagnostic criteria for ADHD unclear, 72% felt that diagnosing ADHD in children was easier compared to adults, and 75% believed that physicians' accuracy in diagnosing ADHD was moderate to poor [30]. Consequently, researchers have made extensive efforts to explore alternative tools for diagnosing ADHD. In recent years, significant progress has been made in utilizing EEGs for early-stage ADHD diagnosis [31]. EEG signal processing has emerged as one of the most popular methods for diagnosing ADHD. This approach is favored due to the accessibility and relatively low cost of collecting EEG data. In 1973, J. Lubar conducted pioneering research on EEG signals in individuals with ADHD. He discovered that children with ADHD exhibited increased theta activity and a sharp decrease in beta activity [32].

Researchers have since delved into various aspects of EEG signals, such as:

- ERP markers [33,34].
- PSD [35].
- Current source density [36].
- nivariate and multivariate EEG measurements [37].
- Visual and auditory continuous performance tests [38].
- Visual short-term memory storage capacity [39].
- omplexity analysis [40].
- Interaction between working memory and attention [40].
- Alpha asymmetry [41].

This case study used the method of extracting theta, alpha, beta, and gamma bands and then injecting these bands into RGB channels as a feature extraction technique.

3.2   Study population and dataset

3.2.1   Subjects

The study included 30 children (22 boys and 8 girls) with a mean age of 9.62 ± 1.75 years who were diagnosed with ADHD by an experienced child and adolescent psychiatrist. Among these children, 25 were diagnosed with the combined subtype of ADHD, 3 with the inattentive subtype, and 2 with the hyperactive subtype. These patients were initially referred to the psychiatric clinic of Roozbeh Hospital in Tehran for ADHD evaluation and were not receiving any medication at the time of the study. The control group consisted of 30 healthy individuals (25 boys and 5 girls) with a mean age of 9.85 ± 1.77 years. The control group was selected from two different sources: 25 boys were chosen from an elementary school, and 5 girls were selected from an all-girls primary school. The control group underwent evaluation by a child and adolescent psychiatrist to ensure the absence of psychiatric problems. Exclusion criteria for both the children with ADHD and the healthy group included a history of major neurological disorders, brain injury (including epilepsy), significant medical illnesses, learning or verbal disabilities, other psychiatric disorders, and the use of benzodiazepine and barbiturate drugs. Additionally, participants who scored above the medium level on the Raven Progressive Matrices Test were included in the study [42].

3.3   Feature extraction

The classification process involved converting the EEG signals into RGB images, which were then fed into two models: CNN and ResNet. The EEG signals were divided into different frequency bands, including delta, theta, alpha, beta, and gamma. The delta wave (0.5−4 Hz) is typically observed during deep sleep. The theta wave (4−8 Hz) is associated with unconscious actions and deep relaxation. Theta waves are particularly important in children and infants. The alpha wave (8−12 Hz) is linked to relaxed states without specific focus. The beta wave (12−40 Hz) corresponds to thinking and concentration, while the gamma wave (above 30 Hz) is involved in long-term memory recall and memory tasks. The gamma wave is strongly influenced by the skull and other brain structures. For the classification process, the delta band was excluded, and the low theta, alpha, beta, and gamma bands (30−40 Hz) were utilized.

In this study, each epoch of data was divided into three parts: theta, alpha, and beta + low gamma. The result of averaging the frequencies within these three parts was injected into one channel of an RGB image. An example of these images can be seen in Fig. 5.11. This process resulted in a total of 16,507 images for further analysis and classification.

The pixels of each image were then normalized. Normalization operations facilitate the learning process of image processing models. The following formula is used for this case:

$$\text{pixel\_normal\_image} = \frac{\text{pixel of image} - \min(\text{pixels of image})}{\max(\text{pixels of image}) - \min(\text{pixels of image})}$$

3.3.1   Data epochs

Deep neural networks require large amounts of data to perform training operations. The reason of needing this large amount of data is the huge space of model hypotheses. The

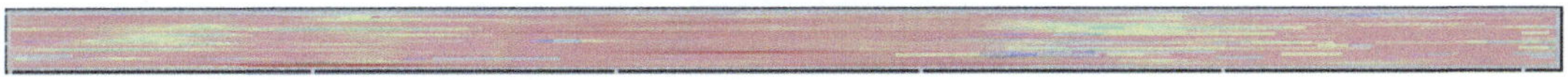

FIGURE 5.11 An example of an extracted image.

database has only 121 samples, which is a small amount. On the other hand, the number of samples available for training is 109, which has dimensions (x * 19), where x is the collection time (seconds) * 128, which is unique to each person according to his performance. Test data also includes 12 individuals with similar conditions to training data. The data is divided into 4-second epochs with 75% overlap at different time intervals.

3.4   Model architecture

3.4.1   Convolutional neural network

The CNN model consists of 13 layers, which include three layers of convolution, three layers of pooling, one layer of flat, two layers of dense, and 4 layers of dropout. Fig. 5.12 shows the network architecture.

3.4.2   Modified ResNet neural network

The architecture of the ResNet network is similar. In this architecture, a residual block is placed between each layer of convolution and dropout. This block consists of three

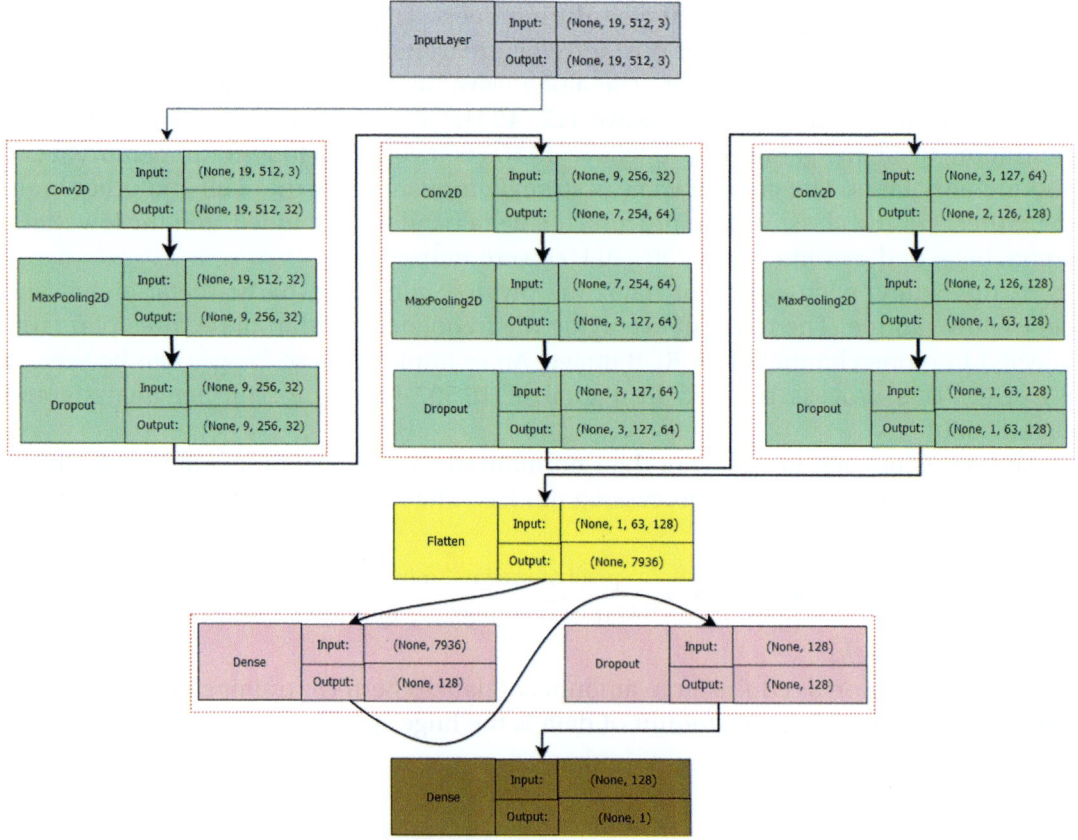

FIGURE 5.12 CNN model architecture [43].

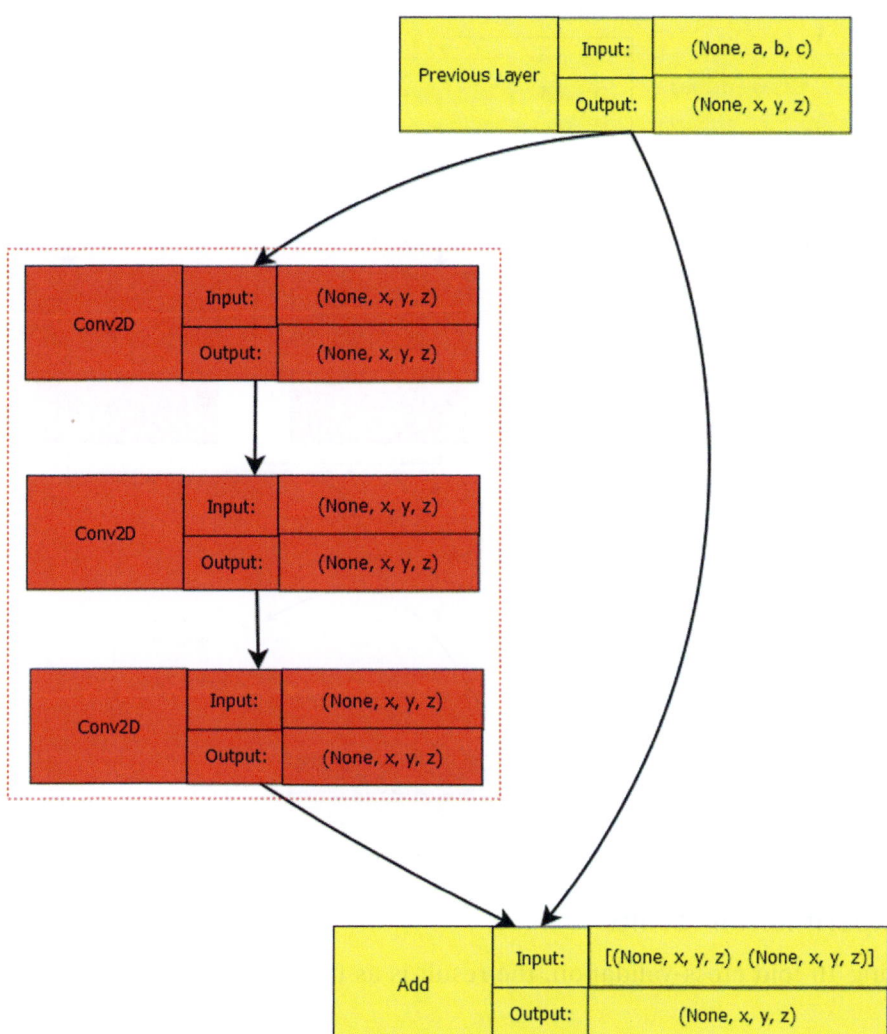

FIGURE 5.13 Residual block for the ResNet model [43].

convolutional layers that perform skip connection operations. Fig. 5.13 displays the residual block used for the ResNet model. Adding these blocks increases the capacity of the model, and more complex structures can be learned from the data. Fig. 5.14 displays the modified ResNet model used in this study [25].

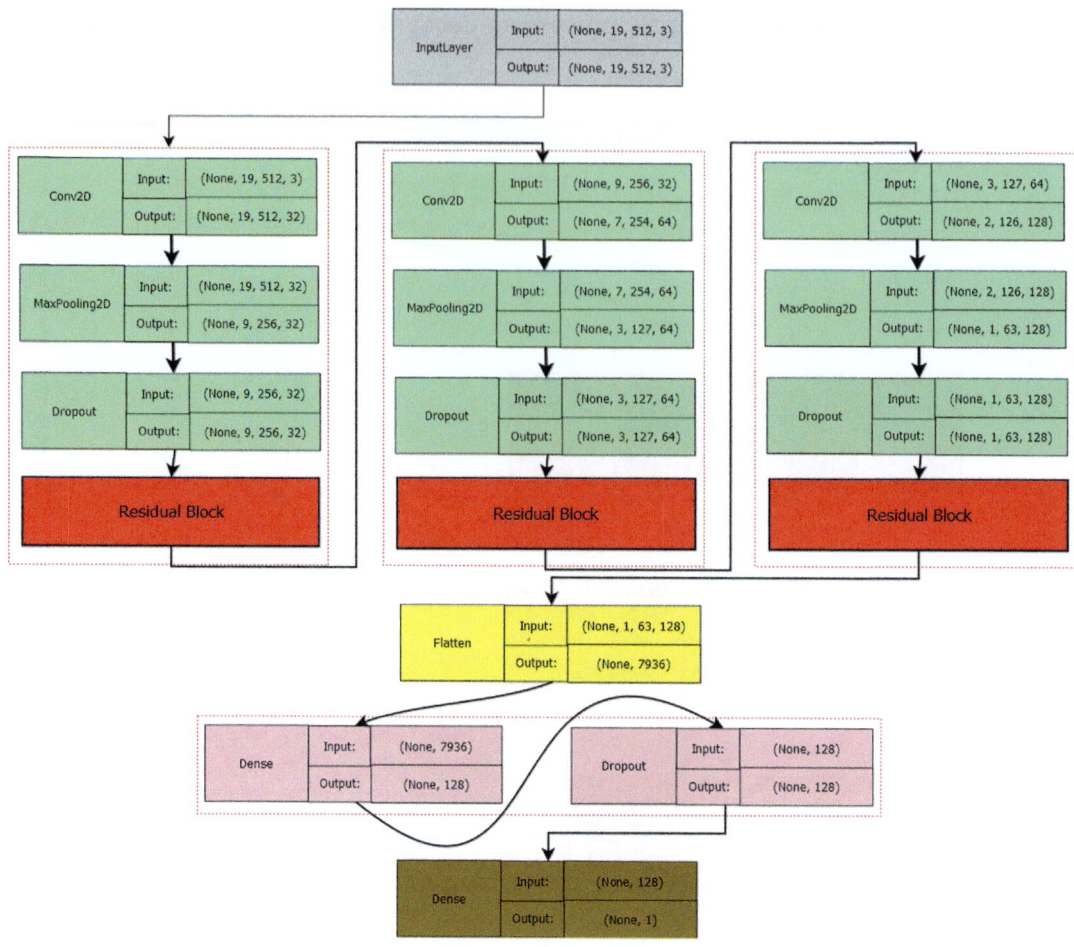

FIGURE 5.14 Modified residual model (ResNet) [43].

3.5 Experimental results

After using 10-fold cross-validation, the result is as follows:

CNN model					ResNet model					
Accuracy	Specificity	Precision	Recall	F1	Accuracy	Specificity	Precision	Recall	F1	
81.1%	57.2%	74%	100%	85.5%	96.3%	94.5%	95.7%	97.7%	96.7%	Fold 1
91.9%	83.1%	88%	98.9%	93.2%	97.1%	96.5%	97.2%	97.6%	97.4%	Fold 2
92.3%	86.5%	90.1%	96.9%	93.4%	98.3%	98.6%	98.9%	98.1%	98.5%	Fold 3
94.6%	90.6%	92.9%	97.8%	95.3%	95.3%	96.8%	97.4%	94.2%	95.8%	Fold 4
94.4%	94.2%	95.4%	94.5%	94.9%	96.3%	97.9%	98.3%	95%	96.6%	Fold 5
95.1%T	95.6%	96.3%	94.9%	95.6%	96.7%	93.8%	95.3%	99%	97.1%	Fold 6
94.8%	93.4%	94.8%	95.9%	95.4%	97.8%	96.8%	97.5%	98.5%	98%	Fold 7
94%	88.5%	91.6%	98.3%	94.8%	96.9%	97.8%	98.2%	96.2%	97.2%	Fold 8
91.7%	83.1%	88%	98.5%	93%	97.3%	96.9%	97.6%	97.7%	97.6%	Fold 9
94.9%	93.9%	95.3%	95.5%	95.4%	95.7%	93.8%	95.2%	97.2%	96.2%	Fold 10
92.52%	86.6%	90.7%	97.1%	93.6%	96.8%	96.3%	97.1%	97.1%	97.1%	Average

According to the results, the average accuracy for the CNN model is 92.52%, and for the ResNet model, the accuracy is 96.8% [43].

4. Discussion

The chapter titled "Neural Signal Processing Using Deep Learning for Diagnosis of Cognitive Disorders" explores the application of deep learning techniques in the field of cognitive disorder diagnosis. This discussion delves into the significance of neural signal processing and its potential to revolutionize the way we diagnose and understand cognitive disorders.

Introduction:

The introduction of the chapter sets the stage by highlighting the growing prevalence of cognitive disorders and the need for accurate and efficient diagnostic methods. It emphasizes the limitations of traditional diagnostic approaches and introduces deep learning as a promising solution.

Neural signals and cognitive disorders:

This section delves into the relationship between neural signals and cognitive disorders. It explains how neural signals, such as EEG, fMRI, and other neuroimaging techniques, provide valuable insights into brain activity and cognitive processes. The discussion highlights the potential of these signals to serve as biomarkers for early detection and monitoring of cognitive disorders.

Deep learning in cognitive disorder diagnosis:

Here, the focus shifts to the application of deep learning algorithms in processing neural signals for cognitive disorders diagnosis. It explores the advantages of deep learning models, such as CNNs, RNNs, and deep belief networks (DBNs), in handling complex and high-dimensional data. The discussion explains how these models can effectively extract meaningful features from neural signals and improve diagnostic accuracy.

Feature extraction and representation learning:

This section dives deeper into the processes of feature extraction and representation learning within deep learning models. It explores various techniques, such as time-frequency analysis, wavelet transforms, and spectrogram representations, used to transform raw neural signals into meaningful features. The discussion highlights the importance of optimizing feature extraction methods to capture relevant information for cognitive disorder diagnosis.

Classification and prediction models:

In this part, different deep learning-based classification and prediction models are examined. The discussion explores how these models can be trained on labeled datasets to differentiate between healthy individuals and those with cognitive disorders. It showcases the potential of these models to predict disease progression and assess treatment efficacy, leading to personalized and targeted interventions.

Challenges and future directions:
The chapter concludes with a discussion on the challenges and future directions of neural signal processing using deep learning for cognitive disorder diagnosis. It addresses limitations such as data scarcity, the interpretability of deep learning models, and ethical considerations. Furthermore, it explores the potential integration of multimodal data, including genetics, imaging, and clinical assessments, to enhance diagnostic accuracy and improve patient outcomes.

5. Conclusion

This chapter presents a comprehensive exploration of the application of deep learning techniques in the field of cognitive disorder diagnosis. The discussion highlights the potential of neural signal processing and deep learning algorithms to revolutionize the way we understand and diagnose cognitive disorders.

The chapter emphasizes the growing prevalence of cognitive disorders and the limitations of traditional diagnostic approaches. It underscores the need for accurate, efficient, and early detection methods to improve patient outcomes and enable timely interventions. By leveraging neural signals, such as EEG and fMRI, deep learning models offer promising solutions for addressing these challenges.

Throughout the chapter, the significance of deep learning in cognitive disorder diagnosis is extensively examined. Deep learning algorithms, including CNNs, RNNs, and DBNs, demonstrate their ability to effectively process complex and high-dimensional neural signals. By extracting meaningful features and patterns, these models enhance diagnostic accuracy and enable the differentiation between healthy individuals and those with cognitive disorders.

Furthermore, the discussion explores the crucial steps of feature extraction and representation learning. Various techniques, such as time-frequency analysis and wavelet transforms, are investigated for optimizing the transformation of raw neural signals into informative features. These techniques play a vital role in capturing relevant information for the accurate diagnosis and monitoring of cognitive disorders.

The chapter also highlights the role of deep learning-based classification and prediction models. By training these models on labeled datasets, they exhibit the potential to predict disease progression and assess treatment efficacy. This personalized approach enables targeted interventions and improves patient care.

While the chapter showcases the transformative impact of deep learning in cognitive disorder diagnosis, it also acknowledges the challenges and future directions of this field. Issues such as data scarcity, the interpretability of deep learning models, and ethical considerations are addressed. Additionally, the integration of multimodal data, including genetics, imaging, and clinical assessments, is explored as a means to enhance diagnostic accuracy further.

In summary, this chapter provides valuable insights into the potential of deep learning techniques in the realm of cognitive disorder diagnosis. By leveraging neural signals and applying advanced deep learning models, we can improve diagnostic accuracy, enable early intervention, and pave the way for personalized treatment strategies. This chapter sets the stage for future research and collaborations to harness the full potential of deep learning in transforming the field of cognitive disorder diagnosis, ultimately improving patient outcomes and quality of life.

References

[1] Uhlhaas PJ, Singer W. Abnormal neural oscillations and synchrony in schizophrenia. Nat Rev Neurosci 2010;11(2):100–13.

[2] Tatum IV WO. Handbook of EEG interpretation. Springer Publishing Company; 2021.

[3] Wichmann T, DeLong MR. Deep brain stimulation for movement disorders of basal ganglia origin: restoring function or functionality? Neurotherapeutics 2016;13:264–83.

[4] Faust O, et al. Deep learning for healthcare applications based on physiological signals: a review. Comput Methods Progr Biomed 2018;161:1–13.

[5] Thomas J, et al. Deep learning-based classification for brain-computer interfaces. In: 2017 IEEE international conference on systems, man, and cybernetics (SMC). IEEE; 2017.

[6] Esteva A, et al. Dermatologist-level classification of skin cancer with deep neural networks. Nature 2017;542(7639):115–8.

[7] McKinney SM, et al. International evaluation of an AI system for breast cancer screening. Nature 2020;577(7788):89–94.

[8] Kadhim YA, Khan MU, Mishra A. Deep learning-based computer-aided diagnosis (CAD): applications for medical image datasets. Sensors 2022;22(22):8999.

[9] Coudray N, et al. Classification and mutation prediction from non–small cell lung cancer histopathology images using deep learning. Nat Med 2018;24(10):1559–67.

[10] Attia ZI, et al. Screening for cardiac contractile dysfunction using an artificial intelligence–enabled electrocardiogram. Nat Med 2019;25(1):70–4.

[11] Poplin R, et al. Prediction of cardiovascular risk factors from retinal fundus photographs via deep learning. Nat Biomed Eng 2018;2(3):158–64.

[12] Hodgkin AL, Huxley AF. A quantitative description of membrane current and its application to conduction and excitation in nerve. J Physiol 1952;117(4):500.

[13] Kandel ER, et al. Principles of neural science, vol 4. New York: McGraw-hill; 2000.

[14] Buzsáki G, Anastassiou CA, Koch C. The origin of extracellular fields and currents—EEG, ECoG, LFP and spikes. Nat Rev Neurosci 2012;13(6):407–20.

[15] Niedermeyer E, da Silva FL. Electroencephalography: basic principles, clinical applications, and related fields. Lippincott Williams & Wilkins; 2005.

[16] Steriade M, Nunez A, Amzica F. A novel slow (< 1 Hz) oscillation of neocortical neurons in vivo: depolarizing and hyperpolarizing components. J Neurosci 1993;13(8):3252–65.

[17] Klimesch W. EEG alpha and theta oscillations reflect cognitive and memory performance: a review and analysis. Brain Res Rev 1999;29(2–3):169–95.

[18] Klimesch W, Sauseng P, Hanslmayr S. EEG alpha oscillations: the inhibition–timing hypothesis. Brain Res Rev 2007;53(1):63–88.

[19] Pfurtscheller G, Da Silva FL. Event-related EEG/MEG synchronization and desynchronization: basic principles. Clin Neurophysiol 1999;110(11):1842–57.

[20] Tallon-Baudry C, et al. Oscillatory γ-band (30–70 Hz) activity induced by a visual search task in humans. J Neurosci 1997;17(2):722–34.

[21] Pesaran B. Spectral analysis for neural signals. Short Course 2008;III:1.

[22] Jiang D, et al. Robust sleep stage classification with single-channel EEG signals using multimodal decomposition and HMM-based refinement. Expert Syst Appl 2019;121:188−203.

[23] Goodfellow I, Bengio Y, Courville A. Deep learning. MIT press; 2016.

[24] Vaswani A, et al. Attention is all you need. Adv Neural Inf Process Syst 2017;30.

[25] He K, et al. Deep residual learning for image recognition. In: Proceedings of the IEEE conference on computer vision and pattern recognition; 2016.

[26] Yang S, et al. Intelligent health care: applications of deep learning in computational medicine. Front Genet 2021;12:607471.

[27] Demontis D, et al. Discovery of the first genome-wide significant risk loci for attention deficit/hyperactivity disorder. Nat Genet 2019;51(1):63−75.

[28] American Psychiatric Association, A. and A.P. Association. Diagnostic and statistical manual of mental disorders: DSM-IV, vol. 4. Washington, DC: American psychiatric association; 1994.

[29] Barkley RA. Attention-deficit hyperactivity disorder: a handbook for diagnosis and treatment. 2006.

[30] Marshall P, Hoelzle J, Nikolas M. Diagnosing Attention-Deficit/Hyperactivity Disorder (ADHD) in young adults: a qualitative review of the utility of assessment measures and recommendations for improving the diagnostic process. Clin Neuropsychol 2021;35(1):165−98.

[31] Adamou M, Fullen T, Jones SL. EEG for diagnosis of adult ADHD: a systematic review with narrative analysis. Front Psychiatr 2020;11:871.

[32] Lubar JF. Discourse on the development of EEG diagnostics and biofeedback for attention-deficit/hyperactivity disorders. Biofeedback Self Regul 1991;16:201−25.

[33] Kaur S, et al. Event-related potential analysis of ADHD and control adults during a sustained attention task. Clin EEG Neurosci 2019;50(6):389−403.

[34] Lau-Zhu A, et al. No evidence of associations between ADHD and event-related brain potentials from a continuous performance task in a population-based sample of adolescent twins. PLoS One 2019;14(10):e0223460.

[35] Lenartowicz A, et al. Alpha modulation during working memory encoding predicts neurocognitive impairment in ADHD. JCPP (J Child Psychol Psychiatry) 2019;60(8):917−26.

[36] Ponomarev VA, et al. Group independent component analysis (gICA) and current source density (CSD) in the study of EEG in ADHD adults. Clin Neurophysiol 2014;125(1):83−97.

[37] González JJ, et al. Performance analysis of univariate and multivariate EEG measurements in the diagnosis of ADHD. Clin Neurophysiol 2013;124(6):1139−50.

[38] Shi T, et al. EEG characteristics and visual cognitive function of children with attention deficit hyperactivity disorder (ADHD). Brain Dev 2012;34(10):806−11.

[39] Wiegand I, et al. EEG correlates of visual short-term memory as neuro-cognitive endophenotypes of ADHD. Neuropsychologia 2016;85:91−9.

[40] Chenxi L, et al. Complexity analysis of brain activity in attention-deficit/hyperactivity disorder: a multiscale entropy analysis. Brain Res Bull 2016;124:12−20.

[41] Alperin BR, et al. The relationship between alpha asymmetry and ADHD depends on negative affect level and parenting practices. J Psychiatr Res 2019;116:138−46.

[42] Mohammadi MR, et al. EEG classification of ADHD and normal children using non-linear features and neural network. Biomed Eng Lett 2016;6:66−73.

[43] Jahani H, Safaei AA. Efficient deep learning approach for diagnosis of attention-deficit/hyperactivity disorder in children based on EEG signals. Cogn Comput 2024. https://doi.org/10.1007/s12559-024-10302-3.

Brain tumor recognition using semisupervised generative adversarial network

Jyotismita Chaki

SCHOOL OF COMPUTER SCIENCE AND ENGINEERING, VELLORE INSTITUTE OF TECHNOLOGY, VELLORE, TAMIL NADU, INDIA

1. Introduction

The human brain is the most important component of the nervous system as it is answerable for delivering proper signals all over the human body. Brain tumor (BT) develops when biological cells split and start to breed abnormally. BTs must be diagnosed immediately as they comprise malignant cells which can cause death. Research undertaken in the United States by the National Brain Tumor Foundation found that around 30,000 individuals are detected with BT every year, and around 43% are dying because of that [1,2]. Patients may be detected with one of several types of BTs based on their symptoms. BTs are characterized as benign or malignant depending on their features [3,4]. The World Health Organization (WHO) divided BTs into four stages, from grade I to grade IV, to define the severity of the malignancy, as seen in Table 6.1.

Radiologists employed numerous procedures for recognizing BTs, such as magnetic resonance imaging (MRI), ultrasonography, and computed tomography (CT) [5]. But brain MRI is the modality which most often used by physicians to detect and locate BTs, which helps the physician to provide the appropriate treatment. Recognizing BTs from MRI scans is a difficult process that necessitates a greater degree of knowledge and expertise, which

Table 6.1 Various grades of BT.

BT type	Explanation
Grade I	Slow rate of growth of BT.
Grade II	The growth of one tissue is influenced by the growth of another.
Grade III	Tumor with cancer, tissues regulate cell reproduction, and tissues influence cells.
Grade IV	Malignant tumor, cell tissue replica happens very fast, significantly affecting the neighboring brain tissue

Signal Processing Strategies. https://doi.org/10.1016/B978-0-323-95437-2.00013-6

may result in misclassification. Misclassification of tumors may cause unsuccessful treatment, deteriorating the health of the patient. The location of a BT is a crucial facet in determining how it disturbs the functionality of a patient and what signs it creates. There are several methods for determining the size of a tumor from a picture, but image segmentation is one of the most prevalent. MRI images include too much data for physical understanding and evaluation. Thus, defining the precise size and location of a BT is a difficult process that plays an important part in BT diagnosis. Because tumors develop indefinitely, they can grow in any location and shape. As a result, a deep learning (DL)-based architecture is required to solve the problem. DL may be used to extract MRI images since they include unique qualities for instance image eigenvalues, texture, and histogram.

DL-based techniques have sparked considerable attention in recent years. Convolutional neural networks (CNNs) and semisupervised models have recently boosted the acceptance of DL-based architectures in the healthcare industry. CNNs are mostly used to extract characteristics from MRI images. CNN starts with raw images and passes them through numerous convolutional and pooling layers to build a hierarchy of increasingly advanced characteristics. For high performance measure in CNN-based systems, massive amounts of properly labeled training data are necessary. Medical imaging data collection and interpretation by skilled physicians are both tough tasks. It is challenging to come up with an automated labeling system for a supervised architecture that is applicable to the image recognition problem. In contrast, transfer learning (TL) has appeared as the favored strategy for various image recognition applications. Let's take the following situation: a small portion of our training data is labeled, but the remainder is not. For the unlabeled component of the training set, TL is unsuccessful. That is what "semi-supervised learning" denotes. A BT that is not discovered early on might have major ramifications for mortality. This study endeavor is centered on recognizing brain MR images that require immediate medical attention. This is the reason for employing a semisupervised learning strategy to distinguish between different BTs as well as no tumors. This method can be used in remote locations where we do not have access to radiology professionals. Because doctors in remote regions are often inexperienced, the recommended model assists them by allowing them to simply upload the brain MRI image to the model, which then recognizes the tumor type.

The authors of Ref. [6] used five DL-based models, namely VGG16, AlexNet, ResNet18, GoogleNet, and ResNet50, as well as five machine learning-based models, namely Naive Bayes, K-nearest neighbors (KNN), decision trees, and linear discrimination with fivefold cross-validation. They used a majority-voting ensemble strategy to improve the overall classification performance of five machine learning and five DL-based models. The authors of Ref. [7] proposed changing the U-Net approach to differentiate and classify the various types of BTs. In the encoder part, the network employs a pretrained ResNet-34 and alters the U-Net topology. To anticipate a group segmentation feature map, the model integrates spatial data from the down sampling path with contextual data from the up-sampling path. Ref. [8] presents research that categorizes BTs (for instance meningioma, glioma tumor, tumor, and pituitary) and autoimmune disease lesions (such as multiple sclerosis) by means of MRI of BTs and individuals with multiple sclerosis. The technique includes four steps. During the preprocessing stage, the tumor and lesion ROIs, Lloyd-max quantization, and Collewet normalization are applied. Within the base learner, a support vector machine (SVM) classifier and prediction model with majority voting is developed. The research project's [9] objective is to increase the accuracy of brain MR image identification using DL algorithms and a TL technique. The authors analyze five various DL models, including GoogLeNet, AlexNet, ResNet50, ResNet101, and SqueezeNet, using an MRI of a BT, and then apply TL techniques to a dataset of BTs. The training results of numerous pretrained DL networks show that the kind of optimizer used has a substantial influence on a pretrained network's performance. It affects the network's accuracy and, more importantly, training time. Next, it is noticed that, of all the pretrained networks, GoogLeNet needed the most time to train—112 minutes—using all of the available optimizers. The RMSProp optimizer takes an average of 74 minutes of training time, whereas the SGDM optimizer requires an average of 63 minutes of training time. The authors of the proposed framework in Ref. [10] perform three research utilizing three CNN models (GoogLeNet, VGGNet, and AlexNet) to categories brain malignancies. Every research then analyses TL mechanisms such as fine-tuning and freezing utilizing MRI slices from a BT dataset. Image augmentation methods are applied to MRIs to decrease the chance of overfitting by increasing the number of dataset samples. Ref. [11] proposes categorizing MRI BTs using a saliency-driven picture representation and CNN-based classification with optimization. The canny edge detection method is used to preprocess MRI images before moving on to saliency-driven picture representation using modified minimum barrier distance and nonlinear diffusion at many layers. Finally, CNN does feature extraction and image classification, while the ADAM optimizer performs optimization. A characteristic of the work reported in Ref. [12] is the classification of BTs into glioma, meningioma, and pituitary using a hierarchical DL system. The preprocessor for BT imaging employs data normalization, data augmentation, and image scaling. Convolution is the following step of CNN, during which the descriptors of the input picture are recovered. Because convolution is a linear process, yet pictures are usually nonlinear, the ReLU layer is used to enhance network disconnection. CNNs use convolutional and pooling layers to learn attributes from preprocessed images. The model classified the input into four categories: glioma, meningioma, pituitary, and no-tumor. The algorithm in Ref. [13] automatically segments the MRI brain image of the BT and classifies the findings. Because this technique combines CNN with gray level cooccurrence matrix (GLCM) descriptors, it provides a highly effective BT detection and classification system. The proposed technique consists of four steps: preprocessing, picture segmentation, feature extraction, optimization, and classification. Noise reduction is the initial step in preprocessing brain MR pictures. Following the classification technique, the segmentation component receives irregular brain MR images and utilizes the fuzzy c-means algorithm to detect tumors and segments. Then there's GLCM and Ant colony optimization (ACO), whose descriptors are obtained from these noise-free brain MRIs. Based on ACO, a large number of descriptors are minimized. To identify

MRI brain pictures as abnormal or normal, the CNN classifier is fed certain features from the brain images. In Ref. [14], the author suggested a novel CNN-based approach for distinguishing between brain MRI images to determine whether or not they are tumorous. The accuracy of the model was 96.08%, and its f-score was 97.3. To get results in 35 epochs, the model employs a CNN with three layers and only a few preprocessing steps. The study's goal is to highlight the importance of diagnostic machine learning technologies and preventative treatment. The authors of Ref. [15] emphasis on creating precise network that can be effectively trained with a smaller number of dataset samples. To extract descriptors from brain MRI, Siamese Neural Network (SNN) is created. The SNN is implemented utilizing a three-layer network. The created SNN is simpler and has a smaller number of parameters as compared to deep CNN. A closest neighbor investigation based on Mahalanobis and Euclidean distances is directed on the SNN-encoded descriptor. Because the encoded descriptor is two-dimensional, the neighborhood investigation is having less computational complexity. Neighborhood analysis is carried out using a KNN model. Three publicly available datasets are utilized to assess the proposed approach, including those from the Harvard, Figshare, and Radiopaedia repositories. The authors of Ref. [16] suggest a customized mask region-based CNN (Mask RCNN) with a densenet-41 backbone network to reliably diagnose and separate BTs. The approach is evaluated using a range of quantitative indicators on two distinct benchmark datasets. According to comparison studies, the unique Mask-RCNN can detect tumor sites more precisely by applying bounding boxes to offer exact tumor areas. In the work in Ref. [17], a novel hybrid CNN-based architecture was built to classify three different forms of BTs using MRI data. The method advocated in this study employs two hybrid deep learning classification algorithms based on CNN. The first method combines an SVM for pattern recognition with a CNN pretrained Google Net model for feature extraction. The second method employs a soft-max classifier in conjunction with a carefully tuned Google Net. The targeted region initialization step of the multistage image segmentation technique described in Ref. [18] is carried out using low-level data from key point features. A series of linear filters is utilized to get high level output from the low-level visual descriptions. A collection of descriptors and filter training data are completed to locate the area of the tumor. For the boundary-refining procedure, the authors employ a probabilistic model and a disparity map. The Fisher vector and autoencoder are also utilized to extract the descriptors. A set of manually constructed descriptors is also collected utilizing a region of interest to train and assess the multilayer perceptron and SVM classifiers. The technique described in Ref. [19] has three basic steps: feature extraction, feature selection, and combination. According to the authors, the original technique was the CNN classification method with no preprocessing of the input photographs. When a CNN adds descriptors derived from feature extraction techniques, the original strategy is enhanced through a series of phases. The production of input descriptors for the network is investigated utilizing a variety of famous feature extraction approaches. After examining the results, a set of feature extraction approaches (local thresholds, Otsu algorithms, watershed algorithms, cluster algorithms, k-means algorithms, and edge detection algorithms) with sufficient performance is chosen for the next stage. Furthermore, the authors assigned weight factors depending on the accuracy of each of the selected methodologies. According to Ref. [20], the system extracts descriptors from brain MRI images using the discrete wavelet transform's strong energy compactness characteristic. The input MRI picture is then classified using a CNN and wavelet descriptors. The author of Ref. [21] proposed a DL architecture based on a 16-layer CNN to recognize various types of BTs using two publicly available resources. A preprocessing stage is accomplished before the images are fed into the proposed structure. To decrease calculations and dimensionality and to enable the network to provide better results faster and with higher clarity, the actual image must first be moved from $512 \times 512 \times 3$ pixels to $128 \times 128 \times 3$ pixels. The architecture is used to detect three types of BTs (pituitary, glioma, and meningioma) as well as the grade of the tumor. Ref. [22] proposed a lightweight attention-guided CNN (AG-CNN) for BT recognition in MRI. The proposed architecture employs channel-attention blocks to focus attention to sections of the image that are critical for tumor recognition. Furthermore, AG-CNN blends feature from different stages via skip connections and global-average pooling. The authors use Gaussian smoothing to the original brain MR images using filters with a kernel size of 7×7 to denoise the input image. The smoothed image is subsequently converted into binary images using the Otsu threshold process. The outer edge contours are discovered in the next preprocessing stage. The authors use the largest contour to define the binary image's four extreme points (top, right, left, and bottom). Finally, the authors crop the images using these four locations. The authors of Ref. [23] describe two rapid and effective deep CNN-based BT identification algorithms for the accurate detection and categorization of various types of BTs. The authors apply conditional random fields to two publicly available datasets to screen out fraudulent results while considering geographical information for the segmentation tasks. The first design recognizes three categories of BT (pituitary, gliomas, and meningiomas). The second architecture distinguishes between low- and high-grade gliomas (LGG and HGG, correspondingly). A brightness enhancement strategy is being investigated as part of the preprocessing phase. When combined with data augmentation approaches, this strategy is quite effective in detecting and classifying BTs. The proposed technique employs three networks, namely vgg16, Alexnet, and vgg19, as well as their deep descriptors [24]. Deep fusion is being employed again to increase classification model performance. Principal component analysis (PCA), which is used to reduce the dimensionality of feature vectors, is also used to evaluate classifier efficacy. In Ref. [25], a DL technology known as faster R-CNN was used to detect tumors in MRI brain images. Three types of tumors can be seen in 2452 images used to train a model. The author of Ref. [26] introduced Parametric Flatten-p Mish (PFpM) to increase performance as one of the most important modules in CNN. The primary limitations of preexisting activation functions may be addressed by PFpM. PFpM also allows the model to acquire composite patterns from data more precisely.

Based on the aforementioned findings and in order to avoid the reasoning methods, this chapter proposes a DL-based network, semisupervised generative adversarial network (SSGAN), to recognize brain MRI in two folds. As the dataset contains fewer brain MRIs, an image augmentation approach is used prior to feed the image to the DL network. To increase the number of samples in the dataset, a mixture of four distinct image augmentation techniques, as well as the brightness correction approach, were applied. The final stage is to recognize images of BTs as well as no tumor using SSGAN.

The following is a summary of SSGAN's two folds and the primary contributions of this chapter:

i. Image augmentation and brightness enhancement: This step can enhance the performance of the DL model. This step helps the network to train and test with variety of images, permitting the network to achieve better more correct results. Data augmentation approaches lower operating expenses by including transformation into datasets. All of the images in the collection had four different geometric alterations applied to them. The options are vertical flipping, horizontal flipping, rotation, orientation, scale-down, scale-up, and a combination of 1 to 5. The combination of scale-down and scale-up is neglected as the final result is dependent on the amount of scale-up or scale-down. A fuzzy inference system (FIS) is developed for brightness adjustment.

ii. SSGAN for the recognition of brain MR images: Semisupervised learning corresponds to the difficult job of training a network using a dataset with a small number of labeled samples and a large number of unlabeled observations. GAN is a DL-based model that uses massive, unlabeled dataset samples to train an image generator framework via an image discriminator framework. The discriminator framework may be utilized to construct a classifier network. The SSGAN model is a GAN architectural expansion that combines training an unsupervised discriminator, a supervised discriminator, and a generator framework all at the same time. As a result, the proposed network structure has a supervised classification framework that handles previously unrecognized samples, as well as a generator framework that generates potential samples. External GPU resources are used to enhance the network application. The model is calibrated so that the chance of a false positive (where the real tumor class is recognized to be nontumor class) is as low as feasible.

The purpose of this endeavor is to improve generalization using SSGAN to enable even more real-world applications. The study's goal is to investigate the application of SSGAN to improve procedures utilized in the detection of brain MRI.

The remainder of the article is structured as follows: Section 2 outlines the proposed methodology, Section 3 reports on the experiments and findings, Section 4 compares the proposed methodology to the method utilized in the by the previous researchers, and Section 5 concludes the chapter.

2. Proposed methodology

The block diagram of the proposed approach is visualized in Fig. 6.1.

2.1 Preprocessing

The brain MRI images are resized to 128×128 pixels to reduce the computational difficulties of the approaches involved. This spatial resize permits the visualization of unwanted characteristics while maintaining the clarity of the image. In this study, resize of brain MRIs are done for two reasons. The first logic is that it can assist us improve the model findings. It can be ensured that the training data is of constant size and form by resizing images, which can help the models learn more successfully. The second reason is that it has the potential to improve the efficiency of the classification models. By resizing brain MRIs, it can be ensured that the model only deal with relevant data and do not spend time analyzing irrelevant data. Thus, this image's size balance enhances the utility of the method.

The purpose of this preprocessing section is to develop a novel FIS to increase the brightness of the brain MRI, therefore overcoming the limits of traditional approaches. FIS is a technique that interprets the values in the input vector and assigns values to the output vector based on a set of rules. The veracity of every assertion becomes a question of degree in fuzzy logic. FIS is the process of utilizing fuzzy logic to create a mapping from a given input to an output. The mapping then serves as a foundation for making decisions and identifying trends. Some components, namely membership functions (MFs), fuzzy logic operators, and if-then rules, are used in the FIS. Finally, a defuzzified value is formed by applying the output MFs.

The proposed FIS includes seven input MFs, which includes trap and Gaussian MFs, to increase the recognition accuracy, which characterize the pixel values as linguistic variables such as very bright, bright, medium, dark, and very dark as shown in Fig. 6.2. These MFs are chosen after examining various brain MRI, and accordingly the range of pixel values are identified.

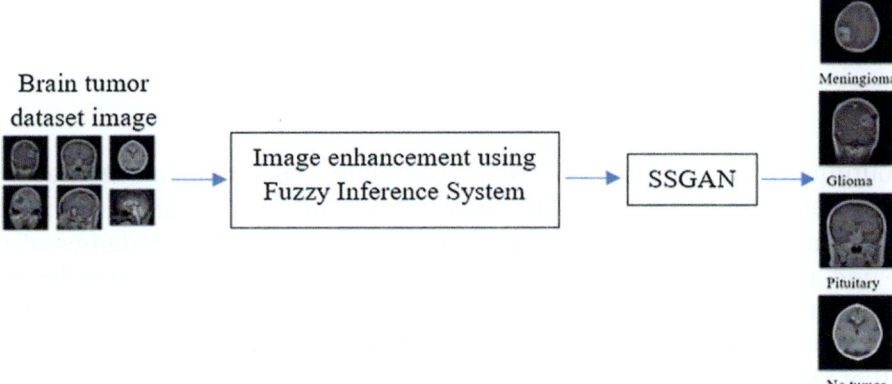

FIGURE 6.1 The block diagram of the proposed approach.

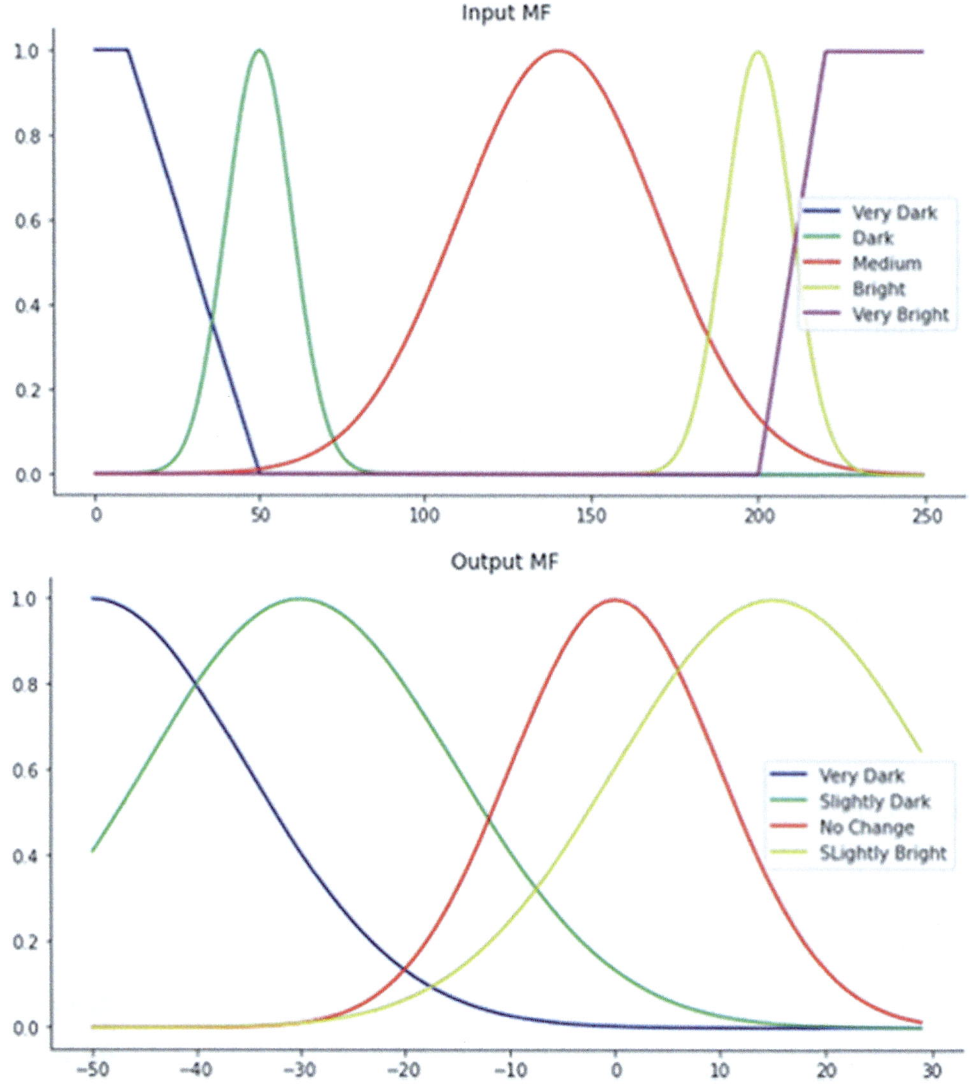

FIGURE 6.2 Input and output MFs.

The output MFs describe the change in pixel value in accordance with the knowledge base principles. From the output MFs, the enhanced crisp pixel value is extracted from via defuzzification. The output MFs are shown in Fig. 6.2. The fundamental intuitive premise underlying the proposed FIS is that if a pixel is bright, it should be brighter, and if it is dark, it should be darker. This notion is used to develop an intuitive FIS. The Mamdani FIS is used in the algorithm. The FIS comprises a knowledge base built by experts that incorporates IF-THEN rules. The following rule base is used in the proposed FIS.

Input pixel brightness	Output pixel brightness
Very dark	Slightly dark
Dark	Very dark
Medium	Slightly bright
Bright	Slightly bright
Very bright	No change

Peak signal-to-noise ratio (PSNR) is calculated using Eq. (6.1) to measure the efficacy of the proposed fuzzy logic-based image improvement technique.

$$\text{PSNR} = 10 \log_{10}\left(\frac{G^2}{\text{RMSE}}\right)$$

$$\text{RMSE} = \frac{\sum_{M,N}\left(\text{RGB}(M,N) - \text{RGB}_E(M,N)\right)^2}{M \times N}$$

(6.1)

where G represents the maximum picture intensity, (M, N) represents image size, and RGB and RGB_E represent the original and enhanced images, respectively.

2.2   Classification using semi supervised generative adversarial network

Diaz-Pinto et al. and Odena et al. [27,28] have discussed the semisupervised GANs (SSGANs) design in detail. The standard GANs utilize two networks, a generator and a discriminator, included in a minimax game to calculate the Nash equilibrium of these two networks [29].

$$\underset{G,D}{\text{minimax}} \, V(G, D) = E_{x \sim p_{\text{data}}(x)}[\log D(x)] + E_{z \sim p_z(z)}[\log(1 - D(G(z)))] \qquad (6.2)$$

In Eq. (6.2), $E_{x \sim p_{\text{data}}(x)}$ denotes the expectation based on the training data, $\left(\text{maximize } \log(D(x))\right.$ and $E_{z \sim p_z(z)}$ denotes the expectation over the data generated by the generator $\left(\text{minimize } \log(1 - D(G(z)))\right).$

SSGANs are the architectural enhancement of GANs. The discriminator is turned into K number of classes rather than binary classification (actual vs. false in traditional GANs) [30].

$$L_{\text{semi-supervised}} = L_{\text{supervised}} + L_{\text{unsupervised}} \qquad (6.3)$$

The semisupervised loss function is represented by $L_{\text{semi-supervised}}$ in Eq. (6.3) and the total loss is employed in optimization.

$$L_{\text{supervised}} = E_{x,y \sim p_{\text{data}}(x,y)} \log\left(p_{\text{model}}(y|x, y < K+1)\right) \qquad (6.4)$$

$$L_{\text{unsupervised}} = -\left\{ E_{x \sim p_{\text{data}}(x)} \log D(x) + E_{z \sim p_z(z)}[\log(1 - D(G(z)))] \right\} \qquad (6.5)$$

$L_{\text{supervised}}$ in Eq. (6.4) denotes the supervised loss demarcated by the cross-entropy loss function in a supervised learning environment with K classes (glioma tumor,

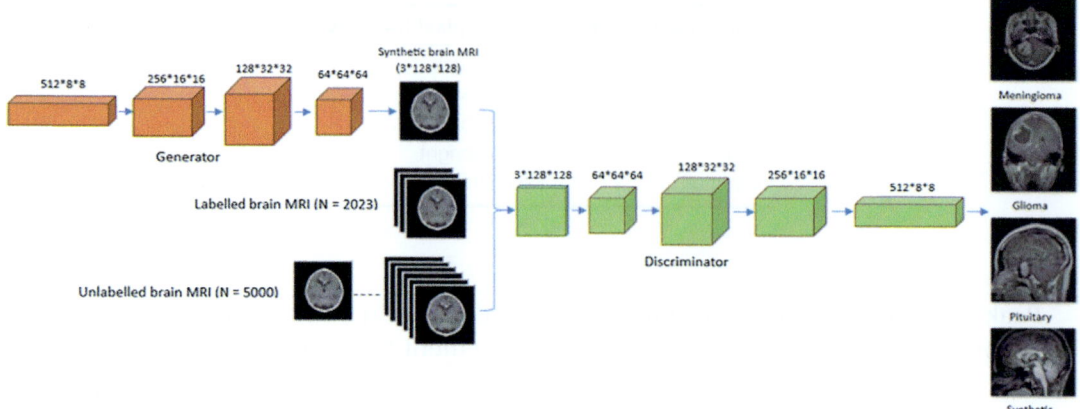

FIGURE 6.3 Schema of SSGAN.

meningioma tumor, pituitary tumor and no tumor in the current research). The loss function of typical GANs is the unsupervised loss function ($L_{unsupervised}$). As a consequence, this design enables unlabeled real-world data to take the part in learning, lowering the amount of labeling work necessary to attain a given level of accuracy.

Fig. 6.3 depicts the SSGANs architectural schema.

Brownlee et al. [31] provided directions for building the generator and discriminator, which we followed. The generator in this design is fed a 100×1 input vector (noise). The brain MRI is then scaled to the required 128×128 image size using four transpose convolutional layers (stride = 2, kernel size = 5) with rectified linear unit activation. A comparable network with a sequence of convolution layers (stride = 2, kernel size = 5) is the discriminator network. In the proposed SSGAN, one neuron is assigned for the unsupervised classification (real vs. synthetic) and four neurons for supervised classification (glioma tumor, meningioma tumor, pituitary tumor, and no tumor) in the discriminator's final output layer.

3. Experimentations and results

To train SSGAN, HP Omen Ryzen 5 Hexa Core 4600H processor, NVIDIA GeForce GTX 1650 Ti graphics, 12 GB SSD and 16 GB 2667 MHz DDR4 RAM, is utilized which minimizes the time of training. TensorFlow 2.9.2 is utilized for the implementation of the proposed method.

3.1   Dataset

This investigation's brain MRI database comprised 7023 images gathered from Ref. [32]. Images in the collection are divided into four categories: meningioma tumor (1645 images), glioma tumor (1621 images), pituitary tumor (1757 images), and no tumor (2000 images). According to the study, the abovesaid dataset is divided into 70% for training and 30% for testing. Fig. 6.4 shows various examples from the dataset.

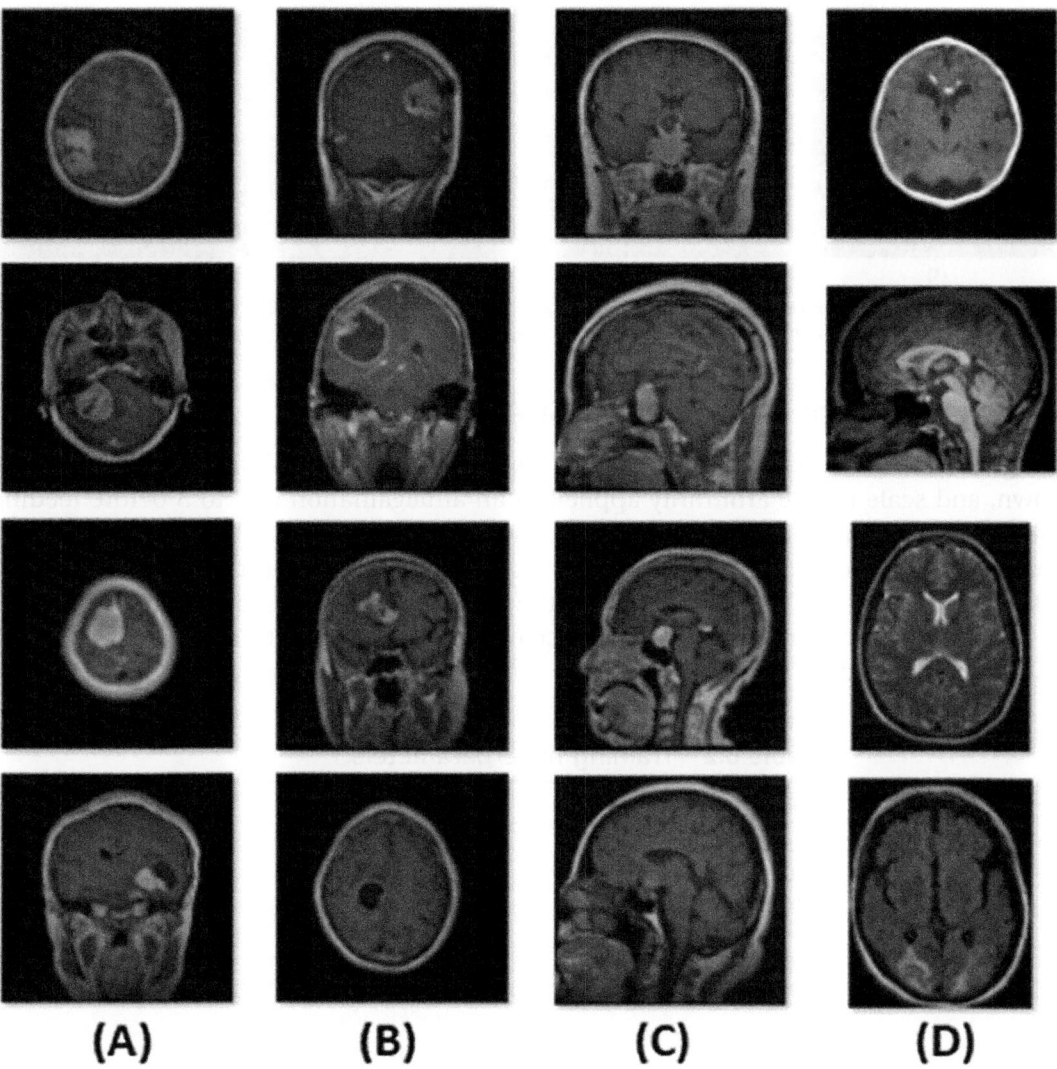

FIGURE 6.4 Sample dataset images; (A) meningioma tumor, (B) glioma tumor, (C) pituitary tumor, (D) no tumor.

3.2 Preprocessing

The suggested fuzzy image brightness enhancement methodology is compared to the output of two existing well-known image enrichment methods, HE and CLAHE. Fig. 6.5 compares the outcomes of the HE, CLAHE, and proposed fuzzy image brightness improvement methods.

PSNR is employed to evaluate the efficacy of the proposed FIS for the image enhancement. The average PSNR for HE, CLAHE, and the proposed FIS enrichment approach is 37.4, 37.2, and 43.6, respectively.

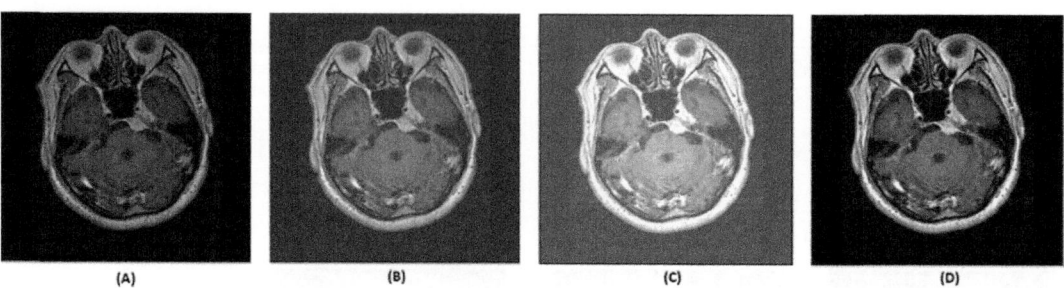

FIGURE 6.5 Image enhancement using fuzzy logic: (A) original image, (B) enhancement using HE, (C) enhancement using CLAHE, (D) enhancement using the proposed approach.

3.3   Performance evaluation of SSGAN

Image augmentation is done in this stage. Horizontal flip, vertical flip, orientation, scale down, and scale up are arbitrarily applied in an amalgamation of 1 to 5 before feeding the image to SSGAN to make the model robust from scale down, scale up, horizontal flip, vertical flip, and orientation. Classification of brain MRI is done using SSGAN as discussed in Section 2.2. Table 6.2 discusses the hyperparameters used to train the system.

The training and validation performance on the brain MRI dataset achieved by using SSGAN is reported in Table 6.3.

Table 6.2   Training hyperparameters.

Parameter	Value
Learning rate	0.001
Execution environment	GPU
Mini batch size	64
Optimizer	Adam
Loss	Categorical cross-entropy

Table 6.3   Training and validation performance using SSGAN.

No. of epoch	Training accuracy (in %)	Training loss	Validation accuracy (in %)	Validation loss	Time (in hrs)
30	78.1	0.704	71.8	0.697	5
50	82.7	0.455	79.3	0.828	7
100	85.6	0.435	82.1	2.1	10
150	90.3	0.354	84.7	0.451	11
200	94.7	0.167	86.4	0.487	12
250	98.4	0.109	89.1	0.31	15
300	98.3	0.111	88.7	0.321	19

Table 6.4 Testing accuracy.

Class	Precision (in %)	Recall (in %)	Specificity (in %)	F1 score (in %)	Accuracy (in %)
Meningioma tumor	92.7	93.3	97.3	90.2	96.2
Glioma tumor	88.9	93.7	95.7	90.3	97.7
Pituitary tumor	88.3	94.3	93.1	92.7	93.4
No tumor	96.4	100	99.3	98.3	100

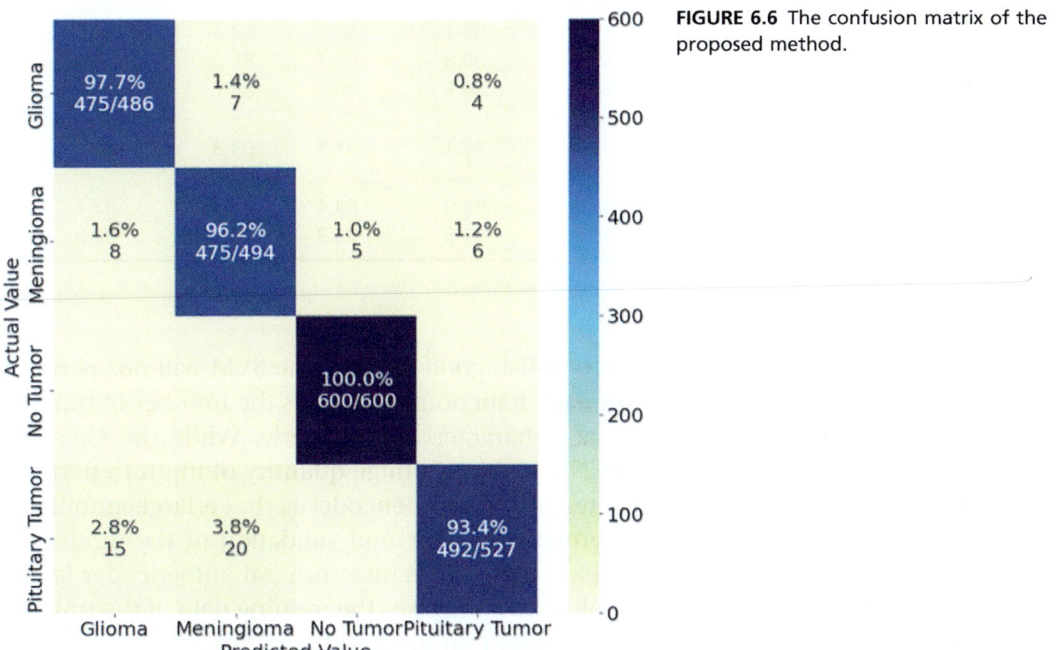

FIGURE 6.6 The confusion matrix of the proposed method.

From Table 6.3, we can see that we attained around 98.4% and 89.1% training and validation accuracy after training for 250 epochs.

Table 6.4 shows the testing accuracy using SSGAN and the brain MRI dataset.

Fig. 6.6 shows the confusion matrix produced by the proposed method.

4. Analysis

The proposed technique is analyzed by contrasting it with the existing method. The comparison is shown in Table 6.5.

Table 6.5 shows that the suggested technique has the best accuracy among the methods employed in this investigation. Some plausible explanations are mentioned below. Big datasets aren't appropriate for the SVM technique. If the dataset is noisy, that

Table 6.5   Comparison of the proposed method with other methods used by the researchers for the recognition of BT.

Model	Accuracy (in %)	Precision (in %)	Recall (in %)	Specificity (in %)	F1 score (in %)
VGGNet [6,10,24]	78.1	80.4	87.8	80.6	82.3
AlexNet [6,9,10,24]	88.7	87.3	87.4	87.7	86.6
GoogleNet [6,9,10,17]	89.3	88.2	89.2	87.4	88.1
SVM [8]	79.8	75.7	76.1	77.2	74.3
SqueezeNet [9]	87.2	85.9	86.4	89.1	85.7
CNN [11,14,19,21]	85.1	85.6	85.8	84.3	85.6
CNN + GLCM + ACO [13]	85.3	83.1	83.7	82.4	83.3
SNN [15]	89.1	88.8	88.3	87.2	88.1
Deep convolutional generative adversarial network (DCGAN) [33]	88.2	86.3	86.4	86.7	85.6
Auxiliary classifier generative adversarial network (ACGAN) [34]	93.7	88.1	92.7	93.3	89.6
Autoencoder [35]	85.2	83.9	84.4	87.1	83.7
The proposed method (SSGAN)	**96.8**	**91.5**	**95.3**	**96.3**	**92.8**

is, target classes overlap, SVM don't accomplish good results. The SVM will not perform well when the number of descriptors for each data point surpasses the number of training data samples. It is also unable to extract characteristics properly. While the ConvNet structure seeks to reduce overfitting, a CNN requires a huge quantity of input to perform successfully. The fundamental disadvantage of the autoencoder is that a large amount of data, adjustment of hyperparameters, processing time, and validation of the model are required prior to even beginning to develop the true architecture. An autoencoder learns how to proficiently characterize a manifold that comprises the training data. If the training data are not typical of the testing data, one may end up concealing rather than clarifying information. An autoencoder learns to record as much data as feasible rather than as much useful data as possible. Furthermore, other forms of GANs could not provide the degree of accuracy provided by SSGAN. For example, our model has a 91.4% accuracy, but the DCGAN has a lower accuracy (88.2%). ACGAN (93.7% accuracy) gets close to accurate prediction but falls short of SSGAN. Comparing these models, we may conclude that our SSGAN model is a better match for this particular application.

5. Conclusions and future scopes

This study proposes an innovative twofold brain MRI recognition method. Data augmentation and recognition are carried out in these two phases. Four distinct transformations are utilized in the order of 1 to 5: horizontal flip, vertical flip, orientation, scale-down, and scale-up. This variety of training images allows the system to account for changes in brain MRI patterns such as orientation, size, and flipping, ensuing in better generalization. To improve the image's brightness, an FIS is constructed. To address the limits of classic brightness enhancement approaches, an FIS is proposed. An SSGAN is utilized in the second fold to detect brain MRI images. In this case, a semisupervised approach is employed to compensate for the limitations of a fully supervised technique. All labeled images are required in a fully supervised approach, whereas SSGAN may handle both labeled and unlabeled images. The model's overall accuracy is 96.8%, and it properly recognizes the BT types as well as no tumor images. As a result, the system contributed to the body of knowledge on brain MRI recognition approaches. The suggested solution is easily adaptable to a wide range of computer vision systems.

We focused mostly on datasets of BTs in this work. In the future, we may apply the model to various brain related diseases to get the same level of accuracy.

References

[1] Díaz-Pernas FJ, Martínez-Zarzuela M, Antón-Rodríguez M, González-Ortega D. A deep learning approach for brain tumor classification and segmentation using a multiscale convolutional neural network. Healthcare 2021;9:153. https://doi.org/10.3390/healthcare9020153.

[2] Mazurowski MA, Buda M, Saha A, Bashir MR. Deep learning in radiology: an overview of the concepts and a survey of the state of the art with focus on MRI. J Magn Reson Imag 2019;49:939–54. https://doi.org/10.1002/jmri.26534.

[3] Hussain UN, Khan MA, Lali IU, Javed K, Ashraf I, Tariq J, Ali H, Din A. A unified design of ACO and skewness based brain tumor segmentation and classification from MRI scans. J Contr Eng Appl Inform 2020;22:43–55.

[4] Thaha MM, Kumar KPM, Murugan BS, Dhanasekeran S, Vijayakarthick P, Selvi AS. Brain tumor segmentation using convolutional neural networks in MRI images. J Med Syst 2019;43:1–10. https://doi.org/10.1007/s10916-019-1416-0.

[5] Bansal S, Kaur S, Kaur N. Enhancement in brain image segmentation using swarm ant lion algorithm. IJITEE 2019;8:1623–8.

[6] Tandel GS, Tiwari A, Kakde OG. Performance optimisation of deep learning models using majority voting algorithm for brain tumour classification. Comput Biol Med 2021;135:104564. https://doi.org/10.1016/j.compbiomed.2021.104564.

[7] Pedada KR, Rao B, Patro KK, Allam JP, Jamjoom MM, Samee NA. A novel approach for brain tumour detection using deep learning based technique. Biomed Signal Proc Control 2023;82:104549. https://doi.org/10.1016/j.bspc.2022.104549.

[8] Shafi ASM, Rahman MB, Anwar T, Halder RS, Kays HE. Classification of brain tumors and autoimmune disease using ensemble learning. Inform Med Unlocked 2021;24:100608. https://doi.org/10.1016/j.imu.2021.100608.

[9] Mehrotra R, Ansari MA, Agrawal R, Anand RS. A transfer learning approach for AI-based classification of brain tumors. Mach Learn Applic 2020;2:100003. https://doi.org/10.1016/j.mlwa.2020.100003.

[10] Rehman A, Naz S, Razzak MI, Akram F, Imran M. A deep learning-based framework for automatic brain tumors classification using transfer learning. Circ Syst Signal Process 2020;39:757–75. https://doi.org/10.1007/s00034-019-01246-3.

[11] Uthra Devi K, Gomathi R. Convolutional neural network based brain tumor classification using robust background saliency detection. J Med Imaging Health Inform 2021;11:2610−7. https://doi.org/10.1166/jmihi.2021.3849.

[12] Khan AH, Abbas S, Khan MA, Farooq U, Khan WA, Siddiqui SY, Ahmad A. Intelligent model for brain tumor identification using deep learning. Appl Computat Intellig Soft Comput 2022;2022:1−10. https://doi.org/10.1155/2022/8104054.

[13] Srilatha K, Chitra P, Sumathi M, Sanju MS, Jayasudha FV. Automated MRI brain tumour segmentation and classification based on deep learning techniques. In: 2022 second international conference on advances in electrical, computing, communication and sustainable technologies (ICAECT); 2022. p. 1−6. https://doi.org/10.1109/ICAECT54875.2022.9807965.

[14] Choudhury CL, Mahanty C, Kumar R, Mishra BK. Brain tumor detection and classification using convolutional neural network and deep neural network. In: 2020 international conference on computer science, engineering and applications (ICCSEA); 2020. p. 1−4. https://doi.org/10.1109/ICCSEA49143.2020.9132874.

[15] Deepak S, Ameer PM. Brain tumour classification using siamese neural network and neighbourhood analysis in embedded feature space. Int J Imag Syst Technol 2021;31:1655−69. https://doi.org/10.1002/ima.22543.

[16] Masood M, Nazir T, Nawaz M, Mehmood A, Rashid J, Kwon HY, Mahmood T, Hussain A. A novel deep learning method for recognition and classification of brain tumors from MRI images. Diagnostics 2021;11:744. https://doi.org/10.3390/diagnostics11050744.

[17] Rasool M, Ismail NA, Boulila W, Ammar A, Samma H, Yafooz WM, Emara AHM. A hybrid deep learning model for brain tumour classification. Entropy 2022;24:799. https://doi.org/10.3390/e24060799.

[18] Kurmi Y, Chaurasia V. Classification of magnetic resonance images for brain tumour detection. IET Image Process 2020;14:2808−18. https://doi.org/10.1049/iet-ipr.2019.1631.

[19] Siar M, Teshnehlab M. A combination of feature extraction methods and deep learning for brain tumour classification. IET Image Proc 2022;16:416−41. https://doi.org/10.1049/ipr2.12358.

[20] Sarhan AM. Brain tumor classification in magnetic resonance images using deep learning and wavelet transform. J Biomed Sci Eng 2020;13:102. https://doi.org/10.4236/jbise.2020.136010.

[21] Nagaraj P, Muneeswaran V, Reddy LV, Upendra P, Reddy MVV. Programmed multi-classification of brain tumor images using deep neural network. In: 2020 4th international conference on intelligent computing and control systems (ICICCS); 2020. p. 865−70. https://doi.org/10.1109/ICICCS48265.2020.9121016.

[22] Saurav S, Sharma A, Saini R, Singh S. An attention-guided convolutional neural network for automated classification of brain tumor from MRI. Neural Comput Appl 2023;35:2541−60. https://doi.org/10.1007/s00521-022-07742-z.

[23] Haq EU, Jianjun H, Li K, Haq HU, Zhang T. An MRI-based deep learning approach for efficient classification of brain tumors. J Ambient Intell Hum Comput 2023;14:6697−718. https://doi.org/10.1007/s12652-021-03535-9.

[24] Sethy PK, Behera SK. A data constrained approach for brain tumour detection using fused deep features and SVM. Multimed Tool Appl 2021;80:28745−60. https://doi.org/10.1007/s11042-021-11098-2.

[25] Lin K, Yang HF, Hsiao JH, Chen CS. Deep learning of binary hash codes for fast image retrieval. In: Proceedings of the IEEE conference on computer vision and pattern recognition workshops; 2015. p. 27−35.

[26] Mondal A, Shrivastava VK. A novel Parametric Flatten-p Mish activation function based deep CNN model for brain tumor classification. Comput Biol Med 2022;150:106183. https://doi.org/10.1016/j.compbiomed.2022.106183.

[27] Odena A. Semi-supervised learning with generative adversarial networks. arXiv Preprint arXiv:1606. 01583 2016. https://doi.org/10.48550/arXiv.1606.01583.

[28] Diaz-Pinto A, Colomer A, Naranjo V, Morales S, Xu Y, Frangi AF. Retinal image synthesis and semi-supervised learning for glaucoma assessment. IEEE Trans Med Imag 2019;38:2211—8. https://doi.org/10.1109/TMI.2019.2903434.

[29] Creswell A, White T, Dumoulin V, Arulkumaran K, Sengupta B, Bharath AA. Generative adversarial networks: an overview. IEEE Signal Proc Mag 2018;35:53—65. https://doi.org/10.1109/MSP.2017.2765202.

[30] Salimans T, Goodfellow I, Zaremba W, Cheung V, Radford A, Chen X. Improved techniques for training gans. Adv Neural Inf Proc Syst 2016;29.

[31] Brownlee J. How to implement a semi-supervised GAN (SGAN) from scratch in keras. Machine Learning Mastery; 2019.

[32] Brain tumor dataset. https://www.kaggle.com/datasets/masoudnickparvar/brain-tumor-mri-dataset.

[33] Ghassemi N, Shoeibi A, Rouhani M. Deep neural network with generative adversarial networks pre-training for brain tumor classification based on MR images. Biomed Signal Proc Control 2020;57: 101678. https://doi.org/10.1016/j.bspc.2019.101678.

[34] Kumaar MA, Samiayya D, Rajinikanth V, Raj Vincent PMD, Kadry S. Brain tumor classification using a pre-trained auxiliary classifying style-based generative adversarial network. 2023. https://doi.org/10.9781/ijimai.2023.02.008.

[35] Nayak DR, Padhy N, Mallick PK, Singh A. A deep autoencoder approach for detection of brain tumor images. Comput Electr Eng 2022;102:108238. https://doi.org/10.1016/j.compeleceng.2022.108238.

7

Multivariate adaptive signal decomposition techniques and their applications to EEG signal processing: An introduction

Kritiprasanna Das, Achinta Mondal, Nabasmita Phukan and Ram Bilas Pachori

DEPARTMENT OF ELECTRICAL ENGINEERING, INDIAN INSTITUTE OF TECHNOLOGY INDORE, INDORE, MADHYA PRADESH, INDIA

1. Introduction

Adaptive signal decomposition techniques are highly useful in a wide range of applications, as they provide flexible and adaptive methods for separating complex signals into their constituent components. These techniques have gained significant attention due to their ability to handle nonstationary and time-varying signals, where traditional methods may be limited. For example, the Fourier transform represents any signal using sinusoidal basis functions of infinite duration. Almost all the signals of interest are of finite duration, and representing a finite duration signal using an infinite duration basis function is not meaningful. Moreover, the time-varying characteristic of the signal cannot be captured appropriately using the Fourier transform. Signals exhibiting time-varying spectral content cannot be adequately characterized by conventional Fourier analysis [1–4]. In order to capture the time-varying spectral content, a short-time Fourier transform (STFT) has been proposed [5,6]. A small sliding window (the duration of the window is much lower as compared to the signal duration) has been used to select the signal corresponding to a particular time, and the spectral content is estimated using the Fourier transform for that time [7]. Choosing the proper window length is a major challenge in STFT; there is no thumb rule for choosing the window length. STFT provides uniform frequency resolution over all frequency components. Additionally, the frequency resolution of STFT-based analysis is limited by Heisenberg's boxes [8]. The wavelet transform has been introduced with the multiresolution property [9,10], which uses a time-localized window function (wavelet) for the analysis of any signal. Though these methods have been applied

Signal Processing Strategies. https://doi.org/10.1016/B978-0-323-95437-2.00011-2

for the analysis of various biomedical signals, predefined basis functions may not be well suited for the representation of real signals of complex nature. Moreover, the time-frequency localization provided by these methods is limited.

The aforementioned problems have been addressed in empirical mode decomposition (EMD), a data-adaptive signal decomposition technique [11]. In biomedical signal processing, adaptive signal decomposition methods play a crucial role in extracting relevant information from complex physiological signals. Biomedical signals often exhibit nonstationary characteristics, making them challenging to analyze using traditional techniques. Adaptive methods such as EMD and its variants allow the decomposition of signals into intrinsic mode functions (IMFs), which capture the underlying oscillatory components at different scales [11]. This enables the identification of specific frequency bands or components associated with physiological processes, such as heart rate variability in electrocardiogram signals or sleep stages in electroencephalogram (EEG) signals. Adaptive signal decomposition techniques provide versatile tools for analyzing, extracting, and manipulating signals in various domains. By adaptively decomposing signals into their constituent components, these methods enable the extraction of relevant information, the removal of noise or interference, and improved analysis, interpretation, and manipulation of complex signals in diverse applications [7]. These methods have been proposed for the analysis of univariate signals.

In recent times, novel approaches have emerged for decomposing multicomponent signals, considering the multivariate/multichannel signal paradigm [12]. The availability of multivariate (multichannel) data has increased significantly due to advancements in sensor technology. To assess the interdependence among multichannel signals through joint time-frequency analysis, the concepts of modulated bivariate and trivariate data oscillations have been introduced initially, followed by the generalization of these concepts to accommodate an arbitrary number of channels [13−15].

The chapter can be classified into two parts: An introduction to multivariate adaptive signal decomposition techniques and their diverse application to EEG signal processing. In the first part, we have demonstrated several multivariate adaptive decomposition algorithms with examples. In the second part, we have presented various applications of multivariate signal decomposition to EEG analysis.

2. Multivariate time series

The analysis of univariate modulated oscillations has seen more advanced development compared to the multivariate case. In both scenarios, the initial step involves establishing a model for the underlying structure of the signal. Let us define a multivariate time series $x(t)$ as follows [15]:

$$x(t) = \begin{bmatrix} x^1(t) \\ x^2(t) \\ \vdots \\ x^C(t) \end{bmatrix} \tag{7.1}$$

Where C is the number of channels, and $x^c(t)$ is the signal corresponding to c^{th} channel. This C-variate signal can also be termed as multichannel signal [12].

The signal can be represented in terms of amplitude-frequency modulated oscillation $u(t) = a(t)e^{j\varphi(t)}$, where $a(t) \geq 0, \frac{d\varphi(t)}{dt} \geq 0$. Here, $a(t)$ and $\varphi(t)$ denote the time-varying amplitude and phase of $u(t)$. Many approaches have been explored for univariate signals [16–18] in order to represent them using amplitude-frequency-modulated oscillatory components. C-variate monocomponent signal can be written in the form as follows:

$$x(t) = \begin{bmatrix} a^1(t)e^{j\varphi^1(t)} \\ a^2(t)e^{j\varphi^2(t)} \\ \vdots \\ a^C(t)e^{j\varphi^C(t)} \end{bmatrix} \tag{7.2}$$

Where $a^c e^{j\varphi^c(t)}$ is the amplitude-frequency modulated oscillatory component corresponding to c^{th} channel.

In practice, signals commonly have a multicomponent nature, implying that they can be expressed as linear combinations of individual signals or components [19]. A signal $x(t)$ with P-components can be written as,

$$x(t) = \sum_p^P \begin{bmatrix} u_p^1(t) \\ u_p^2(t) \\ \vdots \\ u_p^C(t) \end{bmatrix} = \sum_p^P \begin{bmatrix} a_p^1(t)e^{j\varphi_1(t)} \\ a_p^2(t)e^{j\varphi_2(t)} \\ \vdots \\ a_p^C(t)e^{j\varphi_C(t)} \end{bmatrix} \tag{7.3}$$

Where $a_p^c e^{j\varphi_p^c(t)}$ is the pth amplitude-frequency modulated oscillatory component corresponding to c^{th} channel ($p = 1, 2, ..., P$ and $c = 1, 2, ..., C$).

We have considered multichannel synthetic and real-time EEG signals to demonstrate the presented algorithms. For example, we have considered a three-channel synthetic multicomponent signal mathematically defined as,

$$x_s(t) = \begin{bmatrix} x_{s_1}(t) + 0.4x_{s_2}(t) \\ x_{s_2}(t) + x_{s_3}(t) \\ x_{s_1}(t) + x_{s_3}(t) \end{bmatrix} \tag{7.4}$$

Where $x_{s_1}(t)$, $x_{s_2}(t)$, and $x_{s_3}(t)$ are defined as follows:

$$x_{s_1}(t) = 2\sin(70\pi t + 0.5\pi \sin(2\pi t))$$

$$x_{s_2}(t) = (1 + 0.6\sin(2\pi t))\cos(40\pi t) \tag{7.5}$$

$$x_{s_3}(t) = \begin{cases} 0, 0.67\,\text{s} \leq t \leq 1.35\,\text{s} \\ \sin(10\pi t), \text{otherwise} \end{cases}$$

Each row in Eq. (7.4) represents a signal corresponding to a particular channel. In the above-defined synthetic signal, all channels have two components ($P = 2$); for example, channel 1 ($c = 1$) has components x_{s_1} and x_{s_2}. For simulation purposes, a sampling frequency of 100 Hz has been considered for the synthetic signal $x_s(t)$. Fig. 7.1 is showing the signal $x_s(t)$ and its components $x_{s_1}(t)$, $x_{s_2}(t)$, and $x_{s_3}(t)$. A particular multivariate IMF (MIMF) obtained from multivariate decomposition for a synthetic signal corresponding to all the channels is shown in a single row in the figures.

EEG is a noninvasive technique used to measure and record the electrical activity of the brain. Multiple electrodes are placed on the scalp to record the activity of different regions of the brain [20]. An example of resting-state multichannel EEG is shown in Fig. 7.2 [21], which has been used to for decomposition using multivatite algorithms discussed in this chapter.

3. Multivariate adaptive decomposition

Univariate adaptive decompositions like EMD, empirical wavelet transform (EWT), and iterative filtering have attracted the focus of large research communities from various domains. Though these methods have been used for the analysis of multichannel signals, they face drawbacks like the absence of mode alignment, the loss of mutual information, and the unequal number of oscillatory modes among different channels. One major issue with univariate EMD, mode mixing, has been attempted to be solved using a multivariate extension of EMD [22,23]. The key requirements of multivariate decomposition techniques are as follows:

(1) Mode alignment: The estimation of common frequencies or shared oscillatory patterns across channels provides proper mode alignment to the extracted oscillatory modes.

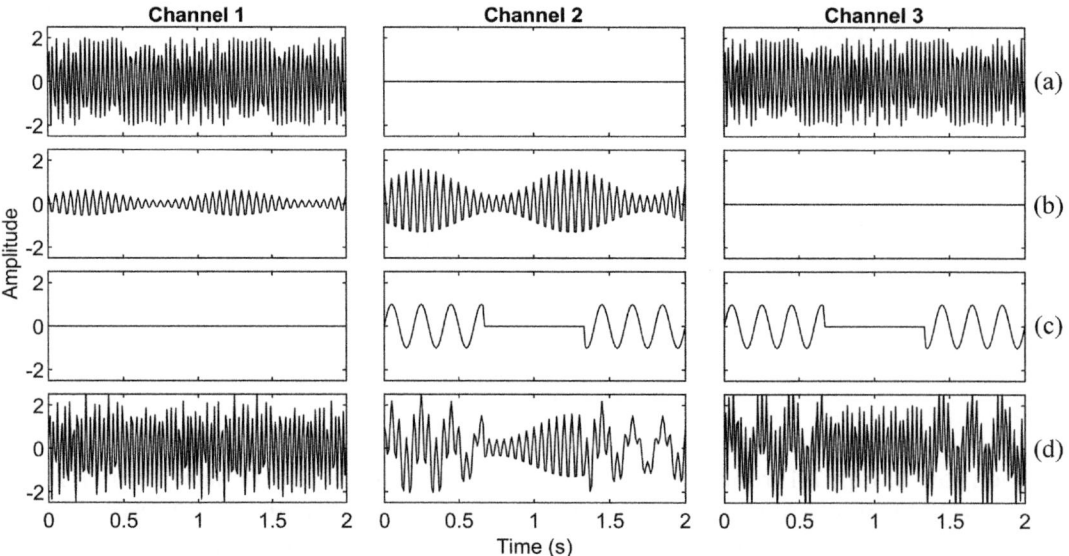

FIGURE   7.1 Amplitude-frequency modulated components (A) $x_{s_1}(t)$, (B) $x_{s_2}(t)$, and (C) $x_{s_3}(t)$, and (D) multicomponent multivariate synthetic signal $x_s(t)$.

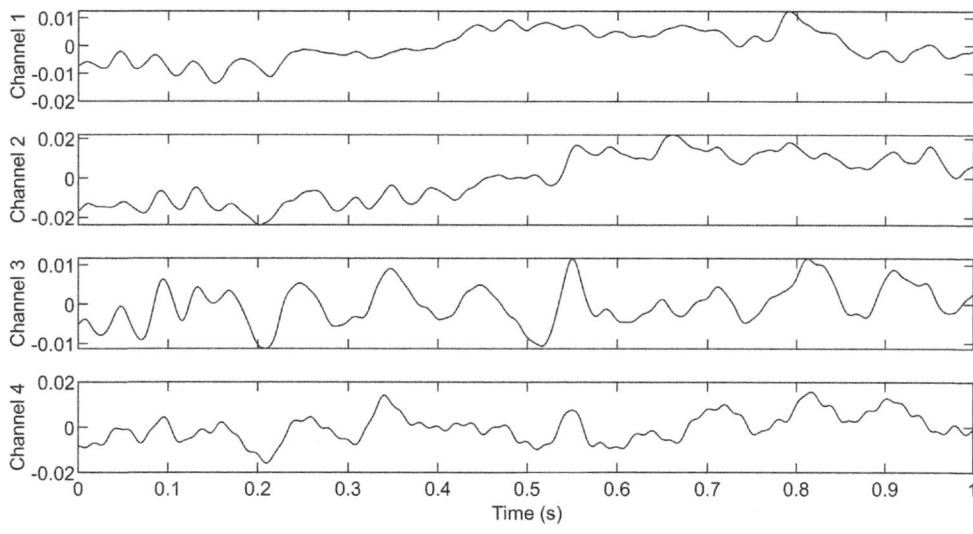

FIGURE 7.2 Four-channel EEG signals.

(2) Equal number of modes: The number of oscillatory components should be the same across different channels.

Proper mode alignment enables the algorithm to identify and characterize the joint instantaneous frequencies in a manner that accounts for variations in different oscillatory modes from different channels. By finding the joint instantaneous frequency, we gain valuable insights into the synchronized oscillatory behavior across multiple channels and obtain a comprehensive picture of the oscillatory patterns present in the multichannel signal [15,24].

The adaptive univariate decomposition techniques like EMD, EWT, variational mode decomposition (VMD), and iterative filtering have been extended to analyze multivariate signals, which have been discussed in the following sections.

3.1   Multivariate empirical mode decomposition

EMD is a data-driven signal processing technique that has gained significant popularity for analyzing nonlinear and nonstationary signals. It is an adaptive method that decomposes a signal into a set of oscillatory components called IMFs. EMD provides an automated and effective way to analyze signals with varying frequencies and time-varying characteristics, making it particularly useful in various fields such as biomedical signal processing, image analysis, and environmental signal processing [11].

The primary goal of EMD is to decompose a signal into its underlying components without making any prior assumptions about its properties. Unlike traditional Fourier-based methods that assume signal components to be sinusoidal or stationary, EMD focuses on capturing the local dynamics and extracting the inherent oscillatory patterns present in the signal. This adaptability makes EMD well-suited for analyzing signals with rapidly changing frequencies or complex nonlinear behavior.

The decomposition process in EMD starts by identifying the local extrema points (maxima and minima) of the signal. These extrema points are then connected by cubic spline interpolation to form upper and lower envelopes, which bound the signal. The mean of these envelopes, referred to as the local mean, represents the slowly varying trend or the low-frequency component of the signal. By subtracting the local mean from the original signal, the high-frequency components are extracted. The process is iteratively repeated on the obtained high-frequency components, treating them as new input signals, until a stopping criterion is met. In each iteration, the local extrema points, envelopes, and local mean are computed specifically for the current input signal. The final result of the decomposition is a set of IMFs, each characterized by a well-defined frequency scale.

The IMFs extracted by EMD satisfy two main criteria: (1) They should have equal numbers of zero crossings and extrema, and (2) the mean value of the envelopes defined by the local maxima and minima should be close to zero. These criteria ensure that the IMFs capture the oscillatory patterns inherent in the signal.

In the literature, several approaches can be found to extend the EMD algorithm for multivariate data, including bivariate EMD [25,26], rotation-invariant EMD [27], trivariate EMD [28], etc. In the context of multivariate signals, the computation of the local mean becomes a critical step since the concept of local extrema is not clearly defined. To address this challenge, Rehman and Mandic [23] proposed a novel approach that involves calculating envelopes and the local mean for multivariate signals by employing real-valued projections along multiple directions on hyperspheres (n-spheres). To extend the concept of EMD, a set of direction vectors has been defined that facilitates the decomposition process in the multivariate domain. A unit hypersphere (n-spheres) has been sampled using both uniform angular sampling methods and quasi-Monte Carlo-based low-discrepancy sequences to obtain these direction vectors. This approach enables the adaptation of EMD to multivariate signals, allowing for effective decomposition and analysis of complex multivariate data.

The multivariate time series $x(t)$ can be represented as a C-dimensional vector $\{x(t)\}_{t=1}^{T} = \{x_1(t), x_2(t), ..., x_C(t)\}$. A set of direction vectors $v^{\gamma_k} = \{v_1^k, v_2^k, ..., v_C^k\}$ along the directions $\gamma_k = \{\gamma_1^k, \gamma_2^k, ..., \gamma_{C-1}^k\}$ is defined on an $(C-1)$-sphere. With the help of this, the MEMD algorithm is described using the following steps [23]:

Step 1: Select an appropriate set of points for sampling on an $(C-1)$-dimensional sphere.

Step 2: Compute the projection, represented as $\{p^{\gamma_k}(t)\}_{t=1}^{T}$, of the input signal $\{x(t)\}_{t=1}^{T}$ along the direction vector v^{γ_k} for all k (the complete set of direction vectors), resulting in the set of projections $p^{\gamma_k}(t)$ for $k = 1$ to K.

Step 3: Identify the time points $\left\{ t_j^{\gamma_k} \right\}$ that correspond to the maxima of the set of projected signals $p^{\gamma_k}(t)$ for k ranging from 1 to K.

Step 4: Perform interpolation on the pairs $\left[t_j^{\gamma_k}, x\left(t_j^{\gamma_k} \right) \right]$ to generate multivariate envelope curves $\Delta^{\gamma_k}(t)$ for k ranging from 1 to K.

Step 5: The mean $m(t)$ of the envelope curves for a set of K direction vectors is computed as follows:

$$m(t) = \frac{1}{K} \sum_{k=1}^{K} \Delta^{\gamma_k}(t)$$

Step 6: Calculate the detail component $d(t)$ as the difference between the original signal $x(t)$ and the mean $m(t)$ of the envelope curves: $d(t) = x(t) - m(t)$. If $d(t)$ satisfies the stopping criterion [11], apply the same procedure described above to $x(t) - d(t)$, treat it as the new input signal. Otherwise, apply the procedure to $d(t)$ instead.

The synthetic signal $x_s(t)$ (defind in Eq. (7.1)) is decomposed using MEMD, and the MIMFs are shown in Fig. 7.3. The first three MIMFs contained two components x_{s_1} and x_{s_2} of the synthetic signal. MEMD fails to separate x_{s_1} and x_{s_2} components properly and suffers from mode mixing. The EEG signal (Fig. 7.2) is decomposed using MEMD, and the MIMFs are presented in Fig. 7.4. The MEMD algorithm is computationally very complex [43,65]. The MEMD algorithm is modified to reduce the computational time in Fast MEMD [29].

3.2   Multivariate empirical wavelet transform

The EWT is a data-driven signal analysis technique that combines the principles of EMD (representing signals with the help of IMFs) and wavelet analysis. It aims to decompose a signal into a set of localized oscillatory components called empirical wavelets, which capture the local dynamics and spectral characteristics of the signal in both time and frequency domains. Unlike traditional wavelet analysis, the EWT does not rely on pre-defined wavelet basis functions but derives the wavelets directly from the data. The decomposition process in EWT involves selecting a set of scales or frequencies for each IMF and applying wavelet transforms at those scales. The wavelet transforms are performed using adaptive basis functions derived from the IMF's local characteristics.

The adaptive wavelet-based bandpass filters generated by univariate EWT result in distinct components for multichannel signals corresponding to different channels. The number of components and their frequency ranges may vary across channels, creating an obstacle for multivariate analysis. To adapt univariate EWT for multichannel signals, a new concept for adaptive boundary detection is proposed to extract MIMFs for each channel [24]. This modified boundary detection ensures the MIMFs have similar

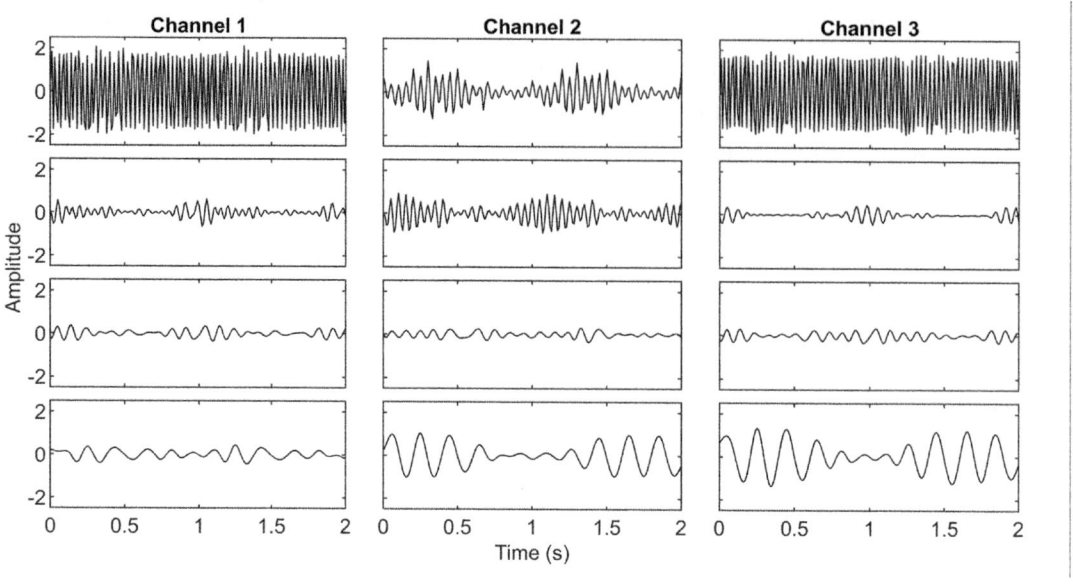

FIGURE 7.3 MIMFs corresponding to the synthetic signal obtained from MEMD.

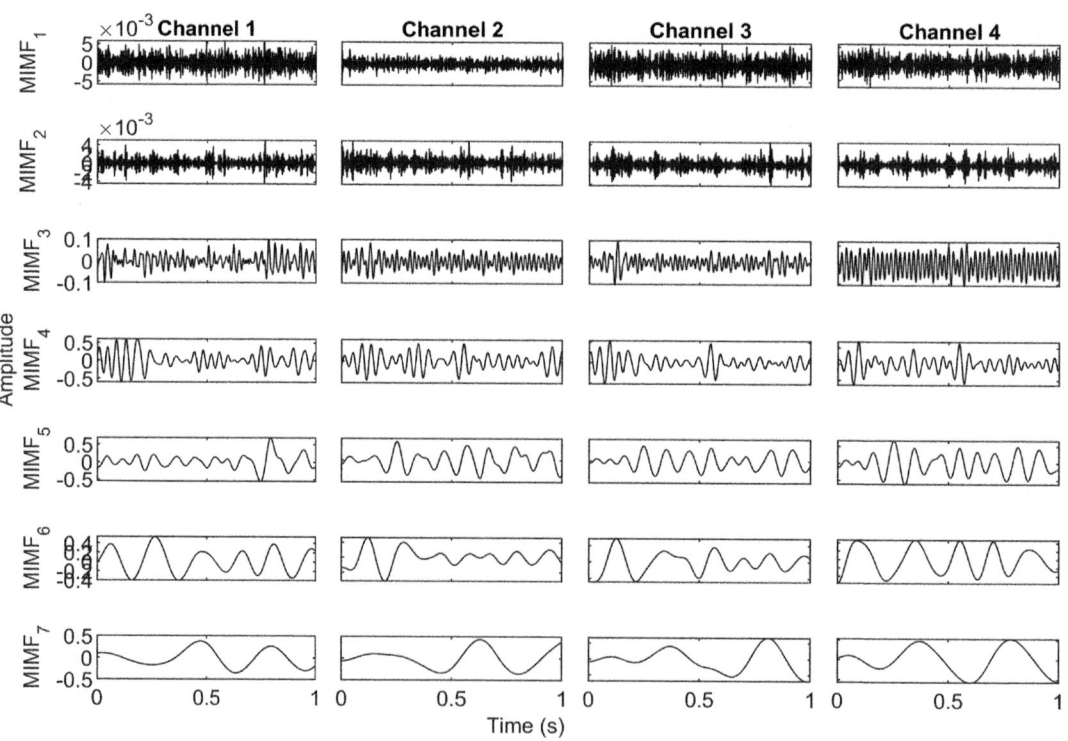

FIGURE 7.4 MIMFs corresponding to the EEG signal obtained from MEMD.

frequency components across different channels. The steps involved in multivariate EWT (MEWT) for multivariate signals $x(t)$ (defined in Eq. (7.1)) are explained below.

Step 1: To obtain a unique set of boundaries for all channels, the authors in Ref. [24] calculated the mean spectrum magnitude of multichannel signals acquired from all channels. The mean spectrum magnitude is defined as follows:

$$\widehat{X}(f) = \frac{1}{C} \sum_{c=1}^{C} |X^c(f)|$$

where the Fourier spectrum of cth channel signal $x^c(t)$ is denoted as $X^c(f)$.

Step 2: The mean Fourier spectrum is segmented into N contiguous segments using the EWT boundary detection method [30]. The set of boundaries can be denoted as $\{\Omega_b\}_{b=0,1,\dots,B}$. It is important to mention that the first boundary frequency (Ω_0) is 0, and the last boundary frequency (Ω_B) is π. Based on these boundaries, the segment can be defined as $[0, \Omega_1], [\Omega_1, \Omega_2], \dots, [\Omega_{B-1}, \Omega_B]$.

Step 3: Empirical wavelets are defined as the bandpass filters applied to each segment. To construct the empirical wavelet-based filter for each segment, we draw inspiration from the concept used in the construction of Littlewood-Paley and Meyer's wavelets [10]. The wavelet and scaling functions are defined empirically as follows [30]:

$$\text{Scaling function}: \chi_b(\Omega) = \begin{cases} 1, & \text{if } |\Omega| \leq (1-\varphi)\Omega_b \\ \cos\left(\dfrac{\pi\Lambda(\varphi,\Omega_b)}{2}\right), & \text{if } (1-\varphi)\Omega_b \leq |\Omega| \leq (1+\varphi)\Omega_b \\ 0, & \text{otherwise} \end{cases}$$

$$\text{Wavelet function}: \Psi_b(\Omega) = \begin{cases} 1, & \text{if } (1+\varphi)\Omega_b \leq |\Omega| \leq (1-\varphi)\Omega_{b+1} \\ \cos\left(\dfrac{\pi\Lambda(\varphi,\Omega_{b+1})}{2}\right), & \text{if } (1-\varphi)\Omega_{b+1} \leq |\Omega| \leq (1+\varphi)\Omega_{b+1} \\ \sin\left(\dfrac{\pi\Lambda(\varphi,\Omega_b)}{2}\right), & \text{if } (1-\varphi)\Omega_b \leq |\Omega| \leq (1+\varphi)\Omega_b \\ 0, & \text{otherwise} \end{cases}$$

where $\Lambda(\varphi,\Omega_b) = \kappa\left(\frac{|\Omega|-(1-\varphi)\Omega_b}{2\varphi\Omega_b}\right)$. Here, the condition on variable φ ensures that the scaling and wavelet functions will have a tight frame, which can be mathematically expressed as $\varphi < \left(\frac{\Omega_{b+1}-\Omega_b}{\Omega_{b+1}+\Omega_b}\right)$. $\kappa(p)$ is an arbitrary function, as defined below.

$$\kappa(p) = \begin{cases} 0, & \text{if } p \leq 0 \\ \kappa(p) + \kappa(1-p) = 1, & \forall p \in [0,1] \\ 1, & \text{if } p \geq 1 \end{cases}$$

Step 4: The approximation and detail coefficients are obtained by taking the inner product of the applied signal $x(t)$ with the scaling and wavelet functions.

The wavelet and scaling functions obtained in the previous step will be the same for all channels; hence, they provide exactly the same number of modes and frequency-aligned modes in all channels. The MIMFs from MEWT for synthetic signals $x_s(t)$ are shown in Fig. 7.5. MEWT has divided the first component x_{s_1} into two modes, which is known as mode splitting problem. The MEWT-based decomposition result for the EEG signal is shown in Fig. 7.6.

3.3   Multivariate Fourier-Bessel series expansion-based empirical wavelet transform

The Fourier-Bessel series possesses orthogonality properties that facilitate signal representation and analysis. This property enables straightforward decomposition and reconstruction of signals using the expansion coefficients [7,31,32]. Since nearly all real-time signals are nonstationary in nature, it is essential to use nonstationary basis functions for their representation. The widely used Fourier transform representation uses sinusoidal functions as basis functions. The Fourier-Bessel representation employs nonstationary Bessel functions as a basis set, which helps to provide a more meaningful representation. The Fourier-Bessel representation of a signal yields unique coefficients, and their length is equal to the length of the signal. In contrast, the Fourier transform provides unique coefficients of length equal to half the length of the signal (for a real signal). As a result, the Fourier-Bessel representation can offer double-frequency

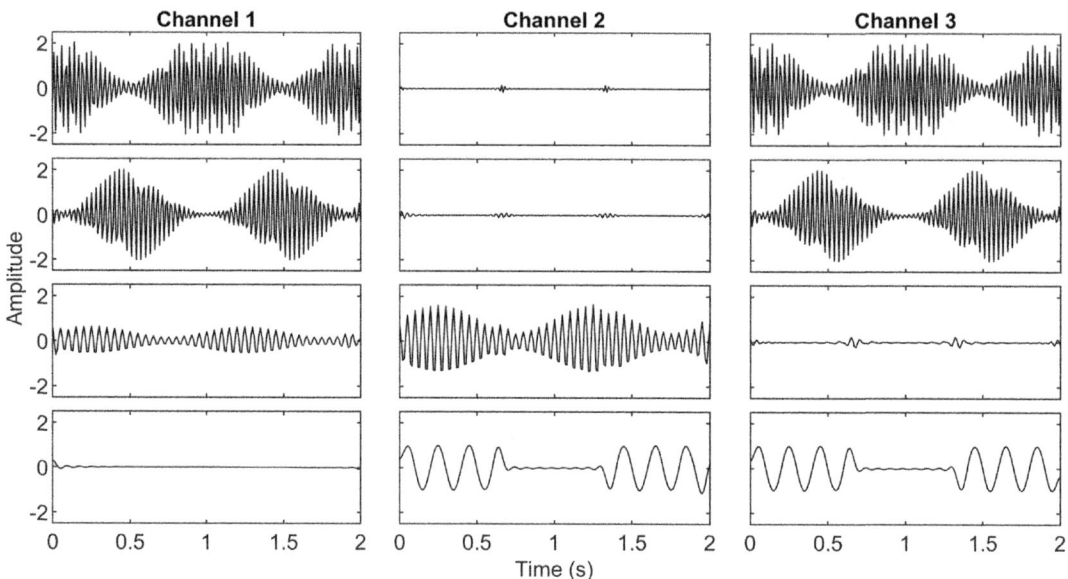

FIGURE 7.5 MIMFs corresponding to the synthetic signal obtained from MEWT.

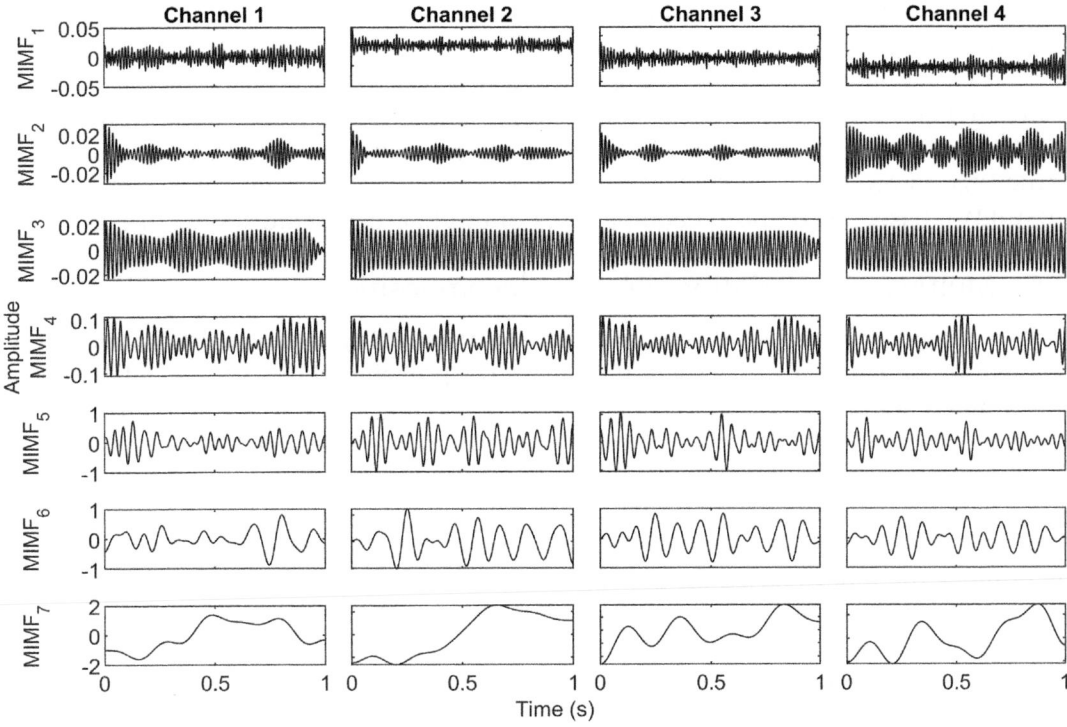

FIGURE 7.6 MIMFs corresponding to the EEG signal obtained from MEWT.

resolution compared to the Fourier transform-based representation. Fourier-Bessel series expansion (FBSE) has been used to represent EEG signals in the literature and has better performance [21,33–35].

By leveraging this advantage, the EWT has been improved by using the FBSE [36]. The univariate FBSE-based empirical wavelet transform (FBSE-EWT) is extended for multichannel signals in Refs. [32,35–37]. The multivariate FBSE-EWT (M-FBSE-EWT) has similar steps to MEWT, described in Section 3.2, except the boundary detection procedure. In M-FBSE-EWT, the FBSE spectrum is used instead of the Fourier transform-based spectrum. Due to the higher resolution of the FBSE spectrum, the signal separation has improved [36]. The FBSE spectrum of the signal ($y(n)$) can be obtained from the FBSE coefficients of the signal [32,37]. The FBSE coefficient can be computed as follows:

$$C_i = \frac{2}{U^2 (J_1(\beta_i))^2} \sum_{n=0}^{U-1} ny(n) J_0\left(\frac{\beta_i n}{U}\right)$$

where, $J_0(\,\cdot\,)$ and $J_1(\,\cdot\,)$ denote zero and first-order Bessel functions, respectively. β_i with $i = 1, 2, \ldots, U$ are the positive roots of the zero-order Bessel function ($J_0(\beta) = 0$) arranged in ascending order. The magnitude of the FBSE coefficient $|C_i|$ will provide the FBSE spectrum [36]. For M-FBSE-EWT, the Fourier spectrum in MEWT is replaced with the

FBSE spectrum for better separation of components, and the other steps are the same as in MEWT.

The separated components using M-FBSE-EWT for the synthetic signal $x_s(t)$ are shown in Fig. 7.7. All three modes present in the synthetic signal are separated into three MIMFs by M-FBSE-EWT. First, MIMF is an insignificant component that can be discarded based on energy-based thresholding or other thresholding techniques. The M-FBSE-EWT-based decomposed components of the EEG signal are shown in Fig. 7.8.

3.4   Multivariate variational mode decomposition

VMD is a signal processing technique that decomposes a signal into a set of modes with distinct frequency bands [38]. It is a data-driven approach that provides an adaptive and self-tuning method for analyzing signals with varying frequencies and time-varying characteristics.

The main idea behind VMD is to find a set of modes that best capture the signal while minimizing cross-mode interference. Unlike traditional Fourier-based methods that assume a fixed set of sinusoidal components can represent the signal, VMD adaptively determines the modes based on the signal's intrinsic characteristics.

The VMD algorithm starts by assuming that the signal can be represented as a sum of K modes, each mode representing a component with a distinct frequency band. The decomposition is obtained by solving an optimization problem that promotes both sparsity and smoothness of the modes. The optimization problem seeks to find the

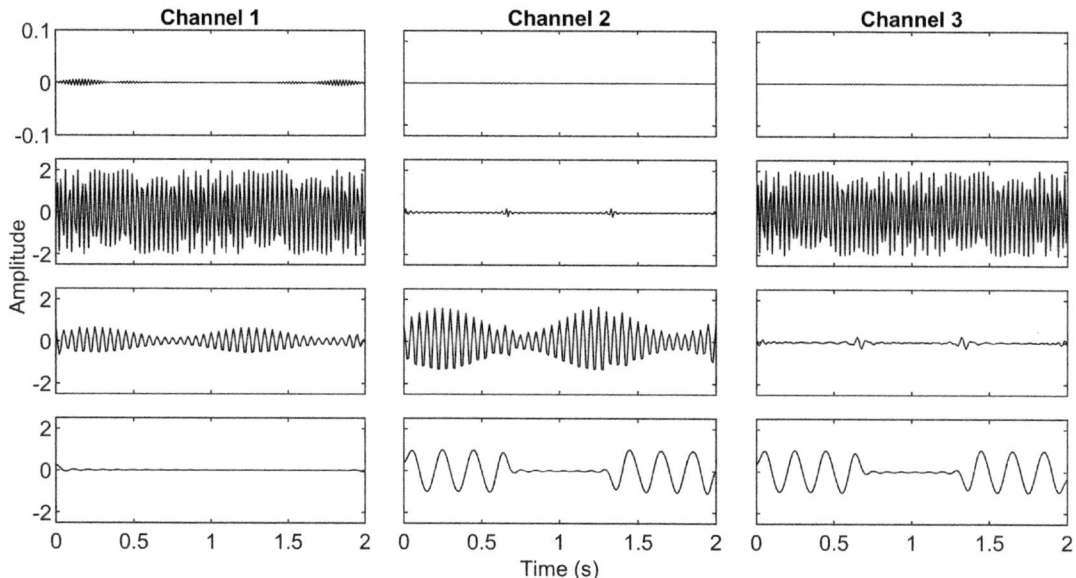

FIGURE 7.7 MIMFs corresponding to the synthetic signal obtained from M-FBSE-EWT.

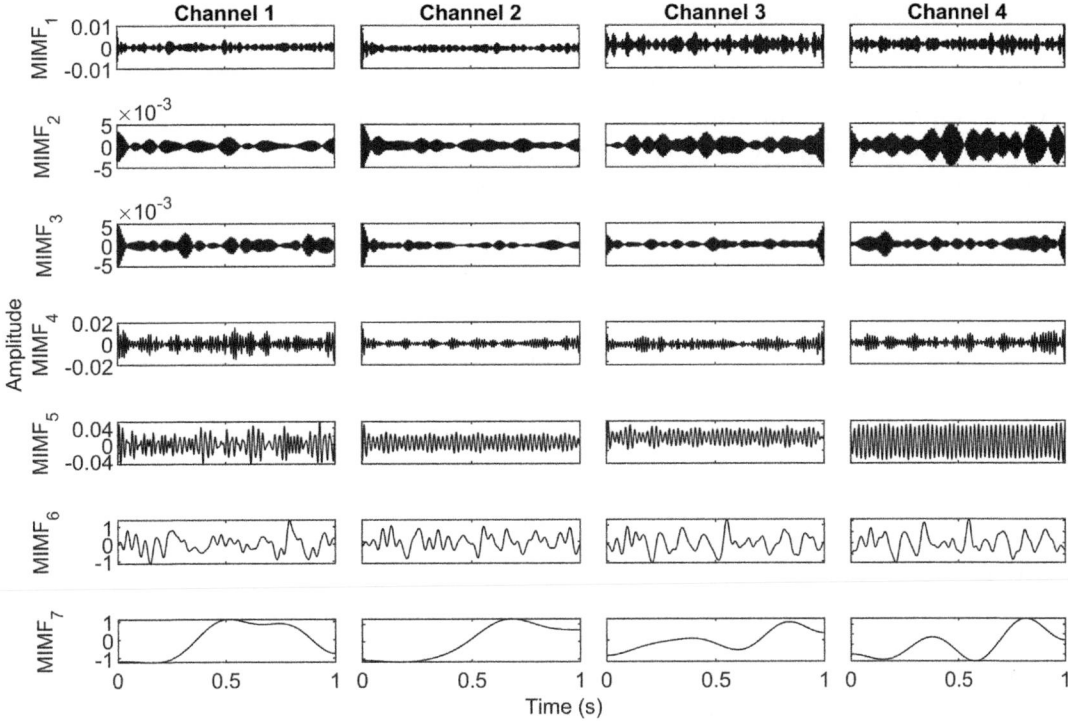

FIGURE 7.8 MIMFs corresponding to the EEG signal obtained from M-FBSE-EWT.

modes that best capture the signal while minimizing the mutual interference between the modes.

To solve the optimization problem, VMD uses a constrained optimization approach. It introduces a constraint on the instantaneous frequency of each mode, which limits the spread of energy in the time-frequency plane. By imposing this constraint, VMD ensures that each mode is localized in both the time and frequency domains.

The VMD algorithm iteratively updates the modes and their associated weights until convergence is achieved. At each iteration, the algorithm estimates the modes by minimizing the objective function, which is a combination of data fidelity and regularization terms. The weights represent the importance or energy distribution of each mode in the signal.

Rehman and Aftab [39] proposed an extension of the VMD algorithm to handle multivariate or multichannel data. The proposed multivariate VMD (MVMD) aims to effectively capture multivariate modulated oscillations (joint or common frequency components shared across all channels) present in the signal. In MVMD, the cost function to be minimized is an extension of the cost function used in standard VMD, specifically tailored for multivariate data. The cost function is defined as the sum of the bandwidths of all signal modes across all input data channels. This modified cost function takes into account the characteristics of multivariate signals and aims to optimize the decomposition by minimizing the collective bandwidths of the modes across all channels. By minimizing this cost function, the MVMD method aims to achieve an effective decomposition of the multivariate data into distinct modes with minimized bandwidths. A C-variate signal $x(t)$ defined in Eq. (7.1) can be represented using K number of multivariate oscillatory components $v_k(t)$ having center frequency ω_k as,

$$x(t) = \sum_{k=1}^{K} v_k(t)$$

where, $v_k(t) = [v_{k,1}(t), v_{k,2}(t), ..., v_{k,C}(t)]$. The steps involved in the MVMD algorithm to find $v_k(t)$ are presented below.

Step 1: Initialize the center frequency ω_k of the multivariate modes, which can be done in various ways, such as by choosing complete random values.

Step 2: Compute the analytic representation $\widehat{v}_k(t)$ of the multivariate signal $v_k(t)$ to obtain the spectrum having a frequency in the positive frequency part only.

Step 3: For each mode obtained in the decomposition, apply a frequency shift to bring the mode's frequency spectrum to the baseband by mixing it with an exponential function tuned to the estimated center frequency of the mode.

Step 4: The estimation of bandwidth is performed by assessing the Gaussian smoothness of the demodulated signal, specifically by considering the squared norm of the gradient. This smoothness measure provides an estimate of the bandwidth of each mode. The resulting constrained variational problem can be formulated as follows:

$$\underset{\{v_{k,c}\}\{\omega_k\}}{\text{minimize}}\left\{\sum_k\sum_c ||\partial_t[e^{-j\omega_k t}\widehat{v}_{k,c}(t)]||_2^2\right\} \text{ subject to } \sum_k \widehat{v}_{k,c}(t) = x^c(t), c = 1, 2, ..., C$$

Step 5: Convert the above-constrained optimization problem into an unconstrained optimization problem by introducing a quadratic multiplier and Lagrangian function, denoted as $\mathscr{L}$. The mathematical expression can be given as follows:

$$\mathscr{L}(\{v_{k,c}(t)\}, \{\omega_k\}, \{\lambda^c\}) = \beta\sum_k\sum_c ||\partial_t[e^{-j\omega_k t}\widehat{v}_{k,c}(t)]||_2^2 + \sum_c ||x^c(t) - \sum_k v_{k,c}(t)||_2^2$$
$$+ \sum_c \langle \lambda^c(t), x^c(t) - \sum_k v_{k,c}(t) \rangle$$

Step 6: The complex optimization problem can be efficiently solved using the alternating direction method of multipliers (ADMMs) algorithm. ADMM is a powerful optimization technique that decomposes the problem into multiple simpler subproblems, making it easier to solve [38,40].

The MVMD-based MIMFs corresponding to the synthetic signal have been shown in Fig. 7.9. It also separates the three components into four MIMFs. The first component x_{s_1} is split into two modes, similar to MEWT. The MVMD-based decomposition of the EEG

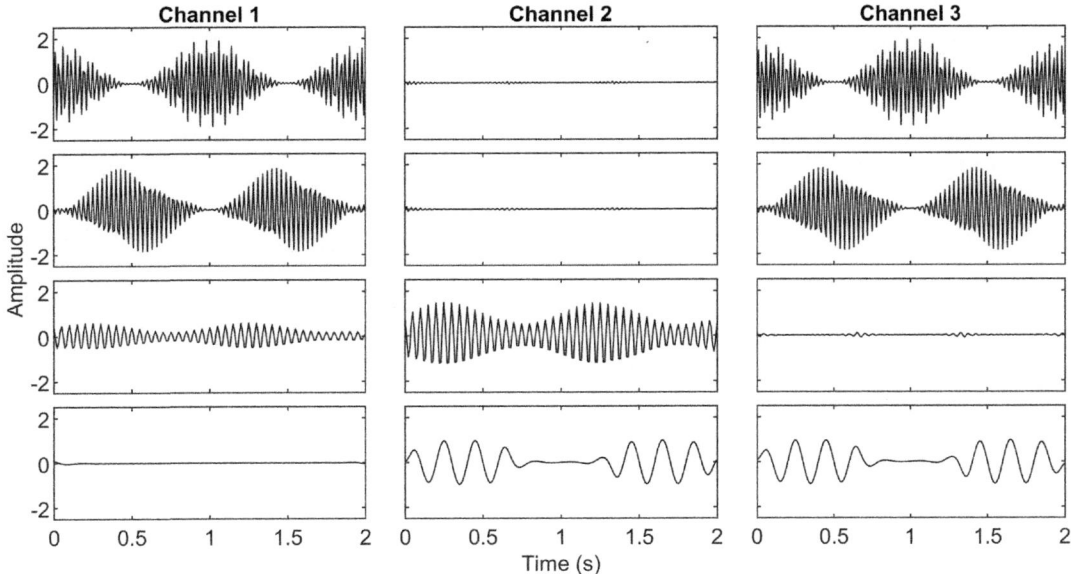

FIGURE 7.9 MIMFs corresponding to the synthetic signal obtained from MVMD.

signal is shown in Fig. 7.10. Liu and Yu have proposed an alternate extension of the VMD algorithm for successive extraction of MIMFs [41].

3.5   Multivariate iterative filtering

An iterative data-driven method has been proposed similar to EMD in Ref. [42], which addresses problems like no proof convergence and the mathematical theory associated with EMD. Iterative filtering extracts the oscillatory components by using a moving average filter designed based on the extrema present in the signal. The univariate iterative filtering has been extended for multivariate signals by choosing a unique moving average filter for all channels [43,44]. Das and Pachori have proposed a simple way of choosing a unique moving average filter by choosing the maximum among the extrema of all channels. The algorithm is defined in discrete signal form. Multivariate iterative filtering (MIF) is described by the following steps:

Step 1: Define a moving average filter $(a(n))$ of length L based on the extrema present in the signal $x(n) \in \mathbb{R}^{N \times C}$, where N is the number of samples [42]. Length L is computed as,

$$L = \left\lfloor \frac{\gamma N}{\max(\mathbf{E})} \right\rfloor$$

Here, γ represent a constant, $\mathbf{E}$ vector hold the number of extrema of all C channels ($\mathbf{E} = [e_1, e_2, ..., e_C]$, where e_c is the number of extrema for c^{th} channel), and $\max(\cdot)$ is an

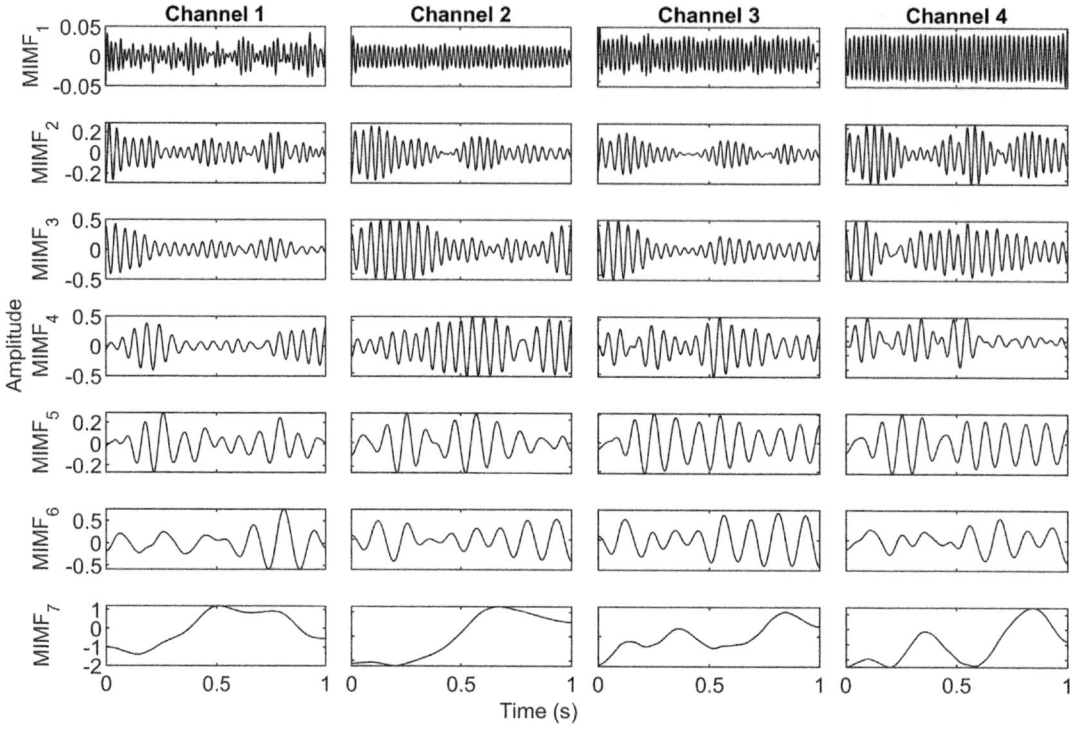

FIGURE 7.10 MIMFs corresponding to the EEG signal obtained from MVMD.

operator to find the maximum value. The moving average operation can be defined by an operator $\mathrm{MA}(\cdot)$ as,

$$\mathrm{MA}(x) = a(n) * x(n)$$

where $*$ denotes the convolution operator.

Step 2: With the help of the moving average operator, the signal $x(n)$ is sifted iteratively, which is defined with an operator $\Xi(\cdot)$ as,

$$\Xi\left(\overset{k}{x}(n)\right) = \overset{k}{x}(n) - \mathrm{MA}\left(\overset{k}{x}(n)\right) = \overset{k+1}{x}(n)$$

The variable k above $x(n)$ denotes the intermediate signal at k^{th} iteration step.

Repeated application of the sifting operator $\Xi(\cdot)$ on the input signal, j times, effectively isolates the fluctuating part of the signal, known as the MIMF. The p-th MIMF I_p, can mathematically expressed as $I_p = \lim_{j \to \infty} \Xi^j(x)$. Here, j represents the number of times the sifting operator operates on the signal $x(n)$, which is ideally infinite. Based on the IMF stopping criterion defined in Ref. [11], the iteration can be stopped after a finite repetition.

Step 3: Upon subtracting the extracted MIMF from the signal $x(n)$, if the resulting signal still contains oscillatory components, proceed to apply steps 1 and 2 iteratively to extract the remaining MIMFs.

Conversely, when there are no oscillatory components present (i.e., the number of extrema is at most one), the remaining signal can be regarded as a trend component $r(n)$. The input signal $x(n)$ can then be expressed as a combination of MIMFs and the trend component as follows:

$$x(n) = \sum_{p=0}^{P-1} I_p(n) + r(n)$$
$$= \sum_{p=0}^{P} I_p(n), r(n) \overset{\triangle}{=} I_P(n)$$

where P is the total number of MIMFs.

The MIF-based decomposed components of the synthetic signal $x_s(t)$ are shown in Fig. 7.11. The first three MIMFs of the MIF algorithm correspond to three components of $x_s(t)$. The MIF-based EEG signal decomposition is shown in Fig. 7.12. Cicone and Pellegrino proposed another extension of univariate iterative filtering in Ref. [44].

3.6   Other multivariate adaptive decomposition techniques

In the literature, several other univariate adaptive data decomposition approaches have been proposed, like ensemble EMD [45], local mean decomposition [46], singular spectrum analysis (SSA) [47–49], local characteristic scale decomposition [50], intrinsic

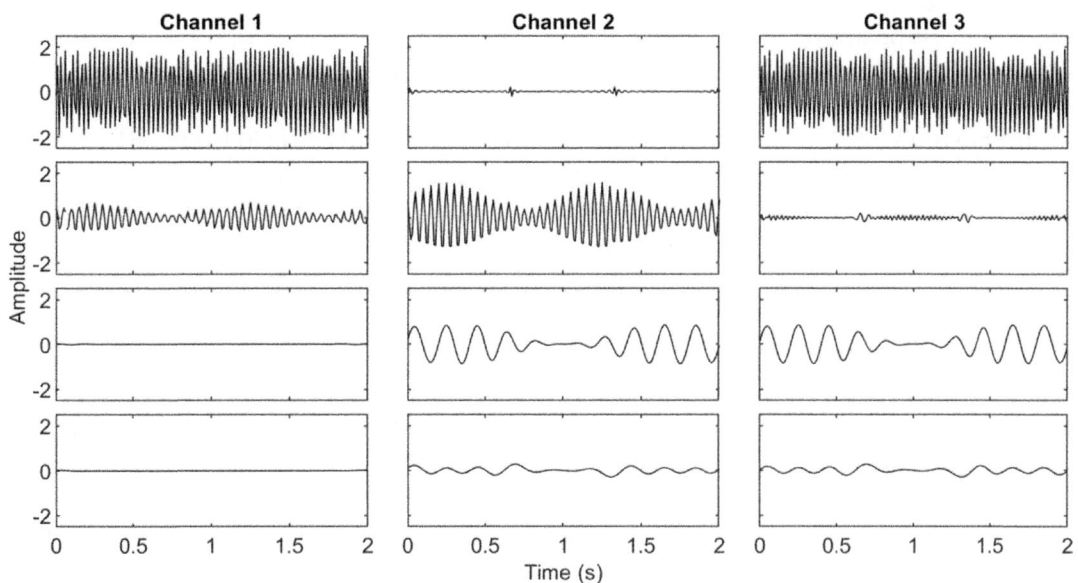

FIGURE 7.11 MIMFs corresponding to the synthetic signal obtained from MIF.

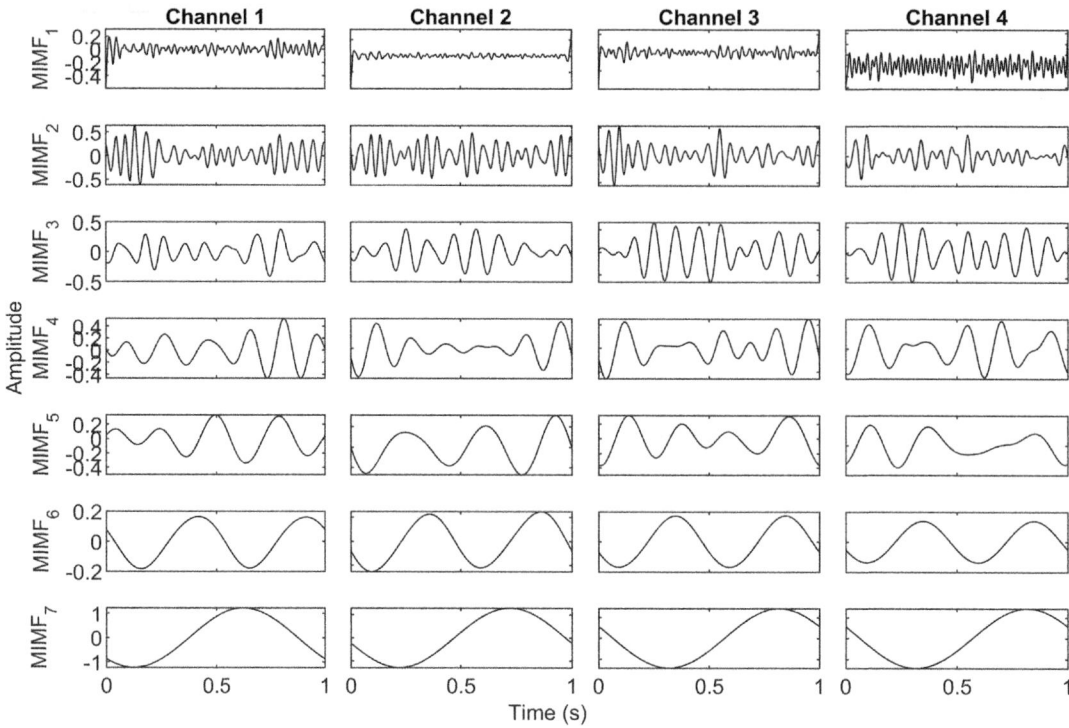

FIGURE 7.12 MIMFs corresponding to the EEG signal obtained from MIF.

time scale decomposition [51], dynamic mode decomposition (DMD) [52], nonlinear mode decomposition [53], adaptive local iterative filtering [54], and the Fourier decomposition method [55]. Several approaches among these techniques have been extended for multivariate data analysis, namely, multivariate nonlinear chirp mode decomposition [56,57], multivariate SSA [58,59], multivariate DMD [60], etc. [61]. Various adaptive signal decomposition and time-frequency analysis frameworks have been discussed in Ref. [7], which can be further extended for multivariate frameworks.

4. Application of multivariate adaptive decomposition to EEG signal processing

The EEG measures and records the electrical activity of the brain produced by the collective firing of neurons in the brain. It is a widely used tool in the fields of neuroscience and clinical diagnostics to study brain function and detect abnormalities [62]. In clinical settings, it is used to diagnose and monitor neurological disorders, such as epilepsy, sleep disorders, and brain tumors. In research, EEG helps to investigate cognitive processes, study brain responses to stimuli, and understand brain disorders and their underlying mechanisms.

Signal processing techniques are applied to EEG data to extract meaningful information. These techniques include filtering to remove noise, spectral analysis to examine frequency content, event-related potential analysis to study specific brain responses, and source localization to identify the origin of brain activity. EEG signals are highly nonstationary and complex in nature. Adaptive signal decomposition techniques adaptively adjust their basis functions to effectively capture the changing features of the signal. By utilizing adaptive methods, these techniques can offer a more accurate and detailed analysis of the complex EEG signals.

EEG signals provide very high temporal resolution but suffer from poor spatial resolution. Spatial resolution can be improved by using more number of electrodes. The processing of these multichannel EEG signals demands multivariate decomposition techniques.

Since the last decades after the introduction of adaptive multivariate decomposition techniques, it has found various applications in EEG signal analysis and classification [33,37,63—66]. The applications of EEG can be broadly classified as brain-computer interfaces (BCIs) and neurological disease diagnosis. The following sections discuss a few representative applications of multivariate adaptive signal decomposition techniques for EEG analysis.

4.1 Brain-computer interface (BCI)

The authors explore the application of MEMD for emotion recognition as high/low arousal and high/low valence states using EEG signals [63]. The study addresses the challenges of nonstationary EEG signals and multichannel analysis. The proposed MEMD-based method extracts MIMFs from multichannel EEG signals and extracts various time and frequency domain features, such as power ratio, power spectral density, entropy, Hjorth parameters, and correlation, for valence and arousal states. The method is evaluated on the 18 channels of a database for emotion analysis using the physiological signals emotional EEG dataset. With MEMD-based features, independent component analysis, and artificial neural networks (ANN), the method obtained an accuracy of $75 \pm 7.48\%$ for high/low arousal and 72.87 ± 4.68 for high/low valence states.

A novel multivariate-multiscale approach for computing EEG spectral and temporal complexity for human emotion recognition was proposed in Ref. [37]. This study introduces a novel multivariate-multiscale approach to compute spectral and temporal entropies from multichannel EEG signals, enabling the recognition of human emotions (positive, neutral, and negative). The proposed method is based on the extension of FBSE-EWT for multichannel signals. The multivariate Hilbert marginal spectrum is derived to compute spectral Shannon and K-nearest neighbor (KNN) entropies. The approach decomposes the multichannel EEG signals into narrow band subbands using M-FBSE-EWT and then computes entropies in both spectral and temporal domains. The extracted entropy features are smoothed and fed into a sparse autoencoder-based random forest classifier for emotion classification. The proposed method achieves an impressive overall classification accuracy of 94.4% using the SJTU emotion EEG dataset and outperforms existing state-of-the-art methods evaluated on the DREAMER emotion EEG public database.

Chang et al. [67] focus on enhancing BCI performance using steady-state visual evoked potential (SSVEP) patterns. To address EEG interference and SSVEP variations across channels and frequency bands, the authors propose a method called MVMD-informed canonical correlation analysis (MVMD-CCA). The approach involves decomposing EEG into inherent mode functions or MIMFs using MVMD to reduce artifact effects. The MIMFs are weighted and reconstructed using the sparrow search algorithm for stability and fast convergence. The canonical correlation analysis is applied for classification. Experimental results with nine subjects using an eight-target SSVEP system demonstrate that MVMD-CCA significantly outperforms canonical correlation analysis, achieving a maximum 14.2% increase in SSVEP decoding accuracy and a 6.5% increase in information transfer rate.

The multivariate adaptive decomposition also helps to improve the performance of several MI-BCI frameworks [64,65,68]. The effectiveness of MEMD in motor imagery (MI) BCI is explored for EEG recordings by addressing the issues relating to EEG signals, i.e., noise and nonstationarity [64]. By applying direct multichannel processing via MEMD, the authors achieved improved localization of frequency information in EEG, particularly through the noise-assisted mode of operation, which offers a highly focused time-frequency representation. The methods are tested on BCI Competition IV dataset I and the Physiobank motor/mental imagery database. The MEMD-based MI BCI method obtained a classification rate of 75.5% using 4-channel MIMFs from an 11-channel MEMD decomposition and 74.1% using 4-channel MIMFs from a 4-channel MEMD decomposition. Sadiq et al. [69] proposed the MI-BCI classification using MVMD. This study proposes a framework for classifying MI EEG signals. MVMD is used to obtain joint frequency-scale modes across all channels in the 18-channel EEG signal from the motor cortex area. Multidomain features, including time domain, frequency domain, nonlinear, and geometrical features, are extracted from each EEG signal. Various feature selection methods (wrapper and filter) are used to enhance classification performance with different channel combinations. The framework is evaluated using subject-independent experiments on two datasets. Remarkably, subject-dependent experiments achieve an average accuracy of 99.8%, while subject-independent experiments achieve 98.3% accuracy. Moreover, for subjects with limited training samples, accuracy exceeds 99.0% using MVMD and the feature selection method, ReliefF. These results demonstrate the effectiveness of the proposed approach for MI EEG signal classification and its potential in BCI research.

4.2 Neurological disease diagnosis

In this section, the application of the multivariate adaptive signal decomposition algorithm for neurological disease diagnosis from EEG signals is discussed. Bhattacharyya and Pachori [24] have explored the multivariate oscillatory nature of EEG using MEWT. The joint instantaneous amplitude and frequency of the multivariate signal are determined on the signal adaptive frequency scale using the MEWT. Features are extracted from each 1 second part of 2 seconds duration from the joint instantaneous amplitude of automatically selected channels of a multivariate EEG signal. Six classifiers are considered to evaluate the 177 h of EEG signal in the Children's Hospital Boston—Massachusetts Institute of Technology database as seizure- and seizure-free. The method has an average sensitivity, specificity, and accuracy of 97.91%, 99.57%, and 99.41%, respectively.

The authors in Ref. [43] proposed a method for detecting schizophrenia based on the analysis of multichannel EEG signals. In the proposed approach, EEG signals are decomposed into their MIMFs. EEG rhythms (delta, theta, alpha, beta, and gamma) are obtained by grouping the MIMFs based on the mean frequency. Hjorth activity, Hjorth mobility, and Hjorth complexity are calculated from each EEG rhythm. Healthy and schizophrenia-related EEG patterns are classified by KNN, linear discriminant analysis, and support vector machines (SVMs) with diffident kernels. The method is evaluated using 19-channel EEG records from 14 healthy and 14 paranoid schizophrenia patients. The SVM achieved the highest classification sensitivity of 99.0% and specificity of 98.8%.

Zahra et al. [70] have decomposed EEG signals into different frequency bands using the MEMD method. Calculated the instantaneous frequency and instantaneous amplitude of the MIMFs by applying the Hilbert transform. The t-test is applied to find the MIMFs with a significant P-value. The instantaneous frequency and instantaneous amplitude from the selected MIMFs are taken as input for ANN to classify seizure and nonseizure EEG signals. The publicly available EEG signals from Bonn University's database are taken to train, validate, and test the ANN-based classifier. The method has an overall classification accuracy of 87.2%.

During surgeries, monitoring the depth of anesthesia is crucial for assessing a patient's level of consciousness. EEG signals have been widely used to assess the depth of anesthesia. However, EEG is susceptible to interference from external sources like electric power, electric knives, and other electrophysiological signals. These interferences can lead to reduced accuracy in determining the depth of anesthesia. MEMD has been used to obtain clean EEG signals based on the selection of specific MIMFs [71]. Various entropy methods, such as spectral entropy, approximate entropy, and sample entropy, have been employed to predict the depth of anesthesia.

The dynamics of EEG changes, evaluated using chaos theory-based embedding reconstruction, have been employed as a biomarker for detecting Parkinson's disease [72]. The MIF is used to decompose the data into multiple narrowband modes. These modes are subsequently represented using phase-space representations in a higher-dimensional space. From these bands, a set of features is extracted, and a classifier is developed to classify the EEG segment based on these extracted features.

5. Conclusion

This chapter has introduced the concept of multivariate adaptive decomposition and its usefulness. The mathematical definition of the multivariate time series is provided. The amplitude-frequency modulated component-based representation of the multivariate signal is mathematically defined. Adaptive signal decomposition techniques can be used to extract these amplitude-frequency modulated components, or MIMFs. Various multivariate adaptive decomposition techniques, including MEMD, MEWT, M-FBSE-EWT, MVMD, and MIF, are described with steps. The decomposition results for both synthetic and real-time EEG signals are presented in the chapter. The area of adaptive multivariate adaptive decomposition is an emerging research area that has applications in almost all areas of science and engineering. In the last few decades, it has drawn the attention of researchers from various fields. More research efforts are required to improve various problems like mode mixing and mode splitting in multivariate adaptive decomposition.

References

[1] Yang Y, Dong X, Peng Z, Zhang W, Meng G. Component extraction for non-stationary multi-component signal using parameterized de-chirping and band-pass filter. IEEE Signal Process Lett 2014;22(9):1373−7.

[2] Orovi I, Stanković S, Draganić A. Time-frequency analysis and singular value decomposition applied to the highly multicomponent musical signals. Acta Acustica United Acustica 2014;100(1):93−101.

[3] Stankovic L, Daković M, Thayaparan T. Time-frequency signal analysis with applications. Artech House; 2014.

[4] Katkovnik V, Stankovic L. Instantaneous frequency estimation using the Wigner distribution with varying and data-driven window length. IEEE Trans Signal Process 1998;46(9):2315−25.

[5] Allen JB, Rabiner LR. A unified approach to short-time Fourier analysis and synthesis. Proc IEEE 1977;65(11):1558−64.

[6] Griffin D, Lim J. Signal estimation from modified short-time Fourier transform. IEEE Trans Acoustics Speech Signal Process 1984;32(2):236−43.

[7] Pachori RB. Time-frequency analysis techniques and their applications. CRC Press; 2023.

[8] Cohen L. Time-frequency analysis. 1st. New Jersey: Prentice Hall; 1995.

[9] Mallat S. A wavelet tour of signal processing. Elsevier; 1999.

[10] Daubechies I. Ten lectures on wavelets. SIAM; 1992.

[11] Huang NE, Shen Z, Long SR, Wu MC, Shih HH, Zheng Q, Yen N-C, Tung CC, Liu HH. The empirical mode decomposition and the Hilbert spectrum for nonlinear and non-stationary time series analysis. Proc Royal Soc A Mat Phys Eng Sci 1971;454:903−95. 1998.

[12] Stanković L, Mandić D, Daković M, Brajović M. Time-frequency decomposition of multivariate multicomponent signals. Signal Process 2018;142:468−79.

[13] Ahrabian A, Looney D, Stanković L, Mandic DP. Synchrosqueezing-based time-frequency analysis of multivariate data. Signal Process 2015;106:331−41.

[14] Lilly JM, Olhede SC. Bivariate instantaneous frequency and bandwidth. IEEE Trans Signal Process 2009;58(2):591−603.

[15] Lilly JM, Olhede SC. Analysis of modulated multivariate oscillations. IEEE Trans Signal Process 2011;60(2):600−12.

[16] Gabor D. Theory of communication. Proc Inst Elec Eng 1946;93:429−57.

[17] Vakman D. On the analytic signal, the Teager-Kaiser energy algorithm, and other methods for defining amplitude and frequency. IEEE Trans Signal Process 1996;44(4):791−7.

[18] Picinbono B. On instantaneous amplitude and phase of signals. IEEE Trans Signal Process 1997; 45(3):552−60.

[19] Stanković L, Brajović M, Daković M, Mandic D. On the decomposition of multichannel nonstationary multicomponent signals. Signal Process 2020;167:107261.

[20] Schomer DL, Da Silva FL. Niedermeyer's electroencephalography: basic principles, clinical applications, and related fields. Lippincott Williams & Wilkins; 2012.

[21] Das K, Verma P, Pachori RB. Assessment of chanting effects using EEG signals. In: 2022 24th international conference on digital signal processing and its applications (DSPA). IEEE; 2022. p. 1−5.

[22] Park C, Looney D, Kidmose P, Ungstrup M, Mandic DP. Time-frequency analysis of EEG asymmetry using bivariate empirical mode decomposition. IEEE Trans Neural Syst Rehabil Eng 2011;19(4): 366−73.

[23] Rehman N, Mandic DP. Multivariate empirical mode decomposition. Proc Royal Soc A Mat Phys Eng Sci 2010;466(2117):1291−302.

[24] Bhattacharyya A, Pachori RB. A multivariate approach for patient-specific EEG seizure detection using empirical wavelet transform. IEEE Trans Biomed Eng 2017;64(9):2003−15.

[25] Tanaka T, Mandic DP. Complex empirical mode decomposition. IEEE Signal Process Lett 2007; 14(2):101−4.

[26] Rilling G, Flandrin P, Gonçalves P, Lilly JM. Bivariate empirical mode decomposition. IEEE Signal Process Lett 2007;14(12):936−9.

[27] Umair Bin Altaf M, Gautama T, Tanaka T, Mandic DP. Rotation invariant complex empirical mode decomposition. In: 2007 IEEE international conference on acoustics, speech and signal processing-ICASSP'07, vol 3. IEEE; 2007. III−1009.

[28] Naveed ur Rehman and Danilo P Mandic. Empirical mode decomposition for trivariate signals. IEEE Trans Signal Process 2009;58(3):1059−68.

[29] Lang X, Zheng Q, Zhang Z, Lu S, Xie L, Horch A, Su H. Fast multivariate empirical mode decomposition. IEEE Access 2018;6:65521−38.

[30] Gilles J. Empirical wavelet transform. IEEE Trans Signal Process 2013;61(16):3999−4010.

[31] Schroeder J. Signal processing via Fourier-Bessel series expansion. Dig Signal Process 1993;3: 112−24.

[32] Chaudhary PK, Gupta V, Pachori RB. Fourier-Bessel representation for signal processing: a review. Dig Signal Process 2023:103938.

[33] Anuragi A, Singh Sisodia D, Pachori RB. EEG-based cross-subject emotion recognition using Fourier-Bessel series expansion based empirical wavelet transform and NCA feature selection method. Inform Sci 2022;610:508−24.

[34] Anuragi A, Singh Sisodia D, Pachori RB. Automated alcoholism detection using Fourier-Bessel series expansion based empirical wavelet transform. IEEE Sensors J 2020;20(9):4914−24.

[35] Pachori RB, Sircar P. EEG signal analysis using FB expansion and second-order linear TVAR process. Signal Process 2008;88(2):415−20.

[36] Bhattacharyya A, Singh L, Pachori RB. Fourier–Bessel series expansion based empirical wavelet transform for analysis of non-stationary signals. Dig Signal Process 2018;78:185–96.

[37] Bhattacharyya A, Tripathy RK, Garg L, Pachori RB. A novel multivariate-multiscale approach for computing EEG spectral and temporal complexity for human emotion recognition. IEEE Sensors J 2020;21(3):3579–91.

[38] Dragomiretskiy K, Zosso D. Variational mode decomposition. IEEE Trans Signal Process 2013;62(3): 531–44.

[39] Rehman NU, Aftab H. Multivariate variational mode decomposition. IEEE Trans Signal Process 2019;67(23):6039–52.

[40] Zosso D, Dragomiretskiy K, Bertozzi AL, Weiss PS. Two-dimensional compact variational mode decomposition: spatially compact and spectrally sparse image decomposition and segmentation. J Mat Imag Vis 2017;58:294–320.

[41] Liu S, Yu K. Successive multivariate variational mode decomposition. Multidimensional Syst Signal Process 2022;33(3):917–43.

[42] Lin L, Wang Y, Zhou H. Iterative filtering as an alternative algorithm for empirical mode decomposition. Adv Adap Data Anal 2009;1(04):543–60.

[43] Das K, Pachori RB. Schizophrenia detection technique using multivariate iterative filtering and multichannel EEG signals. Biomed Signal Process Control 2021;67:102525.

[44] Cicone A, Pellegrino E. Multivariate fast iterative filtering for the decomposition of nonstationary signals. IEEE Trans Signal Process 2022;70:1521–31.

[45] Wu Z, Huang NE. Ensemble empirical mode decomposition: a noise-assisted data analysis method. Adv Adap Data Anal 2009;1(01):1–41.

[46] Smith JS. The local mean decomposition and its application to EEG perception data. J Royal Soc Inter 2005;2(5):443–54.

[47] Vautard R, Ghil M. Singular spectrum analysis in nonlinear dynamics, with applications to paleoclimatic time series. Phys D Nonlinear Phenom 1989;35(3):395–424.

[48] Harmouche J, Fourer D, Auger F, Borgnat P, Flandrin P. The sliding singular spectrum analysis: a data-driven nonstationary signal decomposition tool. IEEE Trans Signal Process 2017;66(1):251–63.

[49] Broomhead DS, King GP. Extracting qualitative dynamics from experimental data. Phys D Nonlinear Phenom 1986;20(2–3):217–36.

[50] Zheng J, Cheng J, Yang Y. A rolling bearing fault diagnosis approach based on LCD and fuzzy entropy. Mech Mach Theory 2013;70:441–53.

[51] Frei MG, Osorio I. Intrinsic time-scale decomposition: time–frequency–energy analysis and real-time filtering of non-stationary signals. Proc R Soc A: Math Phys Eng Sci 2006;463:321–42. 2007.

[52] Schmid PJ. Dynamic mode decomposition of numerical and experimental data. J Fluid Mech 2010; 656:5–28.

[53] Iatsenko D, Ve McClintock P, Stefanovska A. Nonlinear mode decomposition: a noise-robust, adaptive decomposition method. Phys Rev E 2015;92(3):032916.

[54] Cicone A, Liu J, Zhou H. Adaptive local iterative filtering for signal decomposition and instantaneous frequency analysis. Appl Comput Harm Anal 2016;41(2):384–411.

[55] Singh P, Joshi SD, Kumar Patney R, Saha K. The Fourier decomposition method for nonlinear and nonstationary time series analysis. Proc R Soc A: Math Phys Eng Sci 2017;473(2199):20160871.

[56] Chen Q, Xie L, Su H. Multivariate nonlinear chirp mode decomposition. Signal Process 2020;176: 107667.

[57] Huang J, Li C, Xiao X, Yu T, Yuan X, Zhang Y. Adaptive multivariate chirp mode decomposition. Mechan Syst Signal Process 2023;186:109897.

[58] Mahmoudvand R, Rodrigues PC, Yarmohammadi M. Forecasting daily exchange rates: a comparison between SSA and MSSA. 2019.

[59] Rodrigues PC, Mahmoudvand R. The benefits of multivariate singular spectrum analysis over the univariate version. J Franklin Inst 2018;355(1):544–64.

[60] Zhang Q, Yuan R, Lv Y, Li Z, Wu H. Multivariate dynamic mode decomposition and its application to bearing fault diagnosis. IEEE Sensors J 2023;23(7):7514–24.

[61] Matsuda T, Komaki F. Multivariate time series decomposition into oscillation components. Neural Comput 2017;29(8):2055–75.

[62] Cohen MX. Analyzing neural time series data: theory and practice. MIT press; 2014.

[63] Mert A, Akan A. Emotion recognition from EEG signals by using multivariate empirical mode decomposition. Pattern Anal Appl 2018;21:81–9.

[64] Park C, Looney D, ur Rehman N, Ahrabian A, Mandic DP. Classification of motor imagery BCI using multivariate empirical mode decomposition. IEEE Trans Neural Syst Rehabil Eng 2012;21(1):10–22.

[65] Das K, Pachori RB. Electroencephalogram based motor imagery brain computer interface using multivariate iterative filtering and spatial filtering. IEEE Trans Cogn Dev Syst 2022;15(3).

[66] Pranavi Kamaraju S, Das K, Pachori RB. EEG based biometric authentication system using multivariate FBSE entropy. 2023. https://doi.org/10.36227/techrxiv.23244209.v1.

[67] Chang L, Wang R, Zhang Y. Decoding SSVEP patterns from EEG via multivariate variational mode decomposition-informed canonical correlation analysis. Biomed Signal Process Control 2022;71:103209.

[68] Sadiq MT, Yu X, Yuan Z, Zeming F, Rehman AU, Ullah I, Li G, Xiao G. Motor imagery EEG signals decoding by multivariate empirical wavelet transform-based framework for robust brain–computer interfaces. IEEE Access 2019;7:171431–51.

[69] Sadiq MT, Yu X, Yuan Z, Aziz MZ, ur Rehman N, Ding W, Xiao G. Motor imagery BCI classification based on multivariate variational mode decomposition. IEEE Trans Emerg Top Comput Intell 2022;6(5):1177–89.

[70] Zahra A, Kanwal N, ur Rehman N, Ehsan S, McDonald-Maier KD. Seizure detection from EEG signals using multivariate empirical mode decomposition. Comput Biol Med 2017;88:132–41.

[71] QinWei QL, Fan S-Z, Lu C-W, Lin T-Y, Abbod MF, Shieh J-S. Analysis of EEG via multivariate empirical mode decomposition for depth of anesthesia based on sample entropy. Entropy 2013;15(9):3458–70.

[72] Pachori RB, Das K. System and method for predicting Parkinson's disease. 2022. Indian Patent, Application no: 202221027358.

8

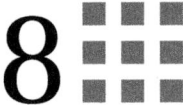

Split learning for human activity recognition

Sandra Pavleska, Valentin Rakovic, Daniel Denkovski and
Hristijan Gjoreski

*SS. CYRIL AND METHODIUS UNIVERSITY IN SKOPJE, FACULTY OF ELECTRICAL ENGINEERING
AND INFORMATION TECHNOLOGIES SKOPJE, MACEDONIA*

1. Introduction

The widespread use of devices like smartphones, smartwatches, fitness trackers, and smart glasses has opened up numerous possibilities for offering new applications to users. This is primarily driven by the valuable contextual information that these devices gather, enabling functionalities such as monitoring patients from a distance [1], identifying and preventing high-risk scenarios like falls [2], enhancing fitness and lifestyle practices [3], spotting cognitive disorders like Parkinson's disease [4,5], and automatically generating logs of activities [6].

Despite the wide array of sensors in modern wearable devices and their ability to collect diverse data, the challenge of data collection persists as a significant issue in HAR. Specifically, considerable privacy concerns emerge, hampering data collection due to the need to share personal and sometimes medical-related information. Moreover, the wearable devices commonly used for HAR are designed as cost-effective solutions that poses low computational capabilities and should foster low energy consumption. As such, the common concept of applying deep centralized learning (DCL) falls short because it does not provide any means for privacy preservation. Federated learning (FL) [7] can address the privacy issues related to DCL; however, it requires that the wearables have substantial computational power and energy budgets, as the learning process is performed locally.

Split learning (SL) [8] represents an alternative approach that can mitigate the setbacks related to DCL and FL. It represents a method that facilitates collaborative model training across diverse distributed data sources without sharing the raw data. It offers several distinct advantages as a powerful approach to machine learning (ML) model training and privacy-preserving data sharing. Unlike other distributed learning concepts, like FL, SL achieves model accuracy that is comparable to DCL. Additionally, the SL design has significant privacy benefits, as it is extremely challenging for malicious actors to reverse engineer the original information from the exchanged learning data. This ensures that the sensitive nature of individual data segments is safeguarded during the model training process. However, effectively implementing and ensuring optimal SL performance in real-world scenarios is a complex task. SL is usually implemented on devices with limited communication and computational resources, demanding a deeper grasp of how different system-level elements influence its dependability and applicability. Analyzing the system-level aspects of SL will facilitate the development of more effective SL-based models, which would advance the practical application of SL in real-world environments. This chapter provides a comprehensive system-level analysis of SL for HAR. It aims to characterize the performance of SL compared to DCL and FL while considering system-level aspects such as computational complexity, communication overhead, and robustness to noisy and error-prone datasets.

The structure of the work is outlined as follows: Section 2 presents the related work related to FL and SL in HAR. Section 3 discusses the HAR dataset employed for training and evaluating the SL model. Section 4 outlines the methodology encompassing the process of feature extraction, the utilized model architecture, and the system architecture. Section 5 presents the evaluation setup, metrics, and experimental specifics. The outcomes of the experiments are showcased and discussed in Section 6. Section 7 concludes the chapter.

2. Related work

The state-of-the-art ML and deep learning (DL) methodologies in the context of HAR require aggregating the available data from all users on a centralized site. However, the centralization of data entails some fundamental drawbacks related to privacy concerns and the challenges associated with data collection. In order to alleviate these drawbacks, several seminal works have investigated the use of FL in the field of HAR. These works focus on enhancing the accuracy and resilience of the FL model [9–12] in the context of different HAR-related scenarios. There has also been an effort to analyze FL from a system-level aspects and evaluate its real-world applicability for HAR [13,14].

However, the application of SL in the context of HAR has been very limited so far. Specifically, the work presented in Ref. [15] focused on using radar imaging to perform indoor activity recognition. Due to the complexity of the proposed problem, using ray tracing and electromagnetic propagation and reflection, the authors propose a large bidirectional recurrent neural network. In order to decrease the complexity of the deployed model, the work proposes to run a small, unidirectional recurrent neural network on a low-power device and make use of intermediate hidden states on another, higher-power device that processes a batch of intermediate computations periodically.

However, the nature of the solution makes it too complex to be deployed in real-world scenarios. A much more effective approach would be to utilize wearable devices that can track different activity-related metrics. To the best of the authors knowledge, this is the first work that investigates the applicability of SL for HAR based on data from wearable devices.

3. Data and preprocessing

For the purpose of training and evaluating our model in this study, we employed the JSI-FOS dataset [14] which contains records of activities of daily living captured using inertial measurement units (IMUs) that were worn by users on various body parts. Specifically, the JSI-FOS dataset encompasses recordings from 10 participants engaged in activities like walking, standing, sitting, running, lying, lying while exercising, kneeling, cycling, allfours_moving, and allfours. Despite the availability of more IMU placements, our analysis is focused solely on data collected from IMUs situated on the dominant hand's wrist and the dominant leg's thigh. Moreover, our consideration was limited to data originating from the accelerometer and gyroscope. During the data collection process, the sensor values were sampled at a frequency of 50 Hz.

Before applying the feature extraction procedure, we executed standard preprocessing steps. Firstly, the continuous data streams were segmented into shorter windows, each lasting 2 seconds and without any overlap. The label assigned to each window corresponded to the most frequently occurring label within that window's readings. Additionally, we computed vector magnitudes from the accelerometer and gyroscope data at each sampling point. Lastly, a band-pass filter was utilized to refine the raw sensor data, effectively eliminating both the gravitational component and inherent noise. Notably, both the filtered and unfiltered versions of the data were retained and utilized. The distribution of data is equitable across the participating subjects in the data collection process. Furthermore, it is evident that the distribution and occurrence frequency of activities remain consistent between the training and testing subsets Fig. 8.1.

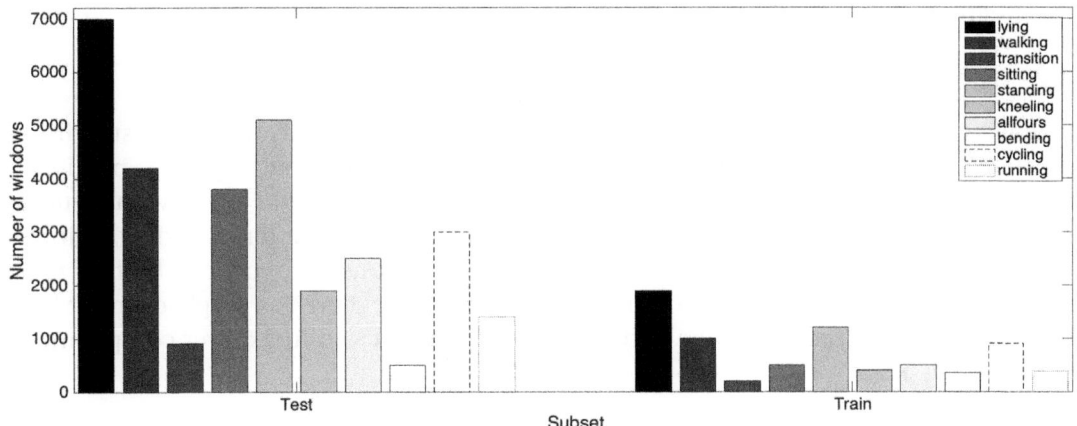

FIGURE 8.1 Dataset distribution.

4. Methodology

This section describes the methodology used in this work. It discusses the feature extraction process, the DL model architecture, and the SL setup.

4.1   Feature extraction

In order to simplify the complexity of both the training and inference phases in our experiments, we opted to employ a straightforward feed-forward neural network (FFNN) that processes extracted features rather than raw sensor data. This approach aligns well with the practical constraints of HAR in the context of Internet-of-Things devices, where computational and energy resources are limited.

After preprocessing the data, the following step in our pipeline entails extracting a comprehensive and diverse array of features, which will empower the FFNN to achieve robust classification accuracy. With this objective in mind, we conducted the extraction of several types of features that have exhibited effectiveness in the analysis of time-series data, particularly data from inertial sensors utilized in HAR. These features are classified into three groups, as follows:

(1) *Generic features*: mean, standard deviation, median, minimum, maximum, range, interquartile range, kurtosis, skewness, and root mean square.
(2) *HAR-specific features*: integral, mean crossing rate, number of peaks, average height of peaks, peak-to-average power ratio, sum, and squared sum.
(3) *Frequency-domain features*: energy, entropy, binned distribution, the three largest PSD magnitudes along with their corresponding frequencies, skewness, and kurtosis.

It's important to highlight that the same feature extraction process was applied to both accelerometer and gyroscope data, as well as to their respective channels, including the magnitude. Additionally, features were extracted from both the filtered and unfiltered data. In total, 1184 features were extracted for each window of data.

4.2   Model architecture

Following the completion of the feature extraction phase, the next and final step in the pipeline involves the learning/inference process, carried out by the previously mentioned FFNN. The FFNN's architecture consists of an input layer with neurons corresponding to the number of features characterizing a singular data window, followed by two fully connected layers containing 64 and 32 neurons, respectively. Both of these layers employ the ReLU activation function. Sequentially, there is a single dropout layer with a rate of 0.2. Concluding the network, the last layer is a softmax layer comprising 10 neurons, Fig. 8.2.

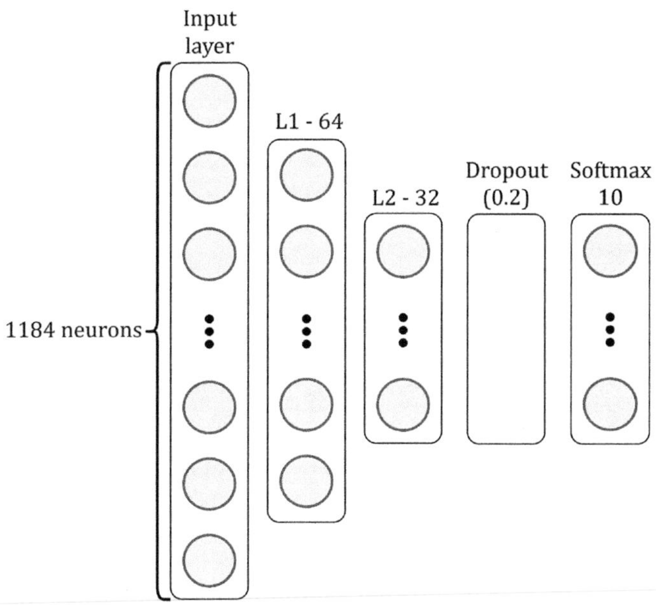

Input
layer

FIGURE 8.2 FFNN model architecture.

L1 - 64

Dropout   Softmax
(0.2)   10

L2 - 32

1184 neurons

It is essential to emphasize that all experiments were conducted using the stochastic gradient descent optimizer with a learning rate of 0.03, the cross-entropy loss function, and a batch size of 64.

4.3   Split learning setup

The SL training approach consists of two primary components. The initial stage involves the segmentation of the neural network into two distinct segments, each of which resides on a separate device (i.e., client device and server device), Fig. 8.3. Virtual workers are created to enable communication between these segments situated on different devices. Both sides initialize the network randomly, after which they progress to the following training steps. The layer at which the model is split is commonly referred to as a split or cut layer.

The training process encompasses several steps. Initially, data undergoes forward propagation through the first layers. The output tensor from these layers (denoted as smashed data) is then transmitted to the server (Fig. 8.3). Subsequently, the server continues the forward propagation by processing the output tensor through its remaining layers. It calculates gradients and propagates them backward. The server-derived gradients from the backward propagation are used to update the model weights on the client. These steps are iteratively executed until convergence. By adhering to this process, it becomes feasible to train the deep neural network without sharing raw

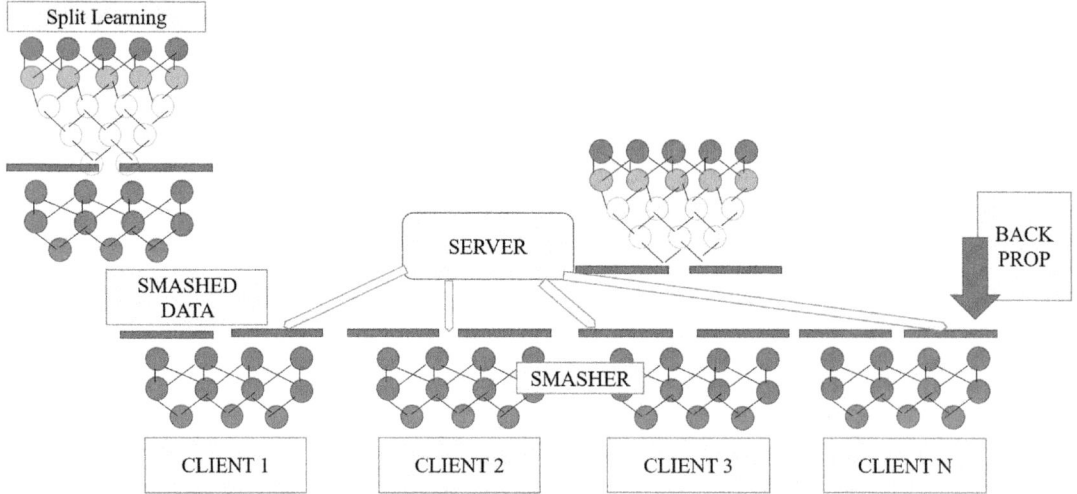

FIGURE 8.3 Split learning generic architecture.

data with the client or revealing any specifics of the model segments held by either the client or the server.

When the scenario consists of multiple clients (i.e., users), then the SL process is executed sequentially, performing the complete training process on a single client at a time. The learned parameters from the client are passed on to the next, and the same process is repeated until all clients have been trained. For example, in Fig. 8.3, the training process starts with Client 1 and lasts for a specific number of epochs. When the training is finished, the client-side model from Client 1 is transferred to Client 2. Then Client 2 starts the training process, but instead of using randomly generated model parameters, it uses the ones learned from Client 1. The process is repeated in the same fashion until the last client, Client N. After the training is completed at Client N, all clients receive the learned model from Client N. In this fashion, all clients participate in building a global model without sharing their local data.

5. Experimental setup

This section presents the evaluation setup of interest, as well as the metrics used in the performance analysis and the experiment scenarios utilized in our study.

5.1   Evaluation setup

The evaluation setup focused more on the personalized approach instead of the leave-one-out approach in order to focus on the SL's capability of building personalized models. Hence, the data of each user is divided into training and test subsets. The training subset consists of approximately 80% of the user's data, i.e., 3100 input data samples per client. The remaining 20% of the user's data, equivalent to around 700 input data samples per client, formed the test set. Validation sets were omitted in this study due to the absence of parameter tuning requirements. The primary emphasis was placed on exposing changes in performance through different configurations exclusively using the test data obtained from individual users.

We implemented precautions during the process of data partitioning in order to address the potential challenge arising from the similarity between data samples (i.e., windows) originating from the same user in close temporal proximity. Our approach involved that data samples corresponding to a continuous execution of a specific activity (termed "activity segments") were exclusively allocated to either the training or test subset, but not both, within each of the two datasets. To achieve this, we followed a sequence of steps: the identification of activity segments within each user's data; the categorization of activity segments based on the nature of the activity performed; and an iterative procedure through the groups of activity segments, where each segment was assigned exclusively to either the training or test subset. For the last step, we allocated activity segments within each group to either the training or test subset while adhering to a distribution where approximately 80% of samples within the group were assigned to the user's training subset, with the remaining 20% attributed to the test subset. By adopting this strategy, we aimed to prevent any unintentional recurrence of similar data during both training and testing phases, facilitating an unbiased evaluation of the model and preserving the integrity of the assessment process.

5.2   Metrics and scenarios

To address the imbalanced distribution of activities within the JSI-FOS dataset, this study used the macro F-score as the primary performance metric. The macro F-score mitigates any inclination toward activities with a larger representation, as it independently calculates the F-score for each activity and computes the average of these individual scores.

The F-score is a harmonic mean of the precision and recall for a specific label. While its interpretability might not be as straightforward as accuracy, higher F-score and macro F-score values (closer to 1.0) signify superior classification performance, whereas lower values (closer to 0.0) indicate poorer performance. Notably, it should be acknowledged that datasets featuring a balanced distribution of activities may yield similar values for both the macro-F-score and accuracy metrics.

In this work, we conveyed several distinct experiments that investigate the performance behaviors of SL and evaluate the system-level advantages and disadvantages of SL-based HAR deployments. Specifically, this work evaluates the SL F-score performance in comparison to DCL as well as FL for system-level aspects such as wearable placement location, communication overhead, client-side computational requirements, as well as robustness to noisy and corrupted data. Table 8.1 depicts the system level and model-related parameters used for the performance analysis.

Table 8.1   Scenario parameters.

System/model parameter	Value
No. of hidden layers	2
Width per hidden layer	[64, 32]
Dropout rate	0.2
Activation function	ReLU
Learning rate	0.03
Loss function	Cross entropy
Batch size	64
No. of users	10
Model optimizer	SGD
FL optimizer	Federated averaging
FL client fraction	0.8
FL no. of local epochs	5
Model parameter size	4B
SL strategy	Vanilla

The following section presents the main results and findings regarding the performance of SL with respect to HAR.

6. Results and discussion

Fig. 8.4 depicts the macro F-score of SL, DCL, and FL with respect to the number of training epochs (for SL and DCL[1]), round (for FL), and for different locations of the wearable device. The wearable location is specifically selected to mimic the most common locations for sensors used in activity tracking. Specifically, the thigh resembles a mobile phone tracker carried in the pocket of the user, whereas the hand location resembles a smart watch or other fitness tracker carried on the wrist.

The results show that SL achieves almost equal performances as DCL and outperforms FL, especially for a lower number of training epochs/rounds. As such, SL represents a better option than FL in situations where privacy preservation is important, since it achieves optimal performance while also preserving the privacy of the local data. It is also evident that SL's performance gain is invariant to the location of the wearable device.

Fig. 8.4 also shows that combining information from multiple wearables provides better model generalization and accuracy. This is a very valuable insight regarding real-world deployments, as it demonstrates that information fusion from different sensors and devices significantly improves the model's performance.

Fig. 8.5 depicts the attained computational load of SL and FL for a specific macro-F-score target. It is evident that SL achieves the required F-score for substantially lower

[1]For clarity and brevity, DCL will be denoted as centralized learning in the remainder of the section.

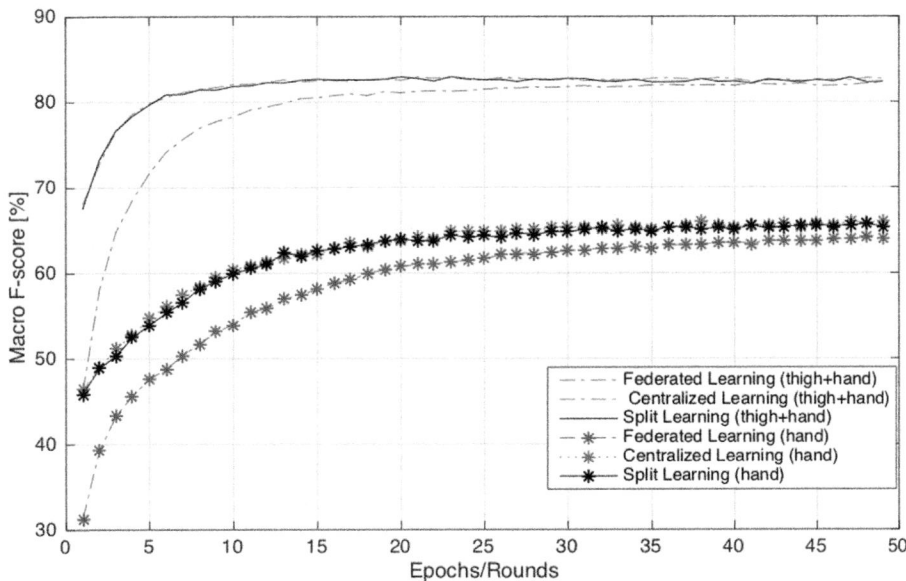

FIGURE 8.4 Macro F-Score versus number of training epoch/rounds.

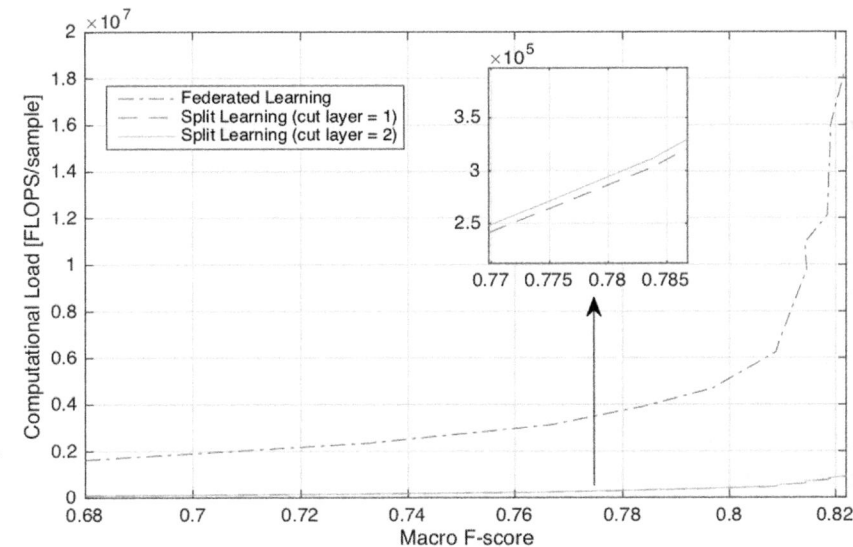

FIGURE 8.5 Computational load versus macro F-score.

computational load, as the model is split and part of it is offloaded on the server. In the case of FL, the complete model is deployed on the device. The figure also shows that in the case of SL, splitting the model at the first layer (cut layer = 1) facilitates a lower load than splitting the model at the second layer (cut layer = 2), although the difference in

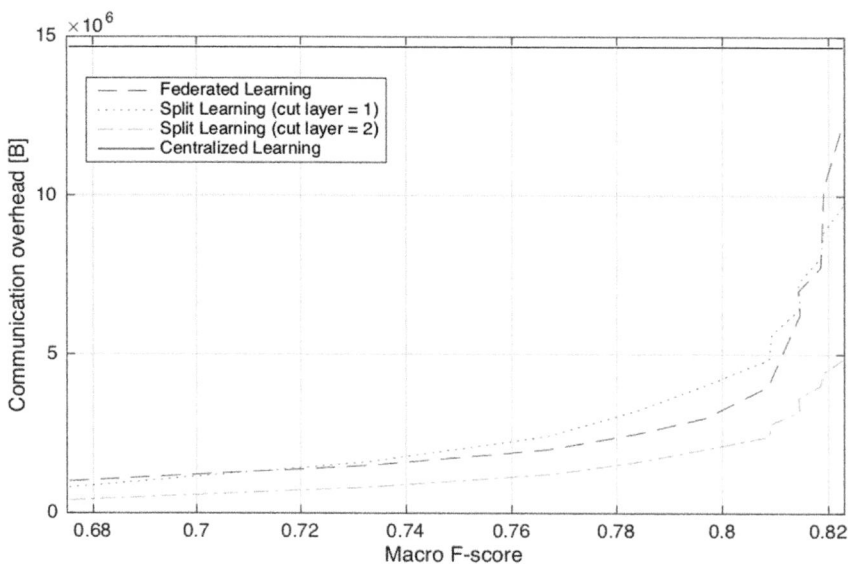

FIGURE 8.6 Communication overhead versus macro F-score.

this case is negligible compared to FL's computational load. This is an intuitive result because splitting the cut layer = 1 results in a smaller part of the model being deployed on the user's device.

Fig. 8.6 depicts the communication overhead, i.e., information exchange requirements for SL, FL, and DCL, in order to attain a target macro F-score. It is clear that the overhead of DCL is constant since the dataset from all users has to be gathered on the central server before the training starts. Moreover, the figure shows that all distributed strategies require lower communication overhead than DCL. It is also evident that FL and SL (cut layer = 1) have very similar overhead footprints. However, SL (cut layer = 2) attains a lower communication overhead since the cut layer (i.e., cut layer = 2) has a lower number of parameters to be exchanged compared to cut layer = 1, as shown in Fig. 8.2.

Fig. 8.7 depicts the generalization performance (i.e., macro F-score) for both SL and FL when the users' dataset has errors in the labels. The figure shows that for a higher percentage of label errors, FL achieves better performances. The primary reason behind this behavior is the principle of averaging weights in the FL strategy [16]. This aspect holds significant value in the FL framework, as real-world data imperfections or inaccuracies could substantially undermine overall performance. During each training round, the global FL model is computed by averaging the local models from a randomly selected subset of clients. Consequently, this process helps in mitigating the impact of erroneous local model weights by combining them with more precise ones, thus reducing the potential for error propagation.

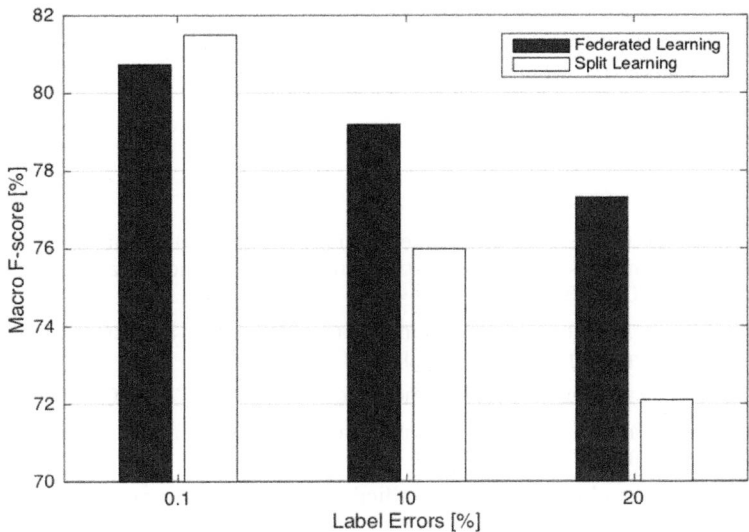

FIGURE 8.7 Macro F-score versus label errors.

By jointly analyzing Figs. 8.48.7, one can conclude that SL is the optimal solution for the application of DL-based HAR. It achieves the same F-score as its DCL counterpart while requiring significantly lower computational and communication requirements. However, if the users' datasets suffer from significant imperfections, then SL may fall short and FL will provide more reliable performances. From the figures, it is also evident that the SL cut layer can play a crucial role with respect to fine-tuning its computational and communication footprint. Specifically, splitting the model closer to the input layer will result in lower computational load and higher communication overhead. In summary, the optimal cut layer should be chosen based on the underlying system-level and deployment requirements.

7. Conclusion

This chapter introduces a comprehensive analysis of the performance of HAR using a novel approach in distributed learning, denoted as SL. The analysis takes a system-level perspective and considers real-world conditions like communication costs, model complexity, and flawed data. It also directly compares SL with DCL, as well as FL. The results clearly demonstrate that SL can be seen as an auspicious technology for DL-based HAR. Moreover, the results show that a wide range of system parameters and setups, including sensor placement, model size, data volume, and the presence of inaccurate data, significantly influence the effectiveness and applicability of SL for HAR scenarios.

Acknowledgment

This work was supported by the WideHealth Project—the European Union's Horizon 2020 Research and Innovation Program under Grant 952279.

References

[1] Luštrek M, et al. A personal health system for self-management of congestive heart failure (HeartMan): development, technical evaluation, and proof-of-concept randomized controlled trial. JMIR Med Informat 2013;9(3):e24501.

[2] Kiprijanovska I, et al. Detection of gait abnormalities for fall risk assessment using wrist-worn inertial sensors and deep learning. Sensors 2020;20(18):5373.

[3] Husain I, Spence D. Can healthy people benefit from health apps? BMJ Clin Res Ed. 2015;350:2520.

[4] Chen Y, et al. FedHealth: a federated transfer learning framework for wearable healthcare. IEEE Intell Syst 2020;35(4):83—93.

[5] Chen Y, et al. PdAssist: objective and quantified symptom assessment of Parkinson's disease via smartphone. In: IEEE International Conference on Bioinformatics and Biomedicine (BIBM); Nov. 2017.

[6] Lee M, et al. A single tri-axial accelerometer-based real-time personal life log system capable of activity classification and exercise information generation. In: Proceedings Annual International Conference of the IEEE Engineering in Medicine & Biology; Aug. 2010.

[7] McMahan B, et al. Communication-efficient learning of deep networks from decentralized data. Proc AISTATS. Apr. 2017;54:273—1282.

[8] Vepakomma P, et al. Split learning for health: distributed deep learning without sharing raw patient data. arXiv 2018;1812:00564.

[9] Ek S, et al. A federated learning aggregation algorithm for pervasive computing: evaluation and comparison. In: IEEE International Conference on Pervasive Computing and Communications; March 2021.

[10] Presotto R, et al. Semi-supervised and personalized federated activity recognition based on active learning and label propagation. Pers Ubiquitous Comput October 2022;26(5):1281—98.

[11] Gudur GK, Perepu SK. Resource-constrained federated learning with heterogeneous labels and models for human activity recognition. In: Proceedings DL-HAR; Jan. 2021.

[12] Zhou X, et al. 2D federated learning for personalized human activity recognition in cyber-physical-social systems. IEEE Trans Netw Sci Eng November 2022;9(6):3934—44.

[13] Cho H, et al. Device or user: rethinking federated learning in personal-scale multi-device environments. In: Proceedings ACM SenSystem; 2021.

[14] Kalabakov S, et al. Federated learning for activity recognition: a system level perspective. IEEE Access 2023;11:64442—57. https://doi.org/10.1109/ACCESS.2023.3289220.

[15] Werthen-Brabants L, Bhavanasi G, Couckuyt I, et al. Split BiRNN for real-time activity recognition using radar and deep learning. Sci Rep 2022;12:7436. https://doi.org/10.1038/s41598-022-08240-x.

[16] Rakovic V, Pejoski S, Hadzi-Velkov Z, Denkovski D. On the double descent effect in federated learning. TechRxiv; 2023. https://doi.org/10.36227/techrxiv.24038145.v1.

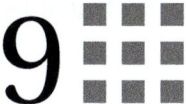

9

Machine learning approaches for epilepsy analysis in current clinical trials

Ishan Ayus and Biswajit Jena

DEPARTMENT OF COMPUTER SCIENCE AND ENGINEERING, INSTITUTE OF TECHNICAL EDUCATION AND RESEARCH, SOA DEEMED TO BE UNIVERSITY, BHUBANESWAR, ODISHA, INDIA

1. Introduction

Epilepsy, a neurological chronic disorder characterized by recurrent seizures, affects millions of people worldwide, making it one of the most prevalent neurological conditions globally. Seizures are caused by abnormal electrical activities in the brain, leading to a wide range of symptoms that can significantly impair an individual's quality of life [1]. These disruptions can manifest as a wide range of symptoms, including convulsions, loss of consciousness, altered sensations, and uncontrolled movements. Despite considerable advancements in medical treatments, a significant proportion of epilepsy patients continue to experience refractory seizures, highlighting the need for novel and more effective approaches to epilepsy analysis and management [2].

Epilepsy is a highly prevalent condition, affecting people of all ages, genders, and ethnic backgrounds. According to the World Health Organization (WHO), approximately 50 million individuals worldwide have epilepsy, making it one of the most common neurological disorders globally [3]. Despite its prevalence, epilepsy remains a complex and challenging condition to manage, with a significant number of patients experiencing treatment-resistant seizures.

Epilepsy inflicts a significant financial and emotional burden on individuals and their families. The financial challenges arise from the high costs of medical care, including diagnostic tests, antiepileptic medications, and potential surgical interventions for drug-resistant cases [3,4]. Emergency room visits and hospitalizations add to the financial strain. The emotional burden of epilepsy is equally profound, with stigma and social isolation affecting the affected individual's mental well-being. Anxiety and depression often accompany the fear of seizures and the uncertainty of managing a chronic condition. The combined financial and emotional cost highlights the significance of offering

Signal Processing Strategies. https://doi.org/10.1016/B978-0-323-95437-2.00008-2

extensive support and resources to more accurately comprehend, identify, and treat epilepsy [5,6].

In recent years, machine and deep learning techniques have emerged as powerful tools with the potential to revolutionize the field of epilepsy research and clinical trials [7−9]. The advent of machine learning and deep learning has dramatically transformed various domains of science and technology, and healthcare is no exception. These approaches leverage the ability of computers to learn patterns and relationships from vast amounts of data, enabling them to make predictions and classifications and assist in decision-making processes. In the context of epilepsy, the potential applications of machines and deep learning are manifold. These approaches can aid in early diagnosis, accurate seizure prediction, and personalized treatment recommendations, ultimately improving the management and outcomes of epilepsy patients [10,11].

The epilepsy analysis pipeline utilizing machine learning begins with data acquisition, capturing EEG signals and patient information. Preprocessing follows, involving noise reduction, feature extraction, and data augmentation. Next, a diverse range of machine learning algorithms like convolutional neural networks (CNNs), recurrent neural networks (RNNs), and support vector machines (SVMs) are employed to train models on these features. Model performance is evaluated through cross-validation and metrics like accuracy, sensitivity, and specificity. Finally, the optimized model is used for real-time seizure prediction and classification, aiding in personalized patient care and treatment strategies. A standard working pipeline of epilepsy analysis employing machine learning paradigms is depicted in Fig. 9.1.

The remaining portions of the chapter have been persuaded like this: Section 2 narrates the background studies necessary for epilepsy analysis, including types of epilepsy and public and private datasets available for analysis. Section 3 describes machine learning techniques for epilepsy analysis. Challenges and considerations in clinical trials

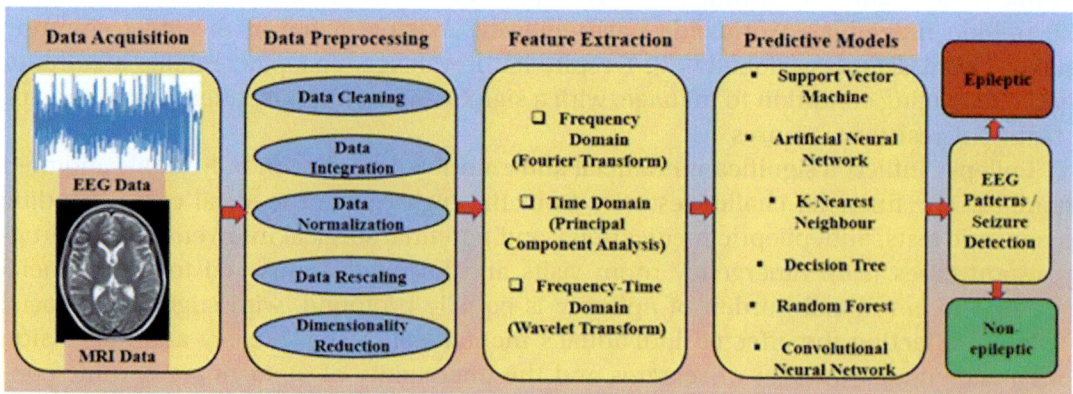

FIGURE 9.1 The working pipeline of epilepsy analysis employing machine learning paradigms.

are available in Section 4. Section 5 provided various case studies considering useful aspects of epilepsy analysis. Section 6 narrates the future directions and implications, while we conclude the chapter theme in Section 7.

2. Background studies

2.1 Types of epilepsy

Epilepsy is a neurological disorder characterized by recurrent seizures. Seizures can manifest in various forms, and there are several types of epilepsy based on their characteristics and underlying causes [12,13]. Each type of epilepsy presents unique symptoms, characteristics, and underlying causes, making it essential for accurate diagnosis and tailored treatment plans for individuals with epilepsy. Some common types of epilepsy include:

1. *Generalized epilepsy*: Absence of seizures (petit mal seizures): Brief lapses of consciousness, often unnoticed, with sudden staring and a lack of response. Tonic-clonic seizures (grand mal seizures): Convulsive seizures involving loss of consciousness and muscle stiffening (tonic phase), followed by rhythmic jerking (clonic phase).

2. *Focal (partial) epilepsy*: Simple focal seizures: Seizures with limited impact on consciousness, characterized by involuntary movements or sensory disturbances. Complex focal seizures: Altered consciousness and repetitive, purposeless movements during the seizure.

3. *Temporal lobe epilepsy*: A specific type of focal epilepsy originating from the temporal lobe of the brain, often characterized by complex partial (CP) seizures.

4. *Frontal lobe epilepsy*: Focal epilepsy originates from the frontal lobes of the brain, leading to various seizure types depending on the affected area.

5. *Juvenile myoclonic epilepsy*: A type of generalized epilepsy usually beginning in adolescence, characterized by myoclonic jerks, tonic-clonic seizures, and absence seizures.

6. *Lennox-Gastaut syndrome*: A severe form of epilepsy characterized by multiple seizure types, intellectual disability, and specific EEG patterns.

7. *Dravet syndrome*: A rare, severe form of epilepsy that begins in infancy, often caused by a genetic mutation, with various seizure types and developmental delays.

8. *West syndrome (infantile spasms)*: Onset in infancy with characteristic spasms, often accompanied by developmental regression.

9. *Progressive myoclonic epilepsies*: A group of rare genetic disorders characterized by progressive myoclonus, tonic-clonic seizures, and neurological deterioration.

10. *Epileptic encephalopathies*: A group of severe epilepsies associated with intellectual and developmental disabilities and ongoing seizures.

Table 9.1 Details of EEG datasets for epilepsy analysis under the machine learning paradigm.

Dataset	Modality	Sampling frequency	Number of channels	Duration/Number	Total seizure data
CHB-MIT	EEG + ECG	256 Hz	18	686 EEG	163
BONN dataset	EEG	173.61 Hz	1	500 EEG	—
Freiburg dataset	EEG	256 Hz	128	557 h	87
Bern dataset	EEG	512 Hz or 1024 Hz	—	10,240 data points per segment	3750
American Epilepsy Society's epileptic seizures detection challenge	EEG	400 Hz	16	627 h	48
TUH EEG corpus	EEG	—	22	200 patients	97
European epilepsy database	EEG	2500 Hz	122	300 patients	—
AIIMS patna	EEG	256 Hz	32	20 patients	105
Zenodo EEG dataset	EEG	256 Hz	19	79 neonates	460

2.2   Dataset description for epilepsy analysis

EEG and neuroimaging datasets comprise brain activity recordings and structural images, respectively, essential for epilepsy research. They help identify seizure patterns, abnormal brain regions, and mechanisms underlying epilepsy, aiding in diagnosis and personalized treatment. These datasets play a critical role in advancing our understanding of epilepsy and improving patient care [13,14]. The summary of all considered datasets has also been provided in Table 9.1.

2.2.1   EEG database

An EEG dataset comprises recordings of the brain's electrical activity captured via electrodes on the scalp. It is vital for epilepsy research as it enables the identification of seizure patterns and interictal abnormalities. Researchers analyze these datasets to understand seizure onset, duration, and brain regions involved in epileptic activity. The EEG data aids in epilepsy diagnosis, treatment optimization, and personalized care. Additionally, it contributes to the development of seizure prediction algorithms, enhancing patient safety and quality of life. EEG datasets offer crucial insights into epilepsy's neurophysiological aspects, advancing our understanding and leading to better management of the condition. An EEG sample for the same is shown in Fig. 9.2.

   CHB-MIT: Children's Hospital Boston–Massachusetts Institute of Technology (CHBMIT) dataset [15] refers to a collection of electroencephalogram (EEG) recordings from pediatric patients. The recordings capture both seizure and non-seizure instances and are provided in the. edf file format. The CHB-MIT dataset contains EEG data recorded from 23 patients, including 5 males aged between 3 and 22 years and the remaining 17 females aged between 1.5 and 19 years. These signals are represented in a

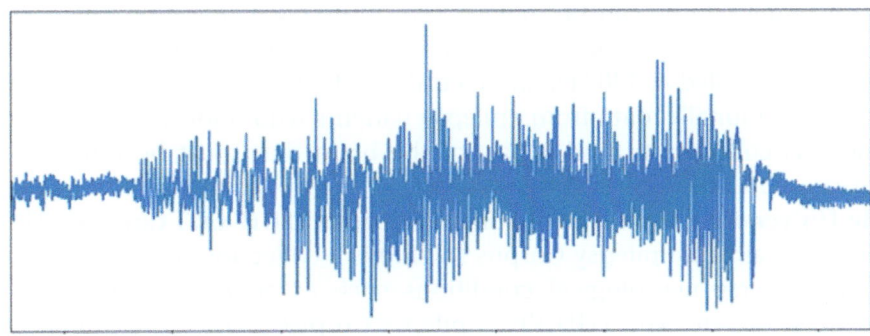

FIGURE 9.2 EEG samples of epilepsy patients.

one-dimensional format and were recorded at a sampling frequency of 256 Hz. The amount of preictal and ictal data extracted is equal. Other signals are occasionally captured in some records, such as an ECG signal in the last 36 files [1,2].

BONN dataset: The Bonn University EEG database is a widely used dataset that was compiled by the Department of Epileptology at the University of Bonn in Germany and contains EEG data from a variety of epileptic conditions, including different seizure types and epileptic syndromes, which are sampled over a duration of 23.6 seconds with a frequency of 173.61 Hz. The study has collected about 100 single-channel data points in five different classes, namely S, F, N, O, and Z. The recordings include both interictal (non-seizure) periods and ictal (seizure) events. The dataset provides valuable insights into the dynamics and patterns of brain activity associated with epilepsy [13,14].

Freiburg dataset: The Freiburg dataset of EEG recordings is a collection of data from 21 patients with physically focal epilepsy. The dataset includes 8 male patients aged between 13 and 47 years and 13 female patients aged between 10 and 50 years. The patients in this dataset were observed and studied before undergoing epilepsy testing at the Epilepsy Checking Center located at the Hospital of Freiburg in Germany. The dataset consists of various data, including CP, simple partial (SP), and epileptic seizure types of tonic-clonic [13,14].

Bern dataset: The Bern dataset is a valuable resource for epilepsy research, specifically focusing on EEG signals. This dataset comprises recordings from patients with epilepsy, where the files are categorized based on the classification of their seizures as "focal" or "non-focal." Each zip file folder contains 750 individual text files, and within each text file, pairs of signals are combined. The signals are represented in a bivariate format, with the X-signal in the first column and the Y-signal in the second column. The dataset adheres to the 10−20 international recording system for electrode placement in EEG recordings [13,14].

The American Epilepsy Society's epileptic seizure detection challenge: The American Epilepsy Society's epileptic seizure detection challenge is a prominent initiative aimed at advancing the field of seizure detection and prediction through machine learning and artificial intelligence. Increasing evidence suggests that the temporal dynamics of brain activity can be divided into four stages: Baseline, postictal (after a seizure), preictal (before a seizure), and ictal (during a seizure). EEG data from 16 electrodes were sampled at 400 Hz, and recorded voltages were normalized to the mean of the group. Additionally, data from epilepsy patients who underwent intracranial EEG monitoring was collected to find a part of the brain that must be removed to avoid further seizures [13,14].

Temple University Hospital EEG Epilepsy Corpus (TUH EEG Corpus): The Temple University Hospital EEG Epilepsy Corpus (TUEP) is a collection of EEG recordings from patients with diverse neurological conditions such as epilepsy, Alzheimer's disease, Parkinson's disease, etc. The TUH EEG Epilepsy Corpus (TUEP) is a publicly available dataset of EEG recordings from 100 subjects with ictal (seizure) and 100 subjects with interictal (non-seizure). The TUEP dataset contains the following information: demographic information, such as age, gender, and race; EEG recordings in EDF file format; manual annotations of seizure events, including the start time, stop time, and channel of the seizure; and manual annotations of artifacts, such as eye movement and muscle activity [13,14].

European epilepsy database: The European epilepsy database (EPILEPSIAE) is a publicly available database of long-term EEG recordings from more than 250 patients with epilepsy supplemented with imaging and clinical data. The information was gathered at European epilepsy centers using scalp and intracranial electrodes, high-quality long-term EEG recordings, as well as derived features and additional clinical and imaging data. In a collaborative effort between epilepsy centers in Portugal (Coimbra), Germany (Freiburg), and France (Paris), high-quality, long-term continuous EEG data are managed in this database [13,14].

AIIMS Patna: The AIIMS Patna Department of Neurology has collected a large dataset of EEG recordings from 20 patients with epilepsy. Following the 10−20 electrode system, 32 channels and a 256 Hz frequency are used for the recording. In this seizure database, a total of 17 male and 3 female patients are taken into account. Ages range from 1 year and 2 months to 38 years. The algorithm is created and tested using continuous multichannel low-noise EEG segments. Marking seizure, non-seizure segments, and types of seizure events as clinician input [13,14].

Zenodo EEG dataset: The Zenodo EEG dataset for epilepsy is a valuable resource for researchers and clinicians alike, aiming to advance our understanding and management of epilepsy. This dataset consists of a diverse collection of EEG recordings from individuals diagnosed with epilepsy, providing a comprehensive view of brain activity during seizure episodes and interictal periods. A total of 79 term neonates were admitted to the neonatal intensive care unit at the esteemed Helsinki University Hospital, and their brain activity was recorded using a multichannel EEG setup. To identify seizures, three experts meticulously annotated the EEG data. After reaching a consensus, it was determined that 39 neonates had seizures, while 22 neonates showed no evidence of seizures [13,14].

2.2.2   *Neuroimaging*

A neuroimaging dataset contains various brain imaging modalities, such as MRI and fMRI scans, used for epilepsy research [16,17]. It provides detailed structural and functional information about the brain, helping to study epilepsy-related abnormalities and neural mechanisms. Researchers analyze these datasets to identify brain regions involved in seizure generation and propagation, aiding in personalized treatment approaches. Neuroimaging data plays a significant role in epilepsy diagnosis, understanding disease progression, and predicting treatment outcomes. Moreover, it contributes to the development of advanced imaging techniques and biomarkers for early detection and intervention. Overall, neuroimaging datasets are crucial tools for advancing epilepsy research, improving patient care, and exploring novel therapeutic strategies. The neuroimaging samples of brain MRI with T1, T2, and Flair modalities [18,19] for epilepsy patients have been shown in Fig. 9.3.

EADC-ADNI HarP dataset: The Alzheimer's disease neuroimaging initiative (ADNI) dataset [20] holds significant relevance beyond Alzheimer's disease, as it has become an essential resource for research in various neurological conditions, including epilepsy diagnosis. The data were gathered at the Hospital of Zunyi Medical University and comprised information from 59 patients (32 males and 27 females) with epilepsy lesions in the hippocampus. All patients were diagnosed using the 2010 version of the diagnostic criteria of the ILAE by intermediate-grade pediatricians or higher. Additionally, 70 healthy volunteers (37 males and 33 females) were selected for the control group, matched in terms of gender, age, and education level.

Open Access Series of Imaging Studies (OASIS) dataset: The OASIS dataset [21] is a valuable resource for epilepsy research, providing researchers with access to

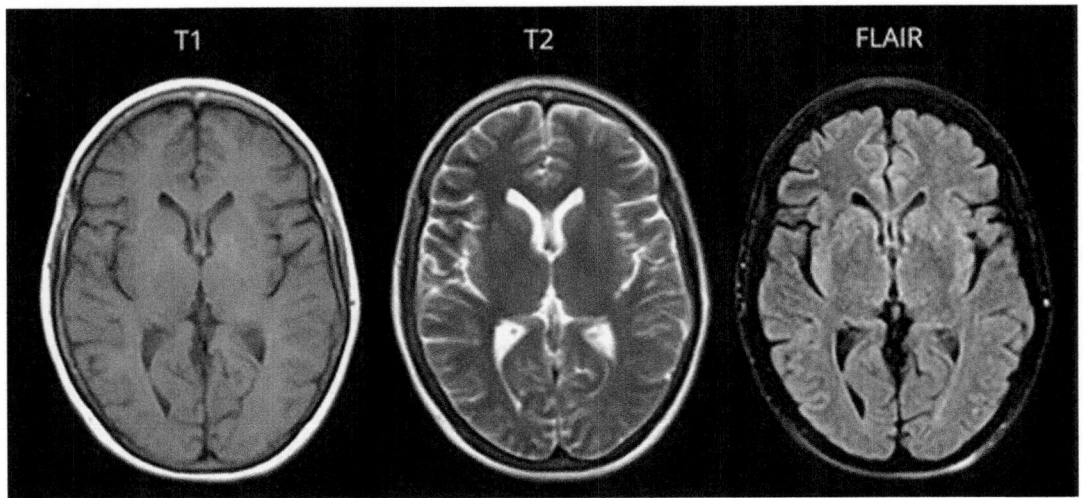

FIGURE 9.3 Neuroimaging samples of brain MRI for epilepsy patients.

neuroimaging data from individuals with epilepsy. OASIS is a collection of openly available brain imaging data that includes structural MRI scans from individuals with various neurological conditions, including epilepsy. The OASIS dataset is used to obtain the 158 MR brain images in sagittal view. The dataset offers an extensive range of brain images, allowing researchers and clinicians to study the brain's structural changes associated with epilepsy and explore potential biomarkers for the condition.

OpenfMRI dataset: OpenfMRI facilitates groundbreaking research on epilepsy through the provision of free and open access to fMRI data. A 3.0 T Signa PET/MRI system (General Electric Healthcare) with an 8-channel high-resolution brain coil was used to acquire the MR images [22]. A standard brain imaging procedure for epilepsy was carried out, which involved the acquisition of axial FLAIR, Fast Field Echo, and T2 weighted images, as well as sagittal T1 weighted images. In addition to the standard procedure, a tractography, a three-dimensional T1-weighted image, and a resting-state functional MRI with closed eyes were obtained. Imaging was done while the patients were awake in this study, and no sedatives were used.

2.3   Preprocessing for epilepsy analysis

Preprocessing of EEG and neuroimaging datasets involves a series of essential steps to clean and enhance the data before analysis. For EEG, preprocessing includes filtering to remove noise, artifact removal, and references to a common average or specific electrode. In neuroimaging, preprocessing involves spatial normalization, intensity normalization, and skull stripping [7]. The importance of preprocessing lies in improving data quality, reducing artifacts, and enhancing the signal-to-noise ratio. In epilepsy research, accurate preprocessing ensures the reliable identification of seizure patterns and abnormal brain regions, leading to more precise diagnosis and treatment planning. Additionally, standardized preprocessing allows for meaningful comparisons across studies, fostering advancements in our understanding of epilepsy's neural basis and facilitating the development of more effective interventions.

2.4   Feature extraction and selection for epilepsy analysis

Feature extraction is a critical step in analyzing EEG and neuroimaging datasets for epilepsy research. In the context of EEG, it involves extracting relevant characteristics from the brain's electrical signals, such as frequency bands, amplitudes, and inter-channel connectivity measures. For neuroimaging, feature extraction includes identifying specific patterns or regions of interest in brain images, like voxel intensities or cortical thickness [6]. The importance of feature extraction lies in transforming complex raw data into meaningful and compact representations that capture essential information about epilepsy-related brain activity. These extracted features serve as input for various machine-learning algorithms and statistical analyses, facilitating the identification of biomarkers, abnormal brain regions, and patterns associated with epilepsy.

Feature extraction plays a crucial role in improving epilepsy diagnosis, predicting treatment outcomes, and advancing our understanding of the disorder's neurophysiological basis, ultimately leading to more personalized and effective management strategies for patients [23].

Feature selection is a critical step in EEG and neuroimaging data analysis for epilepsy research. It involves identifying and choosing the most relevant and informative features from the large pool of extracted data. In EEG, these features may include spectral power, connectivity measures, and time-domain characteristics. In neuroimaging, they could comprise voxel intensities, cortical thickness, or region of interest activations. The importance of feature selection lies in several key aspects of epilepsy research [23]. Firstly, it helps in reducing the dimensionality of the data, leading to more efficient and computationally feasible analyses. Secondly, by focusing only on the most discriminative features, researchers can enhance the accuracy and interpretability of their models, making it easier to identify distinct brain patterns related to seizures and epileptic activity. Moreover, feature selection aids in the discovery of robust biomarkers for epilepsy, which can improve diagnosis, predict treatment outcomes, and potentially lead to more personalized and targeted therapies. Overall, feature selection is crucial in streamlining epilepsy research and paving the way for advancements in understanding, diagnosing, and treating this neurological disorder [15].

2.5 Predictive models for epilepsy analysis

Predictive models using EEG and neuroimaging datasets have been extensively studied in epilepsy research, showcasing their significance in advancing diagnostic and therapeutic approaches. SVMs have demonstrated promising results in classifying seizure patterns with high accuracy [24]. CNNs have been successful in automatically extracting discriminative features from neuroimaging data, aiding in the identification of abnormal brain regions related to epilepsy [25,26]. Additionally, RNNs have shown efficacy in capturing temporal dependencies in EEG recordings, contributing to better seizure prediction [27,28]. Random forests have been utilized to handle high-dimensional EEG and neuroimaging datasets effectively, leading to robust seizure prediction models [23]. These predictive models play a pivotal role in providing early warning systems for seizures, improving patient safety, and tailoring personalized treatments. By delving into the neurophysiological basis of epilepsy, these models enhance our understanding of the disorder, paving the way for innovative research and potential therapeutic interventions [29]. Overall, predictive models using EEG and neuroimaging data are instrumental in transforming epilepsy management and ultimately improving the quality of life for patients [7]. The researchers have devoted their efforts to developing diverse predictive models to gain deeper insights into the neurophysiological basis of epilepsy. These predictive models have been instrumental in driving progress in epilepsy research, fostering innovative approaches for diagnosis, treatment, and more effective management of the condition.

2.6   Performance evaluation for epilepsy analysis

Performance evaluation metrics are essential in assessing the effectiveness of machine and deep learning models in epilepsy research. Some common metrics used include sensitivity (true positive rate, TPR), specificity (true negative rate, TNR), accuracy (ACC), precision (positive predictive value, PPV), and F1-score.

The formulas for these metrics are as follows:

Sensitivity (TPR): TPR = TP/(TP + FN), where TP is the number of true positives (correctly predicted seizures), and FN is the number of false negatives (missed seizures).

Specificity (TNR): TNR = TN/(TN + FP), where TN is the number of true negatives (correctly predicted non-seizures), and FP is the number of false positives (misclassified non-seizures).

Accuracy (ACC): ACC = (TP + TN)/(TP + TN + FP + FN), representing the proportion of correctly classified instances among all predictions.

Precision (PPV): PPV = TP/(TP + FP), measuring the proportion of true positive predictions among all positive predictions.

F1-score: F1 = 2 * (PPV * TPR)/(PPV + TPR), which balances precision and sensitivity, providing a single value to evaluate the model's overall performance.

Additionally, the area under the receiver operating characteristic curve and the area under the precision-recall curve (AUC-PR) are commonly used to evaluate classification models' performance [4]. These performance evaluation metrics allow researchers to objectively assess and compare various machine and deep learning models, ensuring the reliability and efficacy of predictive models for epilepsy management.

3. Machine learning techniques for epilepsy analysis

Epilepsy is a neurological disorder characterized by recurrent and unpredictable seizures, affecting millions of people worldwide. Accurate and timely analysis of epilepsy is crucial for effective treatment and improving the quality of life for patients. Over the years, machine learning techniques have emerged as powerful tools to assist in the analysis of epilepsy, enabling early detection, accurate diagnosis, and personalized treatment plans. This write-up explores some of the key machine-learning techniques applied in epilepsy analysis, highlighting their potential benefits and challenges.

3.1   Predictive and classification modeling

Machine learning techniques enable the development of predictive models to forecast seizures in individual patients. Reinforcement learning (RL) and hidden Markov models (HMMs) have been employed to optimize prediction algorithms and capture state transitions preceding seizures. Classification models are a subset of traditional machine learning algorithms that have proven to be particularly useful in epilepsy analysis tasks. The primary goal of these models is to assign input data to predefined categories or classes based on their features. In the context of epilepsy analysis, classification models can be used to distinguish between different seizure types, classify EEG patterns as normal or abnormal, and even predict seizure occurrences. Commonly used predictive and classification algorithms for epilepsy analysis include:

3.1.1   Logistic regression
Logistic regression is a fundamental machine-learning algorithm that holds significant promise in epilepsy analysis. As a binary classification technique, it plays a vital role in distinguishing between seizure and non-seizure states, making it well-suited for tasks such as seizure detection, classification, and prediction. By modeling the probability of an input belonging to the positive class (seizure) using the sigmoid function, logistic regression can effectively analyze features extracted from EEG signals and other relevant data sources. Its simplicity and interpretability make it an attractive choice for researchers and clinicians, allowing them to gain insights into the importance of different features and facilitating the development of practical and reliable epilepsy analysis solutions. However, challenges such as handling non-linear data and imbalanced datasets require careful consideration, and researchers should combine feature engineering with other machine-learning techniques for more complex analysis tasks in the field of epilepsy research and patient care.

3.1.2   K-nearest neighbors (k-NN)
The k-nearest neighbors (k-NN) algorithm classifies data points based on the majority class among their k-NN. It is particularly useful in scenarios where the decision boundaries are not well defined.

3.1.3   Decision trees
Decision trees recursively split the data based on features to create a treelike structure for classification. They are intuitive and easily interpretable.

3.1.4   Naive Bayes
Naive Bayes is a probabilistic algorithm that relies on Bayes' theorem with the assumption of feature independence. Despite its simplicity, it often performs well in various classification tasks.

3.1.5   Support vector machines (SVM)
SVMs have gained popularity in epilepsy analysis due to their ability to handle both linear and non-linear data separability [24]. SVM is a supervised learning algorithm that finds the optimal hyperplane to maximize the margin between different classes, thereby enhancing its generalization capability. In the context of epilepsy analysis, SVM can be used for seizure detection, the classification of EEG signals, and the localization of

epileptic foci in brain imaging data. The use of various kernel functions enables SVM to efficiently handle complex data representations and capture intricate patterns in the data.

3.1.6 Random forests

Random forests are an ensemble learning technique that leverages the power of decision trees to achieve high accuracy and robustness in epilepsy analysis [27,28]. The algorithm constructs multiple decision trees during the training phase and combines their outputs through voting to make predictions. Random forests can handle a large number of features, are less prone to overfitting, and provide valuable insights into feature importance. In the context of epilepsy analysis, random forests have been utilized for seizure prediction, seizure localization, and automated EEG pattern classification.

3.1.7 Neural networks

Neural networks, especially deep learning architectures, have revolutionized epilepsy analysis in recent years. These models are well-suited for capturing complex relationships in data, learning hierarchical representations, and extracting features automatically. CNNs have shown promising results in analyzing brain images and EEG data, while RNNs are effective in processing sequential data, making them suitable for time-series EEG analysis [9,27,28].

Machine learning techniques have emerged as invaluable tools for epilepsy analysis, offering new avenues for early detection, diagnosis, and personalized treatment. CNNs, RNNs, deep belief networks, and wavelet transforms have shown promise in EEG signal analysis. RL and HMMs have facilitated predictive modeling for seizure forecasting. Despite the challenges, ongoing research and advancements in machine learning hold tremendous potential to revolutionize epilepsy management, ultimately improving the lives of patients and their families. It is vital for interdisciplinary collaborations between machine learning experts, neurologists, and healthcare professionals to drive this field forward and unlock the full potential of machine learning in epilepsy analysis.

4. Challenges and considerations in clinical trials

Machine and deep learning approaches hold great promise for epilepsy analysis in clinical trials, offering the potential to enhance diagnosis, treatment, and patient outcomes. However, their application in this domain also presents several challenges and considerations that need to be carefully addressed [1,3,10,14,17]. In this section, we highlight some of the key challenges and considerations for utilizing machine and deep learning in epilepsy analysis within clinical trials:

Data quality and quantity: Machine and deep learning models heavily rely on large and high-quality datasets for training. In the context of epilepsy, obtaining sufficient and diverse data can be challenging due to the rarity and complexity of seizure events. Ensuring the accuracy and reliability of the data is crucial to avoid biased or misleading results.

Interoperability and data standardization: Clinical trial data often comes from multiple sources and systems, leading to issues with data interoperability and standardization. Integrating data from different sources into machine and deep learning models requires careful preprocessing and harmonization to ensure compatibility and consistency.

Model interpretability: Deep learning models, especially neural networks, are often considered "black-box" models due to their complexity. Understanding the reasoning behind a model's decision is crucial in the medical domain, where transparency and interpretability are critical. Interpretable machine learning models or techniques for explaining deep learning models are essential for gaining trust from clinicians and regulatory authorities.

Overfitting and generalization: Epilepsy datasets may be imbalanced or limited, making overfitting a potential concern. Models trained on one dataset may struggle to generalize to other patient populations or settings. Cross-validation and external validation on diverse datasets are essential to assessing model generalization.

Ethical and privacy concerns: Clinical trial data contain sensitive patient information. Adhering to strict privacy regulations and ensuring patient data confidentiality is crucial. Data anonymization and secure storage methods must be employed to mitigate privacy risks.

Clinical relevance and validity: While machine and deep learning models may achieve high accuracy in analyzing epilepsy data, their clinical relevance and utility must be rigorously validated. The models must demonstrate tangible benefits in improving diagnosis, treatment, or patient outcomes to be meaningful for clinical trials.

Real-time processing: In some clinical trial scenarios, real-time processing of epilepsy data is necessary for timely decision-making or intervention. Deploying computationally intensive deep learning models in real-time applications can be challenging, requiring optimization and hardware considerations.

Regulatory approval and adoption: Machine and deep learning models used in clinical trials need to undergo rigorous validation and gain regulatory approval before widespread adoption. Demonstrating the safety, reliability, and efficacy of these models is crucial for their acceptance in the medical community.

Collaboration and domain expertise: Successful implementation of machine and deep learning approaches in epilepsy analysis requires collaboration between data scientists and domain experts, including neurologists, epileptologists, and clinical researchers. Understanding the clinical context and incorporating expert knowledge is vital for building effective models.

Algorithm bias and fairness: Machine learning models can inadvertently perpetuate existing biases in the data, leading to unfair treatment decisions or recommendations. Ensuring algorithmic fairness and avoiding bias is crucial to providing equitable care to all patients.

Addressing these challenges and considerations is essential for realizing the full potential of the machine and deep learning in epilepsy analysis within clinical trials. By overcoming these hurdles, these advanced approaches can contribute significantly to improving epilepsy management and patient outcomes in a safe and reliable manner.

5. Case studies: Machine learning in current clinical trials

5.1   Case study 1: Real-time seizure prediction using EEG data

The clinical trial focuses on developing deep learning-based seizure prediction models using EEG data to alert patients and healthcare providers before an impending seizure. The trial collects continuous EEG data from participants and utilizes deep learning architectures like RNNs or long short-term memory networks [8,28]. These models analyze EEG patterns and learn to predict seizure events. The deep learning models demonstrate high accuracy in real-time seizure prediction, providing patients with advanced warnings to take precautionary measures or alert healthcare providers. The trial showcases the potential of deep learning for personalized seizure prediction and the development of wearable seizure prediction devices [30].

5.2   Case study 2: Predictive modeling of treatment response in epilepsy patients

The clinical trial aims to integrate machine learning techniques to develop predictive models for assessing treatment responses in epilepsy patients. The trial collects data from patients undergoing different treatments, including patient demographics, medical history, EEG recordings, and treatment outcomes. Machine learning algorithms, such as SVMs or ensemble methods, are utilized to build predictive models [24,26]. The machine learning models show significant predictive power in assessing individual patient responses to specific antiepileptic treatments. The trial demonstrates the potential of machine learning in identifying patients who are likely to benefit from a particular treatment, enabling personalized treatment plans, and improving overall treatment efficacy [31].

5.3   Case study 3: Predicting drug responsiveness in epilepsy patients

The clinical trial aims to utilize machine learning algorithms for patient stratification to predict drug responsiveness in epilepsy patients. The trial collects a comprehensive dataset, including patient demographics, medical history, genetic information, EEG data, and treatment outcomes [3,10]. Machine learning techniques, such as random forests or gradient boosting, are employed to build predictive models. The machine learning models successfully stratify epilepsy patients into different subgroups based on their likelihood of responding to specific antiepileptic drugs. The trial demonstrates that certain patient characteristics and genetic markers can be crucial indicators of drug responsiveness, enabling clinicians to make more informed and personalized treatment decisions [2].

5.4   Case study 4: Deep learning-based brain lesion segmentation in MRI

A clinical trial aims to use deep learning to automatically segment and quantify brain lesions in MRI scans of patients with epilepsy. The trial employs CNNs to analyze MRI images [9,25] and identify regions of interest corresponding to epileptic lesions. The models are trained on a labeled dataset of MRI scans from patients with confirmed lesions. The deep learning approach achieves high accuracy in segmenting brain lesions, providing neurologists with precise and efficient tools for diagnosis and treatment planning [17].

5.5   Case study 5: Closed-loop deep brain stimulation for seizure control

A clinical trial investigates the feasibility of closed-loop deep brain stimulation (DBS) using deep learning algorithms to detect seizure onset and deliver targeted stimulation for seizure control [32]. The trial employs deep learning models to analyze intracranial EEG signals and detect preseizure patterns. The models trigger DBS when seizure onset is predicted. The closed-loop DBS system demonstrates the potential to suppress seizures effectively while minimizing unnecessary stimulation, offering a more precise and adaptive therapeutic approach [32].

6. Future directions and implications

6.1   Current trends and emerging technologies

In recent years, machine learning approaches for epilepsy analysis in current clinical trials have witnessed significant advancements driven by emerging technologies and evolving trends. Deep learning architectures, such as CNNs and RNNs, are increasingly being utilized to process complex and sequential data like EEG signals and medical images. Transfer learning and pretrained models have gained popularity, enabling fine-tuning of models on limited labeled data for specific epilepsy-related tasks. The growing need for model interpretability has led to the development of explainable AI methods, providing insights into model decisions and fostering trust in clinical applications. Multimodal data integration, combining EEG, MRI, genomics, and clinical data, is a prevailing trend, leading to a more comprehensive understanding of epilepsy mechanisms and personalized patient profiles [21,22]. Federated learning addresses privacy concerns and enables collaborative research across multiple sites without sharing raw patient data. The emergence of edge AI and wearable devices allows real-time seizure prediction and monitoring, empowering patients with timely information. Automated biomarker discovery using machine learning aids in improved patient stratification and personalized interventions. As the field progresses, causal inference, counterfactual reasoning, and RL hold promise for evaluating treatment efficacy and developing closed-loop systems for dynamic treatment adjustments. Data augmentation and synthetic data generation techniques address data scarcity challenges, enriching epilepsy datasets for training robust models. These current trends and emerging technologies collectively drive the integration of machine learning into epilepsy analysis in clinical trials, fostering precision medicine and improved patient outcomes [24,32].

6.2   Advancements in data collection and integration

Advancements in data collection and integration have revolutionized the application of machine-learning approaches for epilepsy analysis in current clinical trials. With the increasing availability of electronic health records, wearable devices, and neuroimaging technologies, researchers now have access to vast and diverse datasets encompassing patient demographics, EEG recordings, MRI scans, genetic information, and treatment outcomes [2]. This rich and multimodal data enables the development of more comprehensive and accurate machine-learning models. Moreover, federated learning and edge computing techniques have emerged, allowing decentralized data processing while preserving patient privacy and security. These innovations facilitate seamless collaboration between multiple institutions, enabling the aggregation and integration of data from various sources, ultimately leading to more robust and generalizable models. As machine learning algorithms leverage this wealth of information, they play an instrumental role in patient stratification, seizure prediction, treatment response assessment, and overall personalized care, bringing us closer to transformative advancements in epilepsy management and care [21,22].

6.3   Potential impact on clinical practice

The potential impact of machine learning approaches for epilepsy analysis in current clinical trials on clinical practice is immense. These advanced techniques hold the promise of transforming epilepsy management and care by providing clinicians with powerful tools for more accurate and timely diagnosis, personalized treatment plans, and improved patient outcomes. Machine learning models can aid in patient stratification by identifying distinct subgroups based on clinical and genetic features, leading to targeted interventions and optimized drug selection. Real-time seizure prediction models can empower patients with early warnings, enabling them to take precautionary measures and enhancing their quality of life [30,31]. Moreover, machine learning algorithms can assist in the assessment of treatment responses, identifying patients who are likely to benefit from specific therapies, and guiding treatment adjustments. As these models become more interpretable and trustworthy, they will be seamlessly integrated into clinical decision support systems, providing clinicians with data-driven insights and optimizing treatment decisions. Overall, the adoption of machine learning approaches in epilepsy analysis from current clinical trials has the potential to revolutionize clinical practice, offering more personalized, efficient, and effective care for patients with epilepsy [3,4].

6.4   Ethical considerations and patient-centric approaches

Ethical considerations and patient-centric approaches are paramount when deploying machine-learning approaches for epilepsy analysis in current clinical trials. Patient data privacy and security must be rigorously upheld, adhering to all relevant regulations and obtaining informed consent from participants [13]. Transparent and interpretable models are essential to gaining patient trust and ensuring clinicians can understand and explain model decisions to patients. Researchers should prioritize the fair representation of diverse patient populations in the training data to avoid biased outcomes and ensure equitable care for all. Additionally, machine learning models should not replace clinical expertise but rather augment it, with a focus on providing decision support and improving patient outcomes. Continuous monitoring and evaluation of the algorithms' performance and ethical implications throughout the clinical trial process are necessary. Engaging patients as active stakeholders in the research process can lead to patient-centered solutions that address their needs and preferences. By incorporating patient perspectives, ensuring transparency, and safeguarding patient rights, machine-learning approaches for epilepsy analysis can pave the way for patient-centric and ethically responsible innovations in clinical practice [20].

7. Conclusion

Machine learning approaches have emerged as a transformative force in epilepsy analysis within current clinical trials. These advanced techniques have demonstrated their potential to revolutionize how we understand, diagnose, and treat epilepsy. Deep learning architectures, such as CNNs and RNNs, have enabled more accurate and real-time seizure detection and prediction from EEG signals. Transfer learning and federated learning have addressed data scarcity and privacy concerns, enabling collaborative research while safeguarding patient privacy. Ethical considerations and patient-centric approaches underscore the importance of transparent, interpretable, and fair models that augment clinical expertise and improve patient outcomes. The integration of machine learning in current clinical trials offers the promise of personalized treatments, enhanced patient stratification, and real-time decision support systems, ultimately leading to better care for individuals with epilepsy. As research continues, validation in real-world clinical settings will be vital to harnessing the full potential of machine learning approaches and facilitating their seamless adoption into clinical practice. By combining cutting-edge technology with patient-focused care, machine-learning approaches are poised to play an increasingly critical role in advancing epilepsy management and positively impacting the lives of those affected by this neurological disorder.

References

[1] Rochtus A, et al. Genetic diagnoses in epilepsy: the impact of dynamic exome analysis in a pediatric cohort. Epilepsia 2020;61(2):249—58.

[2] Fattorusso A, et al. The pharmacoresistant epilepsy: an overview on existant and new emerging therapies. Front Neurol 2021;12:674483.

[3] Smits A, Duffau H. Seizures and the natural history of world health organization grade II gliomas: a review. Neurosurgery 2011;68(5):1326—33.

[4] Jena B, et al. Brain tumor characterization using radiogenomics in artificial intelligence framework. Cancers 2022;14(16):4052.

[5] Pradhan R, Dash AK, Jena B. Resource management challenges in IoT based healthcare system. In: Smart healthcare analytics: state of the art; 2022. p. 31—41.

[6] Jena B, Thakar P, Nayak V, Nayak GK, Saxena S. Malaria parasites detection using deep neural network. In: Deep learning applications in medical imaging: IGI global; 2021. p. 209—22.

[7] Das S, Ayus I, Gupta D. A comprehensive review of COVID-19 detection with machine learning and deep learning techniques. Health Technol 2023:1—14.

[8] Gupta D, Kumar V, Ayus I, Vasudevan M, Natarajan N. Short-term prediction of wind power density using convolutional LSTM network. FME Transac 2021;49(3):653—63.

[9] Jena B, Nayak GK, Saxena S. Convolutional neural network and its pretrained models for image classification and object detection: a survey. Concurr Comp Prac Exp 2022;34(6):e6767.

[10] Ramgopal S, et al. Seizure detection, seizure prediction, and closed-loop warning systems in epilepsy. Epilepsy Behav 2014;37:291—307.

[11] Suri JS, et al. Five strategies for bias estimation in artificial intelligence-based hybrid deep learning for acute respiratory distress syndrome COVID-19 lung infected patients using AP (ai) Bias 2.0: a systematic review. IEEE Trans Instrum Meas 2022.

[12] Jokeit H, Schacher M. Neuropsychological aspects of type of epilepsy and etiological factors in adults. Epilepsy Behav 2004;5:14—20.

[13] Natu M, Bachute M, Gite S, Kotecha K, Vidyarthi A. Review on epileptic seizure prediction: machine learning and deep learning approaches. Comput Math Methods Med 2022;2022.

[14] Rahman R, et al. Comprehensive analysis of EEG datasets for epileptic seizure prediction. In: 2021 IEEE international symposium on circuits and systems (ISCAS). IEEE; 2021. p. 1—5.

[15] Jiang Y, Chen W, Li M. Symplectic geometry decomposition-based features for automatic epileptic seizure detection. Comput Biol Med 2020;116:103549.

[16] Yuan J, et al. Machine learning applications on neuroimaging for diagnosis and prognosis of epilepsy: a review. J Neurosci Methods 2022;368:109441.

[17] Shoeibi A, et al. An overview of deep learning techniques for epileptic seizures detection and prediction based on neuroimaging modalities: methods, challenges, and future works. Comput Biol Med 2022:106053.

[18] Saxena S, et al. Prediction of O-6-methylguanine-DNA methyltransferase and overall survival of the patients suffering from glioblastoma using MRI-based hybrid radiomics signatures in machine and deep learning framework. Neural Comput App 2023;35(18):13647−63.

[19] Saxena S, et al. Fused deep learning paradigm for the prediction of o6-methylguanine-DNA methyltransferase genotype in glioblastoma patients: a neuro-oncological investigation. Comput Biol Med 2023;153:106492.

[20] Petersen RC, et al. Alzheimer's disease neuroimaging initiative (ADNI): clinical characterization. Neurology 2010;74(3):201−9.

[21] Marcus DS, Wang TH, Parker J, Csernansky JG, Morris JC, Buckner RL. Open Access Series of Imaging Studies (OASIS): cross-sectional MRI data in young, middle aged, nondemented, and demented older adults. J Cognit Neurosci 2007;19(9):1498−507.

[22] Poldrack RA, Gorgolewski KJ. OpenfMRI: open sharing of task fMRI data. NeuroImage 2017;144:259−61.

[23] Panigrahi N, Ayus I, Jena OP. An expert system-based clinical decision support system for Hepatitis-B prediction and diagnosis. In: Machine learning for healthcare applications; 2021. p. 57−75.

[24] Jena B, Nayak GK, Saxena S. An empirical study of different machine learning techniques for brain tumor classification and subsequent segmentation using hybrid texture feature. Mach Vis Appl 2022;33(1):6.

[25] Jena B, Dash AK, Nayak GK, Mohapatra P, Saxena S. Image classification for binary classes using deep convolutional neural network: an experimental study. In: Trends of data science and applications: theory and practices; 2021. p. 197−209.

[26] Jena B, Digdarshi D, Paul S, Nayak GK, Saxena S. Effect of learning parameters on the performance of the U-Net architecture for cell nuclei segmentation from microscopic cell images. Microscopy 2023;72(3):249−64.

[27] Ayus I, Natarajan N, Gupta D. Comparison of machine learning and deep learning techniques for the prediction of air pollution: a case study from China. Asian J Atmos Environ 2023;17(1):4.

[28] Ayus I, Natarajan N, Gupta D. Prediction of water level using machine learning and deep learning techniques. Iranian J Sci Technol Trans Civil Eng 2023:1−11.

[29] Saxena S, et al. Role of artificial intelligence in radiogenomics for cancers in the era of precision medicine. Cancers 2022;14(12):2860.

[30] Chisci L, et al. Real-time epileptic seizure prediction using AR models and support vector machines. IEEE Trans Biomed Eng 2010;57(5):1124−32.

[31] Delen D, Davazdahemami B, Eryarsoy E, Tomak L, Valluru A. Using predictive analytics to identify drug-resistant epilepsy patients. Health Inform J 2020;26(1):449−60.

[32] Hosain MK, Kouzani A, Tye S. Closed loop deep brain stimulation: an evolving technology. Australas Phys Eng Sci Med 2014;37:619−34.

10

Brainwave and head motion control of a smart home for disabled people

Minoru Dhananjaya Jayakody Arachchige[1] and Marwan Nafea[2]

[1]DEPARTMENT OF AUTOMATIC CONTROL AND SYSTEMS ENGINEERING, FACULTY OF
ENGINEERING, UNIVERSITY OF SHEFFIELD, SHEFFIELD, UNITED KINGDOM; [2]DEPARTMENT OF
ELECTRICAL AND ELECTRONIC ENGINEERING, FACULTY OF SCIENCE AND ENGINEERING,
UNIVERSITY OF NOTTINGHAM MALAYSIA, SELANGOR, MALAYSIA

1. Introduction

Brain−computer interface (BCI) is a rapidly advancing neuroscience technology that facilitates direct communication between the brain and external devices. This enables users to interact with their environment solely through cerebral activity, bypassing the need for involvement from peripheral nerves and muscles [1]. The advent of BCI systems has transformed what were once considered science fiction and fanciful ideas into reality. BCI technology has found profound applications in aiding individuals with disabilities, addressing a considerable global population of approximately 16% who experience varying degrees of disability [2]. As a result, BCI technology has gained widespread adoption in medical contexts to assist disabled patients in recovering lost motor functions [3].

A standard BCI setup encompasses four key elements: signal acquisition, signal processing, application interface, and feedback mechanism [4]. Signal acquisition involves capturing brain activity using three techniques, including invasive, partially invasive, and noninvasive. Signal processing then extracts vital information from neural signals utilizing algorithms for preprocessing, feature extraction, and feature classification. The resulting interpreted signals are transmitted to the application interface, where a feedback mechanism generates real-time output, enabling users to monitor and comprehend the system's actions [5].

Numerous studies investigated the aspects of several BCI systems. For instance, a study introduced a decision tree to classify individual finger movements using electrocorticography (ECoG), achieving classification accuracy exceeding 70% [6]. Another investigation explored the compatibility of different extraction techniques, including

Signal Processing Strategies. https://doi.org/10.1016/B978-0-323-95437-2.00006-9

functional magnetic resonance imaging (fMRI), ECoG, and electroencephalography (EEG) when exposed to natural stimuli [7]. To enhance BCI performance and functionality, integrating BCI with external sensors like ultrasonic sensors, infrared sensors, and electromyography (EMG) has gained prominence. A fusion of EEG and EMG signals was proposed to detect unilateral foot movements, achieving 96% accuracy [8]. Another novel helmet design integrated an inertial measurement unit (IMU) sensor with EEG signals, detecting falling and drowsiness with 98% classification accuracy [9].

While BCI's initial purpose was in the medical realm, yielding assistive devices to enhance motor abilities in disabled patients, its applications have now expanded to nonmedical fields. One such application is the EEG, a noninvasive method that was established by Berger in 1929 [10]. This method assesses the brain's electrical activity by placing electrodes on the scalp. EEGs capture synchronized electrical signals from communicating neurons during specific cognitive tasks. The EEG captures various brainwave patterns, categorized into five frequency bands: delta (δ) waves, theta (θ) waves, alpha (α) waves, beta (β) waves, and gamma (γ) waves. Beta waves are further subdivided into two bands: low beta (β_L) and high beta waves (β_H), while some references add a third band, known as midrange beta (β_M) [11]. These distinct wave patterns are linked to different frequencies and mental states, as outlined in Table 10.1. The amplitude of these bands fluctuates based on internal mental states and external stimuli. Among these, beta waves are commonly utilized in most BCI applications, as they are generated only when the user is in an active mental state.

Signal processing plays a pivotal role in EEG-based BCI systems by extracting and categorizing pertinent features from raw brainwaves. Researchers employ various classification techniques and feature extraction methods during BCI development. For example, an automated unsupervised algorithm was introduced for blink detection using self-learning user-specific brainwave patterns [12]. Another study introduced an attention state detection algorithm using wavelet transform for feature extraction, achieving 80% average accuracy, 25% higher than previous methods [13]. An unsupervised seizure detection algorithm was also developed, identifying rhythmic activities and spikes within EEG signals [14].

Table 10.1 Brainwave frequency bands and their corresponding mental states.

Brainwave type	Frequency range (Hz)	State of mind
Delta (δ) wave	<4	Deep sleep, dreamless sleep, unconscious
Theta (θ) wave	4—7	Meditation, dreaming, recalling
Alpha (α) wave	8—12	Conscious, relaxed, passive attention
Low beta (β_L) wave	12—15	Relaxed yet focused, active thinking
High beta (β_H) wave	16—24	Alertness, cognitive processing
Gamma (γ) wave	25—45	Highly active, information processing

The proliferation of EEG-based BCI research has led to a diverse range of applications beyond medicine, extending to domains like gaming, entertainment, security, education, and smart environments. Numerous EEG-based BCI applications have been highlighted in prior research. For instance, one study introduced an EEG-based human assistance rover, enabling disabled users to control a rover arm with eye blinks and attention levels [15]. Another effort introduced a brainwave-controlled wheelchair system leveraging steady-state visual evoked potential (SSVEP) for control [16]. A distinct endeavor employed a noninvasive BCI system to control quadcopters within a robotic operating system [17]. Additionally, a real-time EEG-based game named *BlinkFruity* allowed users to interact with the game solely through eye blinks, achieving 86% accuracy [18]. Furthermore, an EEG-based user authentication system proposed a two-stage security verification utilizing a mind-generated password for personal device locking and unlocking [19]. EEG-based BCI has expanded beyond these applications, with BCI-based smart home technology gaining traction as an intriguing field. Notable examples include an EEG-controlled smart home using an SSVEP-based system enabling control of six devices through visual stimuli [20]. Another work combined the Internet of things (IoT), augmented reality (AR), and EEG-based BCI for smart home appliance control, effectively employing SSVEPs for real-time control of three appliances [21].

Within a smart home environment, sensors and devices are interconnected to intelligently manage and control devices. Wireless technologies have gained prominence in EEE-based smart systems due to their advantages, including lower transmission capacity requirements and data rates compared to wired counterparts. These wireless technologies include Wireless Fidelity (Wi-Fi), ZigBee, Z-Wave, Bluetooth, and Bluetooth low energy (BLE). Wi-Fi is an Internet protocol (IP)-based wireless communication technology utilized in home networks for broader applications. However, Wi-Fi technology suffers from interference and high-power consumption. ZigBee is a cost-effective and low-power wireless network. Despite having a lower data rate, it offers enhanced coverage and reliability when deployed in a mesh network configuration. Z-Wave is a prevalent wireless technology for home automation and remote-control purposes. However, this technology suffers from limited bandwidth and data rates. Bluetooth is a short-range wireless communication technology commonly employed for personal devices and computer peripherals, offering speed and adaptability. However, it lacks a robust security layer, potentially leading to unauthorized access. BLE offers an attractive solution for the aforementioned issues since it requires lower power consumption and latency while providing better security and interference mitigation.

In this work, we present the development of an EEG-based BCI system tailored for smart home applications due to the promising capabilities of EEG-based BCI technology. The key objective is to assist disabled and elderly individuals grappling with mobility and interaction difficulties. The methodology adopted features a balanced consideration of system complexity, cost, accuracy, power consumption, and portability. The proposed approach encompasses the use of an EEG headset for brainwave extraction, which is then coupled via BLE with an Android application on an Android mobile device, which is also connected to an Arduino Uno board via Bluetooth. An optimized signal processing algorithm translates brain signals into actionable commands. The Android application acts as an interface for users to control different home appliances without requiring physical engagement, as demonstrated in the proof-of-concept implementation. This work aims to provide a step-by-step guide to the process required to develop an EEG-based smart home control system using an EMOTIV Insight headset.

2. Proposed methodology

This research focuses on creating a BCI system that will let users extract and process their brain signals to operate four different devices. The EMOTIV Insight EEG headset, which includes built-in BLE technology, is used to extract the brainwaves. After wirelessly connecting to the headset through BLE, the Android application collects data and uses the EMOTIV Cortex application programming interface (API) to process the brain signals. Following data classification, the mobile application wirelessly transmits the interpreted instruction to the Arduino Uno board through an HC-06 Bluetooth module. Finally, based on the command it receives, the Arduino Uno board controls the electronic appliances. The block diagram of the working principle of this method is illustrated in Fig. 10.1A. In addition, the EMOTIV Insight headset and the locations of the electrodes and their placements when worn by the user are presented in Fig. 10.1B and C.

The involved hardware in this work can be divided into three subsystems, which are the EEG headset, Android application, and control circuit. The design method is divided into five major stages: EEG signal extraction, Android application development, control circuit development, EEG signal processing, and subsystems integration.

2.1   EEG signal extraction

It is essential to connect to and communicate with the Cortex API, which is a WebSocket server that supports the JavaScript object notation remote procedure call (JSON-RPC) protocol, in order to access the features and data from the EMOTIV Insight headset. The Cortex API can be accessed using the WebSocket Secure (*wss*) protocol. Connection to Cortex API can be established by opening a secure WebSocket connection to *wss://localhost:6868*. After the WebSocket connection has been established successfully, the Cortex API is contacted via the JSON-RPC 2.0 protocol. The Cortex returns the outcome of an error and methods can be called with or without parameters.

There are some preliminary steps that must be completed before communicating with the API. The EMOTIV program must first be downloaded to the device since it will be needed to facilitate connections with the Cortex API. For a mobile device, it is necessary to sign up as an EMOTIV-partner developer and request Beta testing. Additionally, in order to receive a client identification (ID) and client secret ID, which are required to verify the identity during the API data exchange, it is required to create an

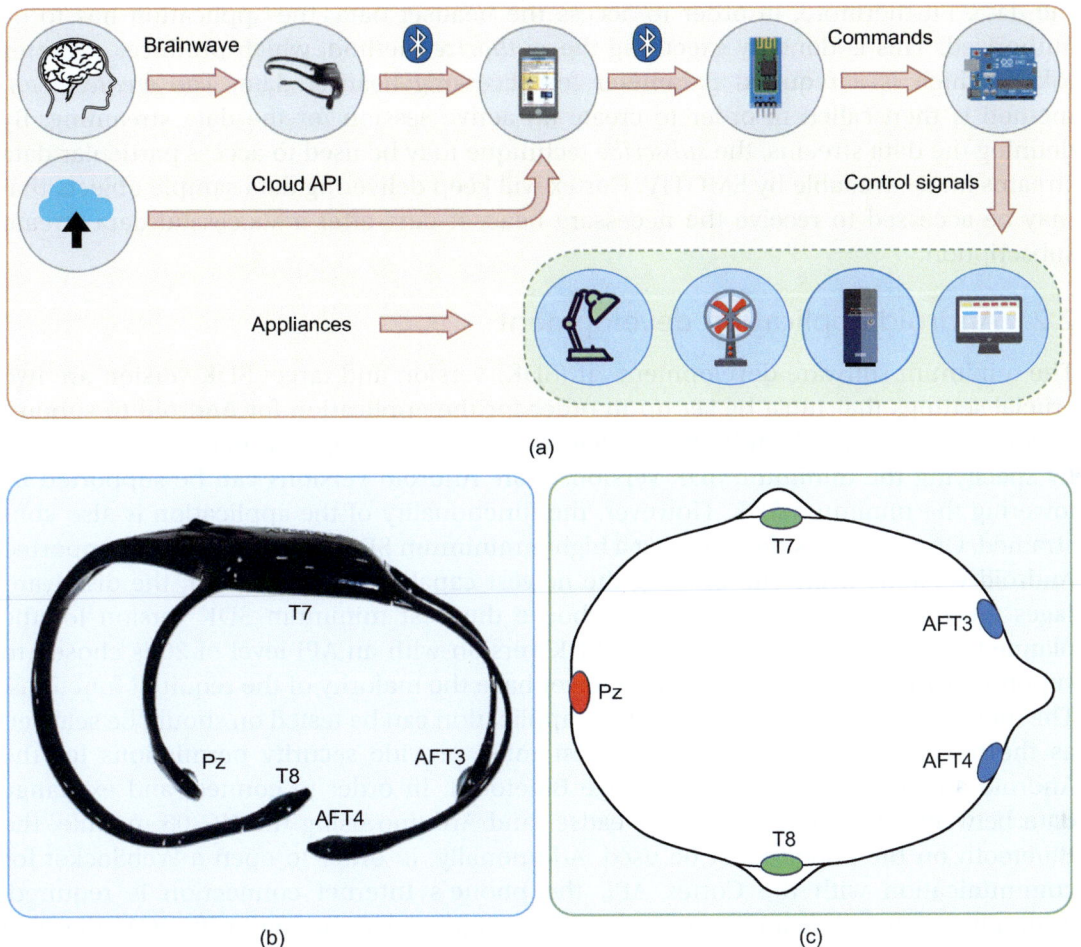

FIGURE 10.1 (A) Working principle of the proposed approach. (B) Top view of EMOTIV Insight headset showing the locations of the electrodes. (C) Top view of a human head with the corresponding locations of the placements of the electrodes.

EMOTIV account with a user-defined EMOTIV ID. Next, the developed software has to request access to the EMOTIV application by using the *requestAccess* method of the API. The EMOTIV application must be granted access manually using the application on the device for the first time. However, once approved, the manual process is no longer necessary. When access is granted, the *queryHeadsets* method is called to find nearby headsets that are available. This method provides a list of headsets that are available, along with their settings and status. Finally, in order to establish a connection with the appropriate headset, the *controlDevice* method is used.

The *queryHeasets* method is continually used until the state of the headset received changes to "connected" in order to verify that the headset is connected before accessing the data. Furthermore, in order to access the headset data, the application has to be authorized. This is done by executing the *authorize* method, which produces a Cortex token, which is a required parameter for accessing headset data. The *createSession* method is then called in order to create an active session for the data streaming. By defining the data streams, the *subscribe* technique may be used to access particular data streams made available by EMOTIV. Cortex will keep delivering data sample objects that may be accessed to receive the necessary headset data after a successful data stream subscription.

2.2   Android application development

The minimum software development kit (SDK) version and target SDK version are two crucial settings that must be set up in order for the application for Android to support various API versions. The minimum API level at which the application may operate is set by specifying the minimum SDK version. More Android versions can be supported by lowering the minimum SDK. However, the functionality of the application is also constrained. On the other side, selecting a higher minimum SDK will result in less supported Android versions while still offering the newest capabilities. As a result, the disadvantages must be traded-off in order to choose the best minimum SDK version for the planned application. In this study, an SDK version with an API level of 26 is chosen to support a variety of Android devices and yet have the majority of the required functions. The most recent Android version that the application can be tested on should be selected as the target SDK. The Android Manifest must provide security permissions for the Android application to use features like Bluetooth. In order to connect and exchange data between the EMOTIV Insight headset and Arduino using the HC-06 module, the Bluetooth on the phone must be used. Additionally, in order to open a WebSocket for communication with the Cortex API, the phone's Internet connection is required. Additionally, the application has to set up a unique certificate authority (CA) that will trust the Cortex WebSocket connection.

A responsive and engaging application with an intuitive visual experience are features of a high-quality graphical user interface (GUI). To make it easy for the user to handle smart home appliances, the application's responsiveness is essential. Therefore, it is necessary to optimize calculations in the application's backend to provide a quick and responsive GUI. The usage of Android GUI widgets, such as buttons, text views, progress bars, and notifications, may make the user interface interactive. However, because the application is designed for disabled people with mobility issues, there must be very little, if any, interaction between the user and the program. As a result, the application was created using just one interactive widget, which must be clicked to launch the application. Additionally, significant data are provided, including the specifications, the application's instructions, and the status of the linked devices.

There are two primary stages that are involved from starting the Android application to operating the smart home appliances: Standby Mode and Selection Mode. Data from the EEG headset is crucial for the user's ability to switch between states and control attributes. The application initially waits for the user to tap a button when it is opened in order to go into the Standby Mode. All Cortex initializations, as well as connection to the EMOTIV Insight headset and Arduino Uno, are carried out consecutively in Standby Mode. The application layout of the Standby Mode displays necessary information that can be of interest to the user. The headset is considered ready by the program once initializations and data extraction have been completed and the signal strength and overall contact quality have surpassed an acceptable threshold.

The program waits until a user blinks twice before switching from Standby Mode to Selection Mode. In the Selection Mode, a visual menu of buttons for various appliances is made available to the user. Until a double blink is observed, which signals the desired appliance to control, the application uses the head motion data to cycle between the appliances. The application then evaluates the user's level of attention after the appliance is defined. Then, the program instructs the Arduino Uno to switch on or off the appliance when it determines that the attention level has risen over the predetermined threshold. The application switches back to Standby Mode after the control process and exits Selection Mode. The application is compelled to go back to Standby Mode at any time throughout the Selection Mode if the user is idle for more than a minute. This additional condition takes into consideration the case in which Selection Mode is accidently activated or in which the user decides to stop controlling the appliances.

2.3   Control circuit development

The control circuit's purpose is to wirelessly receive commands from the Android phone and control electrical appliances accordingly. The controller of this circuit is an Arduino Uno board. The HC-06 Bluetooth module is used for wireless connection between the phone and the Arduino Uno board. The majority of smart home appliances demand greater voltages from the main power supply than Arduino's maximum output voltage of 5 V. Therefore, a 5 V four-channel relay module serves as a link between the Arduino and the appliances.

The RX pin of the HC-06 Bluetooth module was linked to the Arduino digital port, D3. However, the D3 pin produces 5 V, whereas the RX pin only operates at 3.3 V. Therefore, a resistor network was used to scale down the voltage in order to prevent damaging the Bluetooth module pins, with a voltage divider using two resistors of 560 Ω and 1 kΩ.

Following the development of the circuit, the Arduino code is created to control the appliance states in response to commands. The character of the relevant appliance number to be controlled is contained in the command that was received. The Arduino examines the appliance's status, whether it is on or off, and toggles the state appropriately after successfully retrieving the appliance number.

2.4   EEG signal processing

Several characteristics from the EEG headset must be retrieved and categorized in order for the application to function. The user's blink, active attention level, and head motions must be distinguished from the headset to transition to other states. Additionally, the application's responsiveness must be improved by minimizing the amount of backend calculations needed. As a result, the computation for signal processing of raw EEG data is reduced by using the processed classifications offered by detection algorithms from the EMOTIV Cortex wherever possible to identify the required transition circumstances.

Three distinct types of machine learning-based detection algorithms that can categorize various characteristics and circumstances are offered by EMOTIV. Performance metrics, which recognize various cognitive states, mental instructions, which recognize user's mental thinking patterns, and facial expressions are some of these classes. The *met* data stream of performance metrics from EMOTIV is used to determine the user's active focus level. The performance metrics data, which include a detection flag and intensity level that range from 0 to 1, provide information on the level of focus. After the detection flag is verified to be true, it is tested to find out if the intensity of the focus level is higher than a predetermined threshold level in order to determine the user's active focus/ attention level. Multiple attention level tests were conducted, and the ideal normalized threshold was found to be 0.3.

Facial expressions data from EMOTIV, or *fac* data stream, is utilized to identify the user's blinks. The movement of the eyes is shown in the facial expression data, which can recognize both eyes blinking and each eye winking separately. The time difference between each blink is measured and evaluated to find if it falls within a certain time range in order to identify a double blink. The ideal time interval between blinks during double blink was found to be between 0.1 and 0.4 seconds after a number of double-blink testing. To prevent duplicate blink detection when a single blink yields several blink detections in a very short amount of time, the lower limit was not set to 0 seconds.

By collecting raw data in the form of quaternions from the built-in accelerometer, magnetometer, and gyroscope, the EMOTIV Insight headset produces headset motion data. The only permitted movements are turning up, down, left, and right, which restricts the possibilities that may be taken into account. Numerous motion testing revealed that the accelerometer produced the most reliable and distinct results for the aforementioned movement possibilities. Therefore, only the accelerometer data were considered in this work. It should be considered that implementing a straightforward thresholding approach reduces the amount of computing needed. To gather the benchmark average data for the x, y, and z axes of the accelerometer during the idle state, the user is first instructed to stay still for 10 seconds. To differentiate the specific motion involved, the difference between the current sensor reading and the average is evaluated against a threshold after the average has been generated. The best-normalized threshold was determined to be 0.1 after multiple experiments.

2.5   Subsystems integration

The primary software used in the Android application is depicted in Fig. 10.2A. The program launches and establishes a connection with the EMOTIV Cortex server. After putting on the EEG headset and turning on both the headset and the Arduino controller, the user is asked to tap the *Start Application* button after a successful connection. The primary application then alternates between the stages of Selection Mode and Standby Mode. The application provides ongoing control of appliances until either the application is canceled voluntarily by the user, or the application has connectivity issues with the relevant devices, as can be seen from the flowchart. This guarantees that user physical interaction with the program is kept to a minimum, which is essential for impaired users. The detailed subprograms utilized in the application are depicted in Fig. 10.2B. This algorithm logically follows the proposed state transition diagram outlined in the previous sections, facilitating the seamless transition between Standby Mode and Selection Mode. Moreover, an interrupt system is activated within the Selection Mode to respond to the user's inactivity.

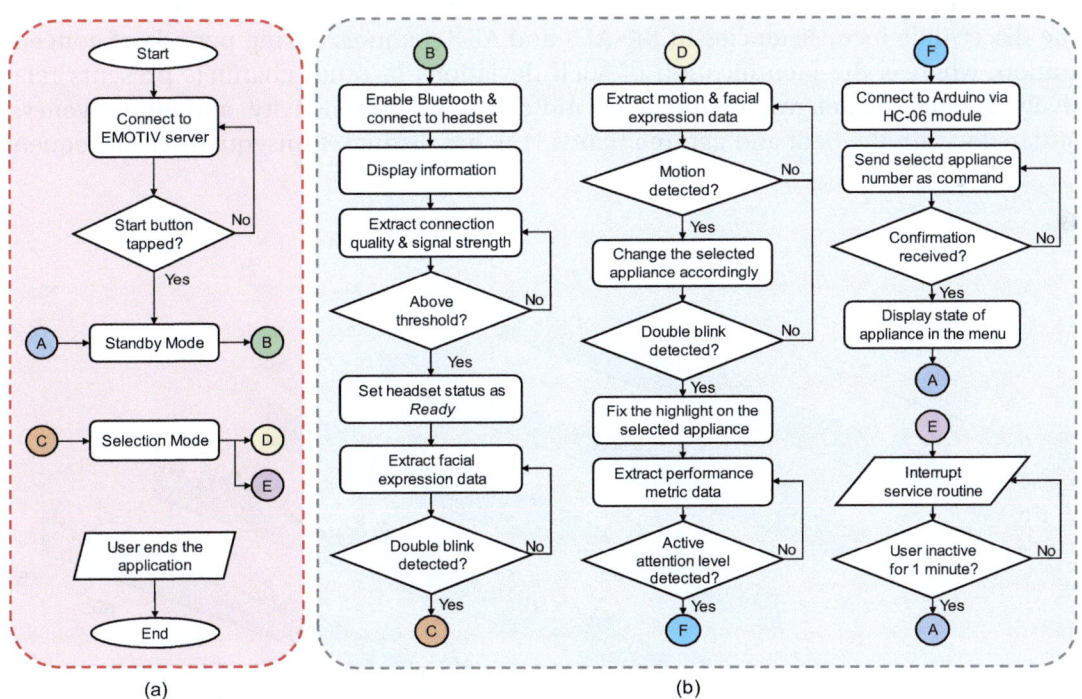

FIGURE 10.2 Android application flowchart. (A) Main program. (B) Subprograms.

3. Results and analysis

The presented methodology was subject to evaluation through the systematic analysis of the influence exerted by attention levels, blinking occurrences, and physical motion on the data derived from the EMOTIV Insight headset. These evaluations subsequently served as the foundation for the conception of a GUI, facilitating user control over four household devices. The explanation of outcomes derived from these sequential stages is presented in the subsequent subsections.

3.1   Attention level data analysis

In order to enable user-directed control over appliances, the application necessitates the detection of an active attention level. Experiments were executed to evaluate the impact of focus or attention on the extracted raw EEG data. This analysis of attention levels, as illustrated in Fig. 10.3, spanned a duration of 60 seconds. Within this interval, users were directed to rest for 10 seconds (marked with green arrows in Fig. 10.3) and then enhance their attention levels for 10 seconds (marked with red arrows in Fig. 10.3). Observations extracted from Fig. 10.3 demonstrate that instances of increased user focus correspond to slight surges or notable deviations in the raw EEG data. Particularly noteworthy are the discernible inconsistencies in the AF3 and AF4 channels during periods of concentration, whereas the identification of such deviations in other channels presents relatively greater challenges. Frequencies indicative of user activity and attentiveness, situated within the beta and gamma bands [11], are distinct. Consequently, subsequent

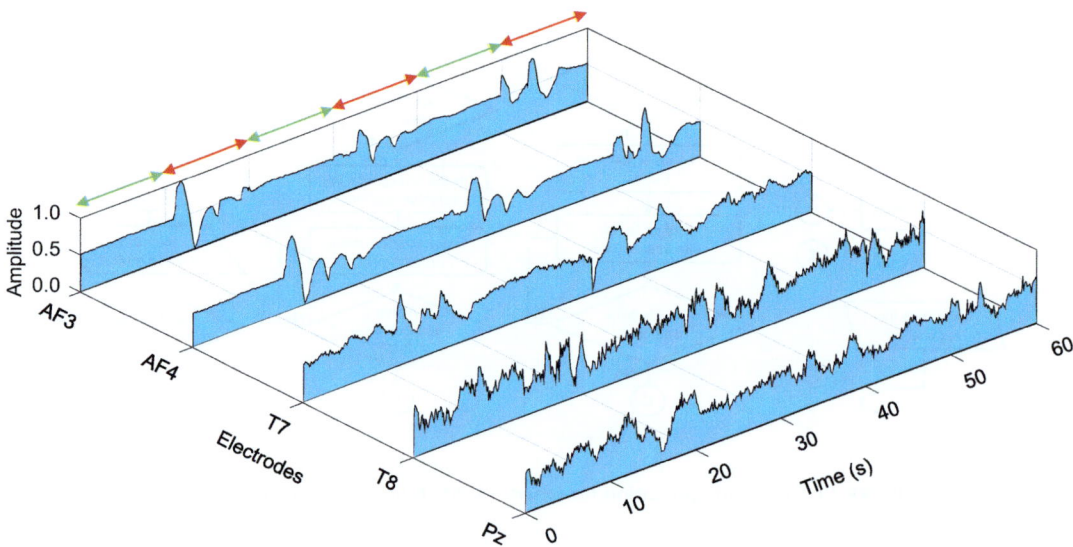

FIGURE 10.3 Raw EEG bands during the active attention experiment. Green arrows represent the rest durations, while red arrows represent the attention durations.

analyses involving the evaluation of frequency components at the levels of low beta, high beta, and gamma bands were executed to delve more profoundly into attention-related patterns for the purpose of detection.

Fig. 10.4 portrays the presence of low and high beta bands during the attention evaluation. The figure examines the perceptible impact of attention, notably centered in the AF3 and AF4 channels. However, the initial attempt by the user to concentrate, occurring between the 10th and 20th seconds, remains somewhat indistinct.

In contrast, an examination of the high beta band within Fig. 10.5 reveals amplitude variations that characterize attention, prominently noticeable in the AF3 and AF4 channels, concurrently with distortions aligned with instances of user-directed focus. Importantly, according to previous studies [22], optimal representation of user attention and focus is expected in the AF3, AF4, and Pz channels. Moreover, subtle yet perceptible deviations are anticipated within the T7 and T8 channels.

Within the context of gamma band analysis, as depicted in Fig. 10.6, discernible characteristics appear during the user's attempts to focus within the time intervals of the attention durations across all channels. This accentuates the suitability of employing gamma band analysis for evaluating attention levels. The visual representation highlights the optimal portrayal of active attention levels within the AF3, AF4, T7, and T8 channels. However, observations indicate that signals derived from the Pz channel exhibit noise during the periods of rest. This phenomenon may stem from factors, such as the user's inability to achieve complete cerebral relaxation during such intervals, interference from muscular noise, or diminished EEG quality in these channels. Notably, the EMOTIV

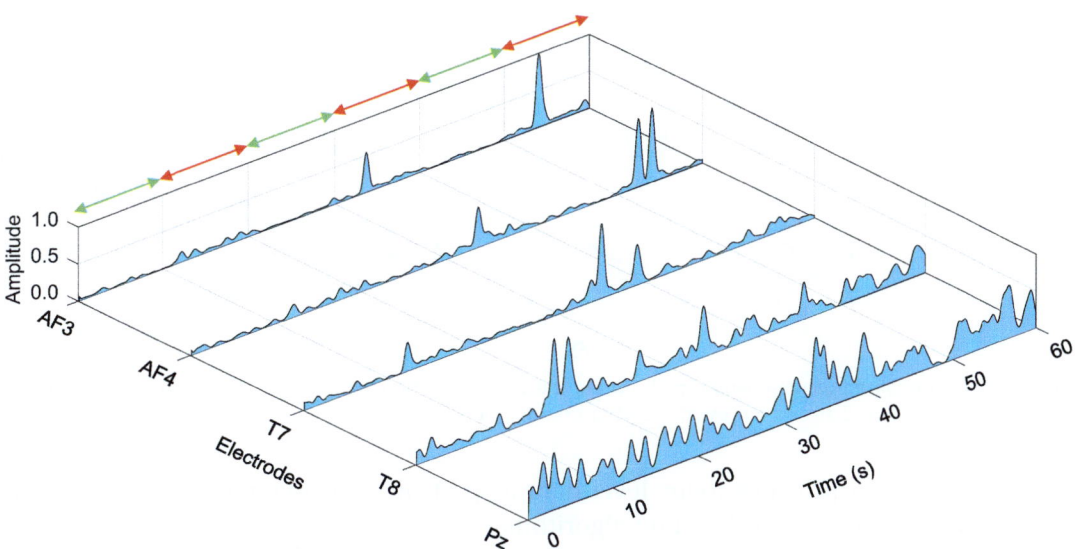

FIGURE 10.4 Normalized low beta band signals during the active attention experiment. Green arrows represent the rest durations, while red arrows represent the attention durations.

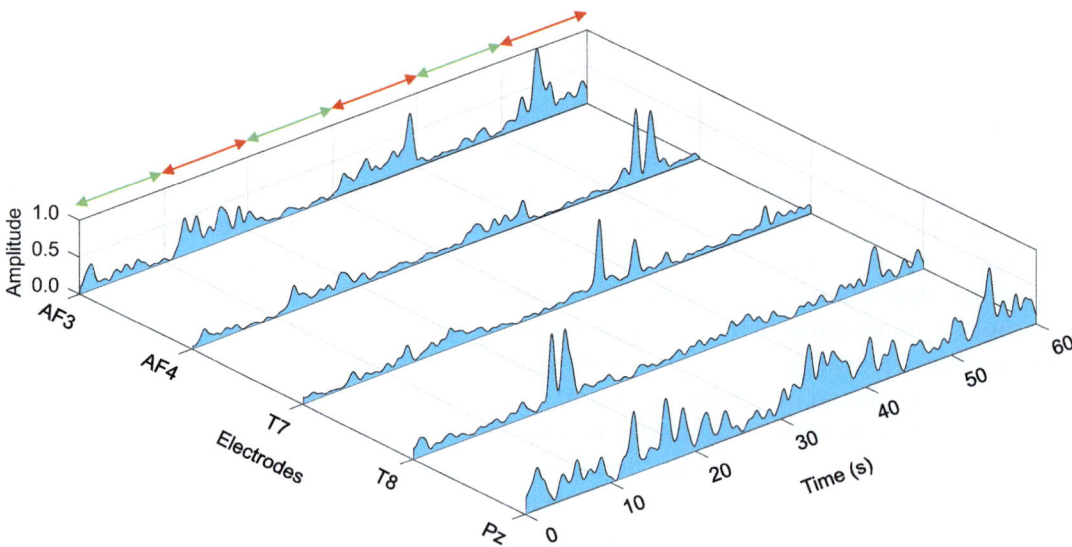

FIGURE 10.5 Normalized high beta band signals during the active attention experiment. Green arrows represent the rest durations, while red arrows represent the attention durations.

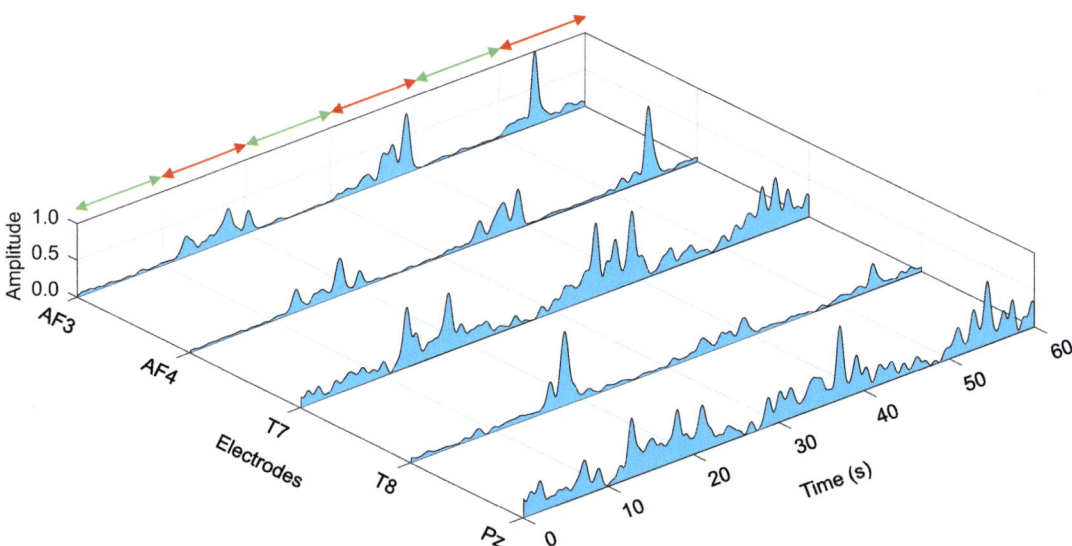

FIGURE 10.6 Normalized gamma band signals during the active attention experiment. Green arrows represent the rest durations, while red arrows represent the attention durations.

Cortex server also provides focus-level data as part of performance metric data, categorized using diverse deep learning algorithms.

The attention levels were extracted from the high beta band and gamma band since they showed more reliable results compared to the low beta band, as demonstrated in

Figs. 10.5 and 10.6. Examination of the graph reveals a consistent elevation in attention level throughout the temporal segments of 10–20 seconds, 30–40 seconds, and 50–60 seconds, aligning with instances where the user was prompted to concentrate. The trigger of active attention level detection occurs when the focus level derived from EMOTIV surpasses the threshold of 0.3. The assessment of active focus level detection's efficacy involved repetitive algorithmic testing (20 iterations) and subsequent comparison to projected outcomes, yielding a 75% accuracy rate for the algorithm governing active attention detection. Instances of detection lapses largely emanate from the algorithm's inaccurate identification of active focus during user-rest intervals subsequent to focused periods. The presence of variance in focus levels across different users and mental states introduces susceptibility to unreliability and inaccuracies in detection algorithms grounded in threshold principles. Elevating the precision of such algorithms is attainable through the integration of methodologies, such as wavelet transform and artificial intelligence-driven classifications [23,24].

3.2   Blinking data analysis

To facilitate the transition from Standby Mode to Selection Mode and fixation of appliance choice, a dual blinking action was implemented. A blink experiment was conducted to scrutinize the influence of blinking on extracted EEG data. In this testing scenario, the user executed blinks at 5-second intervals over a 25-second duration. Specifically, blinks are prompted at the fifth, 10th, 15th, and 20th seconds, as illustrated in Fig. 10.7. To assess the impact of blinking on data, raw EEG data from each headset channel were graphed across time. The outcomes indicate visible data surges in the AF3 and AF4 channels during blink occurrences, whereas the Pz, T7, and T8 channels show a lack of substantial surges. This

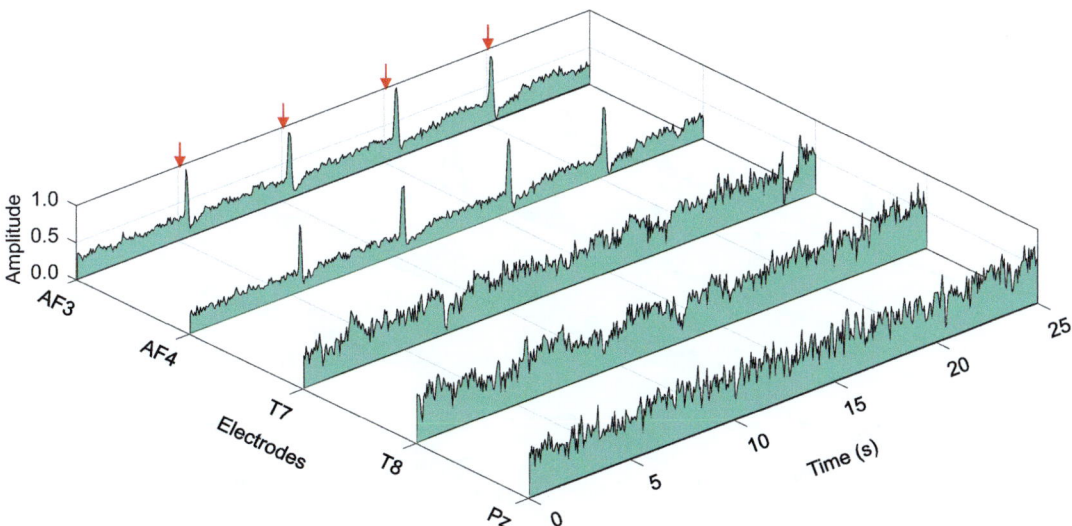

FIGURE 10.7 Raw EEG bands during the blinking test. The arrows correspond to the locations of the actual blinks.

discrepancy could be attributed to the proximity of AF3 and AF4 channels to the eye, being situated on the forehead. As a result, it became evident that for user blink classification, an analysis of AF3 and AF4 channels was imperative.

Therefore, an exploration of blinking's impact on frequency bands in AF3 and AF4 channels was pursued. The averaged band power data for these channels was represented over time in Fig. 10.8. The visual depictions reveal that the theta and alpha bands are most effective in discerning blinks, especially the alpha band, due to its comparatively amplified surge magnitude at the anticipated moment, contrasting with the beta and gamma bands. However, an initial minor peak is observable in the frequency band plots, potentially stemming from considerable startup noise or involuntary blinking.

Blink-level data were also extracted by the EMOTIV Cortex server and classified using diverse deep learning algorithms. Evidently, four surges were identified, signifying blink occurrences at distinct temporal points. A slight lag in blink detection was caused by the blink data's lower sampling rate, set at 32 Hz, in comparison to the raw EEG data's 128 Hz. In this study, the application's functionality is augmented through the utilization of a dual blink, as opposed to a singular blink. This adjustment aims to enhance detection accuracy by eliminating noise-induced blink detections. The proposed algorithm identifies a dual blink when two blinks materialize within a specific time span of 0.1−0.4 seconds. This adjustment enhances EMOTIV's capacity to detect multiple blinks within a brief time frame, even if the user executes a singular blink. The performance of the double blink detection algorithm was evaluated through repetitive testing (20 instances) and subsequent comparison with expected outcomes. Following multiple trials, the double-blink detection algorithm attained an accuracy rate of 90%. Instances of detection failures are attributed to marginally faster or slower double blinks, falling beyond the predetermined time gap range.

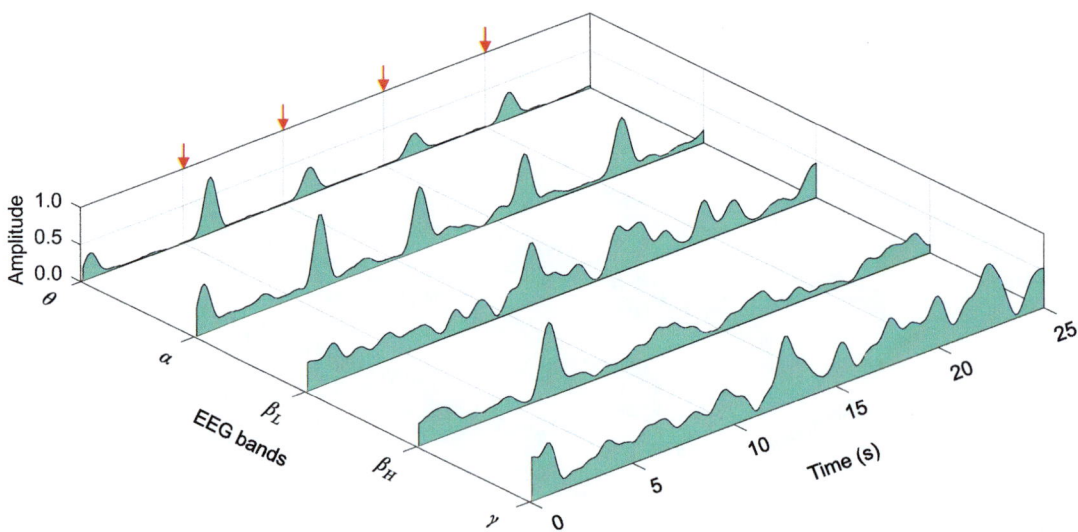

FIGURE 10.8 Averaged frequency bands of AF3 and AF4 channels during the blinking test.

3.3   Motion data analysis

During the phase dedicated to selecting the electrical devices within the application, the user's head movements are utilized as a navigational tool for navigating the appliance menu. In order to optimize the processing efficiency of the application, the scope of user motion is confined to four discrete actions: tilting the head upwards and downwards and turning to the left and to the right. A comprehensive investigation was undertaken using data derived from the gyroscope, magnetometer, and accelerometer sensors to effectively differentiate between these four distinct categories of motion. Despite the potential for achieving high precision through sensor fusion, this approach was avoided to mitigate the computational burden on the backend calculations on the phone. The selection of a sensor capable of distinctly discerning these specific motion types was guided by analytical evaluation.

From Fig. 10.9, it becomes evident that during each instance of motion, one axis reading remains constant, while the other two axes undergo changes. Furthermore, initial readings for each instance of motion demonstrate a relative degree of consistency across the data. The accelerometer readings were selected for analysis due to their ability to effectively differentiate among the different motion types. It is noteworthy that although the accelerometer may not provide readings during continuous speed rotations that involve rotational motion, this limitation was neglected.

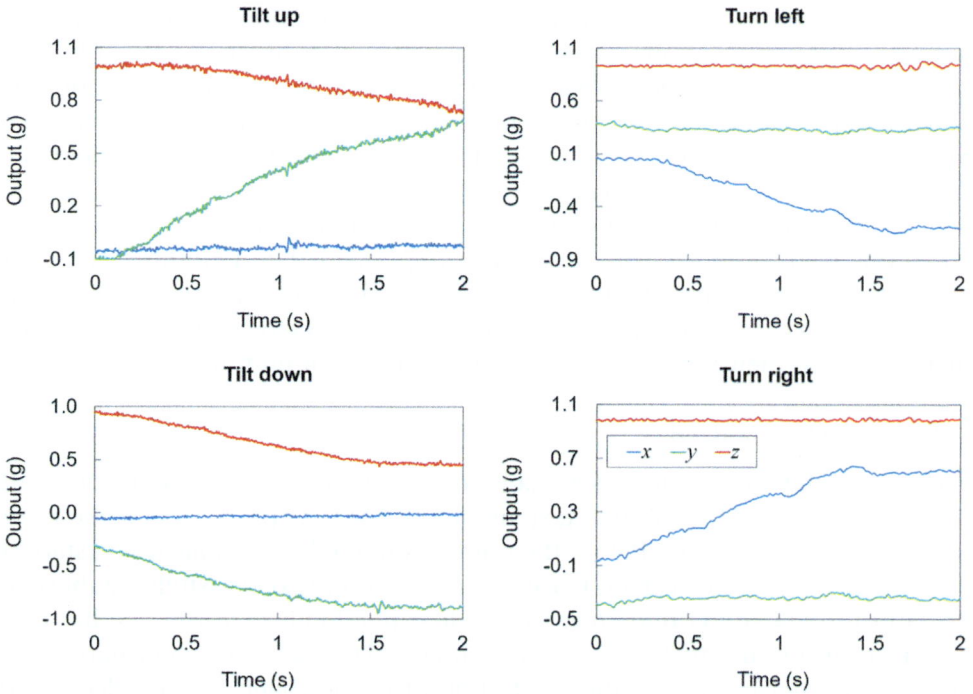

FIGURE 10.9 Head motion data based on the accelerometer's reading during the tilt down, tilt up, turn left, and turn right motions.

The results reveal that during tilt-up and down actions, x-axis readings exhibited constancy. During downward motion, both z-axis and y-axis readings declined below their initial values, while during upward motion, the z-axis readings decreased, and the y-axis readings increased. When it comes to turning left and right, the x-axis readings maintained a relative level of constancy. During leftward motion, both y-axis and x-axis readings increased, whereas, for rightward motion, both readings decreased. Upon collecting data for these distinct motion patterns, it was observed that the constant axis value remained within a range of 0.1. Consequently, the algorithm designed for motion detection was formulated to identify corresponding motions when alterations in two-axis readings surpass the threshold of 0.1, while changes in the other axis remain below 0.1. Additionally, the algorithm mandates that the user maintains a stationary position for a brief period of time (10 seconds) to calibrate the average initial reading. This measure serves to avoid fixed initial values and enhances the adaptability of the system.

By incorporating the initially computed reading as opposed to the reading from the preceding time step, the algorithm was enabled to filter out the motion associated with returning to the original position. This design ensures that the algorithm when presented with scenarios such as a user turning right and subsequently returning to the starting point, detects exclusively the rightward motion owing to its comparison with the originally calibrated reading. This mechanism establishes a high degree of reliability in the algorithm's motion detection capability.

To quantitatively evaluate the performance of the motion detection process, the precision in identifying motions is assessed through a series of repeated algorithmic tests (20 trials), and then compared against the expected outcomes. Each evaluation of motion involves a preliminary 10-second calibration phase, followed by the activation of the motion detection algorithm. After multiple trials, the algorithm attained a perfect success rate of 100% in detecting both tilting upwards and downwards. For turning right and left, it achieved an 85% success rate. Instances of failure in detecting turn-left and turn-right motions were observed, primarily occurring when the algorithm generated dissimilar detection outcomes. The observed decrease in accuracy in the detection of left and right motions could potentially be attributed to the relatively modest deviations in axis readings from their initial values during the course of motion.

3.4   Appliances control

The GUI in Standby Mode comprises a solitary button designed for interaction with the assistant, accompanied by multiple text views that serve as platforms for the presentation of pertinent information. Before the button is tapped, the application establishes a connection with the Cortex server. Once this connection is successfully established, the server status is updated to *Connected*. The guide's function is to lead users through tasks relevant to transitioning to the subsequent phase. The exhibited information is categorized into two device groups: the headset and the controller. In the Selection Mode, the guide section furnishes instructions for users to select and manage appliances. To distinctly indicate the presently selected appliance during the appliance selection stage, a yellow background is applied behind the chosen appliance. The highlighting feature shifts among the appliances, guided by an assessment of the user's head motion. A double blink by the user serves to confirm the intended appliance for control and to terminate the query across appliances. Upon detection of this double blink, a lime-colored background is applied to the respective appliance. This transition is influenced by the user's level of focus.

Fig. 10.10 presents the outcomes of user-conducted experiments employing the integrated system for appliance control. This is accompanied by screenshots from the Android application displayed on a smartphone during the experiment. The control circuit was connected to four appliances: light (appliance 1), fan (appliance 2), computer (appliance 3), and monitor (appliance 4). The objective of the experiment was to successively activate the computer, monitor, light, and finally the fan. The user initiated the EEG headset, consequently activating both the headset and controller. Upon tapping the button, the Standby Mode was initiated, facilitating back-end connections and processes. The application awaited a double-blink to transition to the Selection Mode. Following the double blink, the application follows the algorithmic flow as outlined in Fig. 10.2, utilizing user head motion and double blinks to choose and activate the computer.

Similar steps were undertaken to sequentially activate the monitor, light, and fan, culminating in the final outcomes of both the integrated system and the application. To evaluate the impact of focus and blinks on the extracted brainwave data, as well as the influence of head motion on the motion sensor data, diverse tests were executed. Furthermore, the performance of each developed detection algorithm was assessed, with accuracy determined through repeated trials involving various scenarios. The experiments successfully demonstrated the sequential activation of configured appliances, relying on user blinks, head motion, and attention levels while circumventing the need for physical intervention.

The preceding outcomes clearly demonstrate the appropriateness of the proposed methodology for the elderly population and individuals with disabilities. Furthermore, in

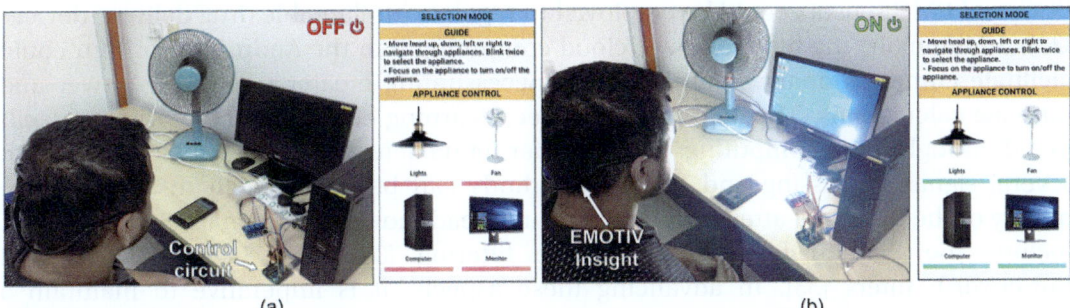

(a)   (b)

FIGURE 10.10 Experiments and screenshots of the developed Android application using the integrated system during the Selection Mode. (A) At the beginning of the experiment when all appliances were switched off. (B) At the end of the experiment after switching on the appliances.

scenarios characterized by complete paralysis, adjustments can be made to the suggested approach to eliminate the requirement of head motion for navigating the mobile application. To achieve this, the methodology could incorporate right and left winks as alternative input gestures, given the headset's capability to detect such actions. Moreover, the proposed strategy adeptly balances considerations of portability, cost-effectiveness, accuracy, and the breadth of features provided. For example, while opting for a budget-friendly headset, such as the NeuroSky MindWave [11], this headset sacrifices EEG sensing and extraction precision. On the other hand, employing more intricate systems like Brain Products LiveAmp 64 [25] or Neuroscan SynAmps [26] amplifies system complexity and cost, thereby constraining portability. As a result, despite potentially offering enhanced EEG sensing and extraction precision, these advanced systems become less pragmatic for smart home control. Such deliberations are pivotal in the development of systems intended for real-world deployment.

Further improvements to the system can be achieved by substituting Bluetooth with Wi-Fi for communication between the application and controller has the potential to expand the connection range and alleviate latency concerns. This transition would facilitate remote control of devices positioned in various areas. Upgrading the relay and controller systems could extend their capabilities to accommodate appliances with higher current demands, such as refrigerators or air conditioners. Further enhancements could extend to the Android application domain, encompassing smart appliances. This extension could facilitate a more comprehensive level of control, enabling actions that surpass the mere activation and deactivation of appliances. For instance, in the context of appliances like fans, users could regulate parameters such as fan speed. The efficacy of such measures could be strengthened through the integration of the BCI system with other sensor modalities like ultrasonic and infrared sensors. This expanded integration would amplify the system's functionality, incorporating environmental sensing and collision avoidance capabilities [27]. The incorporation of wearable sensors like EMG or electrocardiogram (ECG) into the application could also contribute to monitoring user health and generating alerts in response to detected anomalies [28].

This system has the potential to extend into medical treatment contexts, where EEG, ECG, and EMG sensors could be employed to regulate implantable drug delivery devices [29]. Additionally, the integration of cloud computing into the Android algorithm could facilitate computationally intensive tasks, necessitating greater processing power and intricate calculations. The deployment of deep learning classification algorithms, facilitated through cloud computing, could further refine detection accuracy [30]. Moreover, several signal processing and filtering methods can be investigated to enhance the quality of the extracted attention, blinking, and head movements signals. Such methods include the Kalman filter [31], adaptive filtering [32], and Savitzky–Golay and Butterworth filters [33]. In advancing these aspects, it is imperative to maintain a meticulous equilibrium among factors, such as system cost, complexity, portability, and accuracy, ensuring the practicality of the solutions for the elderly and disabled demographic.

4. Conclusion

This work presented an EEG-based BCI system for smart home control, with the goal of assisting elderly or handicapped people in engaging with their surroundings. To operate household appliances, the system utilized double-blink, attention level, and head motion data. During user focus, the attention test indicated identifiable surges at the AF3 and AF4 channels, with gamma band frequency signals displaying the most recognizable characteristics. Tilt up, tilt down, turn left, and turn right actions were detected using accelerometer data. Each detection algorithm's performance was analyzed and assessed. The identification of double blinks showed 90% accuracy, whereas active attention level detection achieved 75% accuracy. Turn-up and turn-down motions were recognized with 100% accuracy, while turn-left and right motions were detected with 85% accuracy. Four distinct household appliances were controlled via an Android application as proof of concept. The developed system is inexpensive, portable, and consumes little power, making it appropriate for a variety of smart home and wireless portable applications.

Acknowledgment

This work was supported by the University of Nottingham Malaysia.

Ethical approval

This study was approved by the Science & Engineering Research Ethics Committee (SEREC), University of Nottingham Malaysia (application ID: MD24920).

References

[1] Tang X, Shen H, Zhao S, Li N, Liu J. Flexible brain–computer interfaces. Nat Electron 2023;6(2): 109−18. https://doi.org/10.1038/s41928-022-00913-9.

[2] World Health Organization. Global report on health equity for persons with disabilities. World Health Organization; 2022.

[3] Rimbert S, Fleck S. Long-term kinesthetic motor imagery practice with a BCI: impacts on user experience, motor cortex oscillations and BCI performances. Comput Hum Behav 2023;146:107789. https://doi.org/10.1016/j.chb.2023.107789.

[4] Rasheed S. A review of the role of machine learning techniques towards brain–computer interface applications. Mach Learn Knowl Extr Oct. 2021;3(4):835−62. https://doi.org/10.3390/make3040042.

[5] Kohli V, Tripathi U, Chamola V, Rout BK, Kanhere SS. A review on virtual reality and augmented reality use-cases of brain computer interface based applications for smart cities. Microproc Microsyst 2022;88:104392. https://doi.org/10.1016/j.micpro.2021.104392.

[6] Yao L, Shoaran M. Enhanced classification of individual finger movements with ECoG. In: Conference record - asilomar conference on signals, systems and computers. IEEE Computer Society; 2019. p. 2063−6. https://doi.org/10.1109/IEEECONF44664.2019.9048649.

[7] Haufe S, et al. Elucidating relations between fMRI, ECoG, and EEG through a common natural stimulus. Neuroimage 2018;179:79−91. https://doi.org/10.1016/j.neuroimage.2018.06.016.

[8] Hooda N, Das R, Kumar N. Fusion of EEG and EMG signals for classification of unilateral foot movements. Biomed Signal Proc Control 2020;60:101990. https://doi.org/10.1016/j.bspc.2020.101990.

[9] Dhole SR, Kashyap A, Dangwal AN, Mohan R. A novel helmet design and implementation for drowsiness and fall detection of workers on-site using EEG and random-forest classifier. Procedia Comput Sci 2019;151:947−52. https://doi.org/10.1016/j.procs.2019.04.132.

[10] Berger H. Über das elektroenkephalogramm des menschen. Arch Psychiatr Nervenkr 1929;87(1):527−70.

[11] Nafea M, Hisham AB, Abdul-Kadir NA, Harun FKC. Brainwave-controlled system for smart home applications. In: 2018 2nd international conference on BioSignal analysis, processing and systems (ICBAPS); 2018. p. 75−80. https://doi.org/10.1109/ICBAPS.2018.8527397.

[12] Agarwal M, Sivakumar R. Blink: a fully automated unsupervised algorithm for eye-blink detection in EEG signals. In: 2019 57th annual allerton conference on communication, control, and computing, allerton 2019. Institute of Electrical and Electronics Engineers Inc.; Sep. 2019. p. 1113−21. https://doi.org/10.1109/ALLERTON.2019.8919795.

[13] Djamal EC, Pangestu DP, Dewi DA. EEG-based recognition of attention state using wavelet and support vector machine. In: Proceeding - 2016 international seminar on intelligent technology and its application, ISITIA 2016: recent trends in intelligent computational technologies for sustainable energy. Institute of Electrical and Electronics Engineers Inc.; 2017. p. 139−44. https://doi.org/10.1109/ISITIA.2016.7828648.

[14] Tsiouris KM, Konitsiotis S, Koutsouris DD, Fotiadis DI. Unsupervised seizure detection based on rhythmical activity and spike detection in EEG signals. In: 2019 IEEE EMBS international conference on biomedical and health informatics, BHI 2019 - proceedings. Institute of Electrical and Electronics Engineers Inc.; 2019. https://doi.org/10.1109/BHI.2019.8834644.

[15] Bin Nasir T, Lalin MAM, Niaz K, Karim MR, Rahman A. EEG based human assistance rover for domestic application. In: ICREST 2021 - 2nd international conference on robotics, electrical and signal processing techniques. Institute of Electrical and Electronics Engineers Inc.; 2021. p. 461−6. https://doi.org/10.1109/ICREST51555.2021.9331224.

[16] Zhang H, Dong E, Zhu L. Brain-controlled wheelchair system based on SSVEP. In: Proceedings - 2020 Chinese automation congress, CAC 2020. Institute of Electrical and Electronics Engineers (IEEE); 2020. p. 2108−12. https://doi.org/10.1109/CAC51589.2020.9327651.

[17] Chhabra K, Mathur P, Baths V. BCI controlled quadcopter using SVM and recursive LSE implemented on ROS. In: IEEE transactions on systems, man, and cybernetics: systems. Institute of Electrical and Electronics Engineers Inc.; 2020. p. 4250−5. https://doi.org/10.1109/SMC42975.2020.9282898.

[18] Parbez RMS, Mamun KA. BlinkFruity: a real-time EEG based neurofeedback game for brain-computer interface. In: 2020 2nd international conference on advanced information and communication technology, ICAICT 2020. Institute of Electrical and Electronics Engineers (IEEE); 2020. p. 404−9. https://doi.org/10.1109/ICAICT51780.2020.9333469.

[19] Hossain T, Rakshit A, Konar A. Brain-computer interface based user authentication system for personal device security. In: 2020 international conference on computer, electrical and communication engineering, ICCECE 2020. Institute of Electrical and Electronics Engineers Inc.; 2020. https://doi.org/10.1109/ICCECE48148.2020.9223069.

[20] Adams M, et al. Towards an SSVEP-BCI controlled smart home. In: 2019 IEEE international conference on systems, man and cybernetics (SMC). IEEE; 2019. p. 2737−42. https://doi.org/10.1109/SMC.2019.8914668.

[21] Park S, Cha HS, Im CH. Development of an online home appliance control system using augmented reality and an SSVEP-based brain-computer interface. IEEE Access 2019;7:163604−14. https://doi.org/10.1109/ACCESS.2019.2952613.

[22] Jayakody Arachchige MD, Nafea M, Nugroho H. A hybrid EEG and head motion system for smart home control for disabled people. J Ambient Intell Hum Comput 2023;14(4):4023−38. https://doi.org/10.1007/s12652-022-04469-6.

[23] al-Qerem A, Kharbat F, Nashwan S, Ashraf S, blaou khairi. General model for best feature extraction of EEG using discrete wavelet transform wavelet family and differential evolution. Int J Distributed Sens Netw 2020;16(3). https://doi.org/10.1177/1550147720911009.

[24] Alshebly YS, Nafea M. Isolation of fetal ECG signals from abdominal ECG using wavelet analysis. IRBM 2019. https://doi.org/10.1016/j.irbm.2019.12.002.

[25] Shao L, et al. EEG-controlled wall-crawling cleaning robot using SSVEP-based brain-computer interface. J Healthc Eng 2020;2020:6968713. https://doi.org/10.1155/2020/6968713.

[26] Chai X, et al. A hybrid BCI-controlled smart home system combining SSVEP and EMG for individuals with paralysis. Biomed Signal Proc Control 2020;56:101687. https://doi.org/10.1016/j.bspc.2019.101687.

[27] Khan MJ, Zafar A, Hong K-S. Hybrid EEG-NIRS based active command generation for quadcopter movement control. In: 2016 international automatic control conference (CACS); 2016. p. 200−5. https://doi.org/10.1109/CACS.2016.7973909.

[28] Sarhan SM, Al-Faiz MZ, Takhakh AM. A review on EMG/EEG based control scheme of upper limb rehabilitation robots for stroke patients. Heliyon 2023;9(8):e18308. https://doi.org/10.1016/j.heliyon.2023.e18308.

[29] Nafea M, Nawabjan A, Mohamed Ali MS. A wirelessly-controlled piezoelectric microvalve for regulated drug delivery. Sens Actuators A Phys 2018;279:191−203. https://doi.org/10.1016/j.sna.2018.06.020.

[30] Mehmood I, et al. Deep learning-based construction equipment operators' mental fatigue classification using wearable EEG sensor data. Adv Eng Inf 2023;56:101978. https://doi.org/10.1016/j.aei.2023.101978.

[31] Yadav S, Saha SK, Kar R. An application of the Kalman filter for EEG/ERP signal enhancement with the autoregressive realisation. Biomed Signal Proc Control 2023;86:105213. https://doi.org/10.1016/j.bspc.2023.105213.

[32] Yadav S, Saha SK, Kar R, Mandal D. EEG/ERP signal enhancement through an optimally tuned adaptive filter based on marine predators algorithm. Biomed Signal Proc Control 2022;73:103427. https://doi.org/10.1016/j.bspc.2021.103427.

[33] Siew HSH, Alshebly YS, Nafea M. Fetal ECG extraction using Savitzky-Golay and Butterworth filters. In: 2022 IEEE international conference on automatic control and intelligent systems (I2CACIS); 2022. p. 215−20. https://doi.org/10.1109/I2CACIS54679.2022.9815469.

Independent component analysis methods for motor imagery-based brain-computer interfaces

Paulo A.A.L. Viana[1,2], Sarah N.C. Leite[3] and Romis Attux[1]

[1]SCHOOL OF ELECTRICAL AND COMPUTER ENGINEERING, UNIVERSITY OF CAMPINAS (UNICAMP), CAMPINAS, BRAZIL; [2]AI R&D LAB, SAMSUNG R&D INSTITUTE BRAZIL, CAMPINAS, BRAZIL; [3]DIVISION OF ELECTRONICS ENGINEERING, AERONAUTICS INSTITUTE OF TECHNOLOGY (ITA), SÃO JOSÉ DOS CAMPOS, BRAZIL

1. Introduction

Electroencephalography (EEG) signals play a key role in several brain-computer interface (BCI) applications. EEG signals are expressions of brain physiology and behavior as evoked by specific stimuli under distinct paradigms [1]. The modus operandi of the motor imagery (MI) paradigm requires that a subject imagine limb or tongue movements while their brain electrical activity is recorded. Subsequently, this recorded activity is classified, and the corresponding motor task labels are associated with the commands of the targeted application [2,3]. Fig. 11.1 shows a diagram with the stages of a classical MI-based BCI.

The literature highlights that common MI instances include imagery of left and right foot movements, along with arm gestures [4]. Importantly, such a setup in a BCI can be effectively employed to control both physical and virtual devices.

Specific brain regions are closely associated with patterns observed during the performance of each task. In the case of MI, the motor cortex exhibits event-related desynchronization or synchronization (ERD or ERS, respectively), primarily in the *mu* (8−13 Hz) and *beta* (13−25 Hz) frequency bands [5]. These bands represent distinct frequency intervals in which brain activity is detected, and the amplitude of the recorded signals is influenced by the task being performed. Furthermore, due to the crossing of the neuron paths at the medulla, the response to left and right-hand imagery can be observed in the contralateral hemispheres of the brain [6].

Various electrical physiological processes, including ocular, muscular, and cardiac activity, can influence EEG recordings. Moreover, nonphysiological sources like line noise and movement artifacts can also contaminate the signals. Algorithms based on

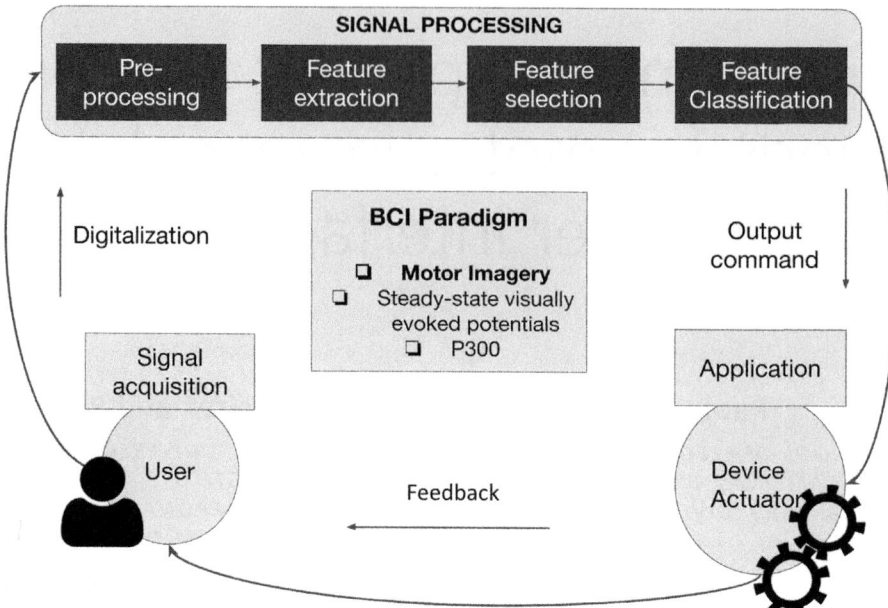

FIGURE 11.1 Stages of an MI-based BCI.

independent component analysis (ICA) are capable of separating statistically independent sources from a set of observed mixtures [7]. In the context of BCI, ICA is extensively used, mainly for artifact removal, spatial filtering, SNR enhancement, and electrode selection [8–18]. Typically, it is assumed that the observed signals result from a linear combination of these sources, allowing their separation by finding the inverse of the mixing linear system. However, different ICA algorithms may rely on distinct assumptions about the characteristics of these sources, leading to potentially diverse outcomes.

In MI applications, we focus on selecting the independent components (ICs)—the sources—that carry the most relevant information about the performed task. Nonetheless, it is important to note that many techniques assume that the sources are stationary, which is not always guaranteed in real EEG data [19]. Hence, the dynamic nature of EEG signals should be carefully considered when employing ICA in BCI setups to ensure accurate and reliable results.

BCI can also benefit from deep learning strategies applied to signal processing and classification, as deep neural networks have shown great applicability over a broad range of scenarios. Many works [20–22] have proposed network architectures that can process raw or almost raw EEG and may perform well even in session-to-session transfer learning and cross-subject protocols. We employ the EEGNet, which originally operated on raw EEG signals, and investigate whether the network can be augmented by using ICA as a preprocessing step.

In our initial experiments, we performed a thorough comparison of several ICA algorithms based on system classification accuracy. Each ICA method was utilized as a preprocessing step in our application, which involves a 4-class MI task performed by subjects while their EEG signals were recorded. Following the source separation stage, the signals underwent further processing, and features were extracted from the separated sources. These features served as characteristic vectors for a downstream classifier.

The general goals of this work are:

1. Comparing how different ICA methods and classification algorithms impact the performance of a motor-imagery BCI system.
2. Evaluating how ICA methods as preprocessing for a deep-learning-based classifier affect the convergence speed and the accuracy of the MI-based BCI.

The specific goals are:

1. To identify, among the tested ICAs and classifiers, which, on average, results in the highest BCI accuracy for the analyzed dataset.
2. For each classifier, identify the ICA method that maximizes the MI-based BCI accuracy.
3. For each ICA, identify the classifier method that maximizes the MI-based BCI accuracy.
4. To find the combination of ICA and classifier that results in the highest MI-based BCI accuracy.
5. To analyze which ICA method leads to the highest accuracy with EEGNet.
6. To analyze the impact of ICA on the number of iterations required to train the EEGNet.

2. Methods

In this section, we describe the data, ICA methods, and machine learning algorithms for data classification. We used two open-access datasets in our experiments, the BCI Competition IV 2a [23] and the OpenBMI [24] dataset, and they are presented in Section 2.1. In Section 2.2, we present the problem of ICA as well as the associated motivation and algorithms. In Section 2.3, we present the feature extraction method used in our experiments as well as the feature subset selection technique. In Section 2.4, the employed classification algorithms are discussed. Finally, Section 2.5 presents the EEGNet, which is a neural network used for MI-based BCI.

2.1 Dataset

The BCI Competition IV dataset 2a (dataset D1) [23] contains two sessions of motor imagery tasks, performed on two different days. Subjects were instructed to imagine the movement of their left hand, right hand, foot, or tongue according to the cues that are presented on a monitor screen. On each day, each subject performed 36 tests of each imagination task, resulting in 288 tests in total. At each trial, the subject first heard a beep indicating attention, then, 2 seconds later, a visual cue was displayed on a monitor for 1.25 seconds, which instructed the subject to perform the imagination task for 4 seconds. EEG was collected using the 10–20 international system with 22 electrodes at 250 Hz. The collecting equipment had a bandpass filter between 0.5 and 100 Hz, with an additional notch filter at 50 Hz to suppress line noise. For our experiments, we use the whole 4s of imagination as our sample windows.

The OpenBMI dataset (dataset D2) [24] contains four sessions of MI tasks, two at each day. Differently from the former dataset, subjects performed only the imagination of moving their right and left hands. At each session, 50 trials of the imagination task were carried out for each hand, yielding 400 sessions in total. The trials started with a fixation cross displayed on a screen for 3 s, then at $t = 0s$, an arrow indicates that the subject needs to imagine the movement for 4 seconds, succeeded by 6 seconds (± 1.5 seconds) of a blank screen. EEG was collected using the 10–20 international system, but using 62 electrodes at 1000 Hz. Those signals were bandpass-filtered between 8 and 30 Hz with a fifth-order IIR Butterworth filter, and we cropped the trial window from $t = 1s$ to $t = 3.5s$ to create the sample windows, reproducing the preprocessing done in Ref. [24].

For both datasets, we use the data from the first day as training and the data from the second day as tests. The classifiers and ICA unmixing matrices are also adjusted or fitted using only the training data, except in the case of ORICA, since it is an adaptive algorithm. All experiments are based on the intersubject protocol, in which the ICA and the model are adjusted and evaluated for each subject individually.

2.2   Source extraction

Blind source separation (BSS) is a family of methods that can be used for unmixing signals that are mixed/superposed and can be used for extracting discriminative information from the EEG signals. Assuming that there are specific brain regions that activate during MI tasks, we can suppose that each EEG electrode captures a superposition of those brain regions. ICA is a subset of algorithms from the BSS family that assumes linearly mixed signals. Although ICA has proven useful for removing eye blink artifacts [25], it has also been used as an unsupervised alternative to common spatial patterns (CSP) filtering for extracting motor-related ICs [10], and many use cases have been proposed and compared in terms of accuracy, robustness, and generalization [26–29].

For solving the unmixing problem, uncorrelatedness is not enough; statistical independence is also necessary. Independence is a stronger condition and also a more difficult one to quantify and ensure for real-world signals. Eq. (11.1) shows the generative model of ICA, where $\mathbf{x}$ is the vector of observed variables, $\mathbf{A}$ is the mixing matrix, and $\mathbf{s}$ is the source vector. Each row of $\mathbf{A}$ can be interpreted as a spatial filter that mixes the sources.

$$x = As \tag{11.1}$$

This model assumes that the observed variables are linear mixtures (weighted sums) of the unobserved source variables, and, in some models, there is also an additional noise term. Generally, the assumptions are that the number of sources is equal to or less than the number of observed variables and that sources are nonGaussian (even though some algorithms can handle up to one Gaussian source) and statistically independent [7]. Independence can be defined using the joint probability density functions (pdf) of the sources, as in Eq. (11.2), where $s = (s_1, s_2, \ldots, s_N)$ is the source vector and N is the number of sources.

$$p(s_1, s_2, \ldots, s_N) = \prod_{n=1}^{N} p(s_n) \tag{11.2}$$

We say that the random variables s_i are independent if and only if their joint pdf can be factorized as the multiplication of the individual pdf's, which means that one variable does not convey information about the others. Fig. 11.2 shows an example of ICA: three source signals (left) are linearly mixed (center), and, after source separation, recovered (right). Note that ICA does not necessarily preserve the order and scale of the sources; this, however, is not a significant limitation in most cases.

If an ICA algorithm estimates the mixing matrix A, it is necessary to invert it to obtain a separation condition of the form $s = A^{-1}x$, where $s = (s_1, \ldots, s_N)$ is the N-dimensional multivariate source vector, $x = (x_1, \ldots, x_N)$ is the N-dimensional multivariate observed vector, and A^{-1} is the unmixing matrix. It should be noticed that some ICA algorithms assume that the input signals are uncorrelated, which means that a whitening step may be required beforehand.

The existing algorithms approach the source separation problem in different ways, but the outcome is always an estimate of the mixing or unmixing matrix. Most methods produce similar results for ideal or synthetic signals, but dealing with real signals poses a different challenge, and each method may yield distinct results as each has algorithmic differences and real EEG signals may only approximately follow ICA assumptions [30]. That is also why different scenarios may require different methods.

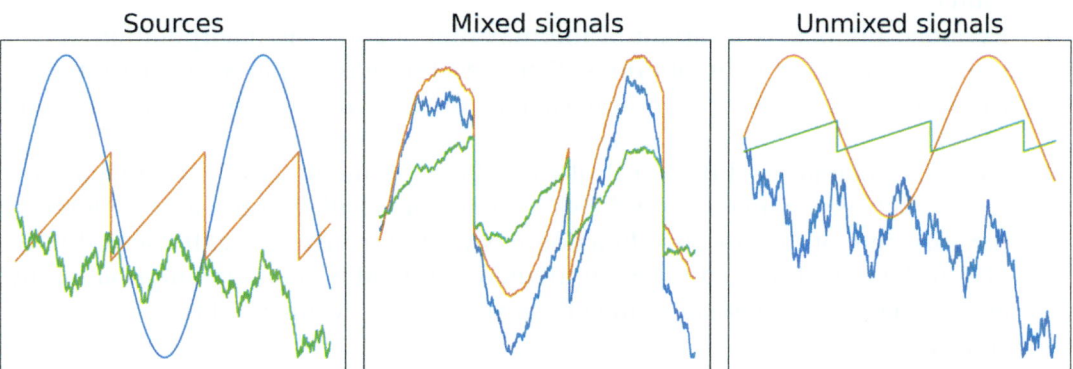

FIGURE 11.2 Example of an application of ICA.

In [26], five ICA methods are compared in terms of their capability of creating valid filters (based on the "maximum projection on the primary electrodes" criteria) and accuracy in a three-class MI task. It was found that not only simplified Infomax (compared to FastICA, Infomax, second-order blind identification (SOBI), and joint approximate diagonalization of eigenmatrices (JADE)) created valid filters more often, but it also greatly improved classification accuracy when using only the IC variances as features and also outperformed the CSP method, which is a classical supervised method for spatial filter design [14]. For our experiments, we chose a set of eight methods that are commonly cited in BCI literature and that are readily available in computing toolkits to evaluate if different approaches have a great impact on BCI accuracy. Next, we briefly present the characteristics of each of them.

i. JADE is an ICA method that diagonalizes the fourth-order cumulant tensors of estimated sources, which may be interpreted as a nonlinear decorrelation procedure [31]. Aside from assuming that there is at most one Gaussian source, JADE assumes that the mixing filters are linearly independent, the sources are independent at each time point, the sources' covariance and cumulants have consistent estimates, the additive noise is Gaussian and independent of the sources, and it is spatially white. In this algorithm, the fourth-order cumulant tensor is calculated and decomposed into orthogonal eigenmatrices, which are then jointly diagonalized. It has the goal of canceling all third- and fourth-order cross-cumulants, since the second-order cumulant is canceled by the initial whitening, and this can be seen as a higher-order decorrelation.

ii. Infomax [32] is an ICA algorithm based on the goal of minimizing the mutual information among the separated sources. The method uses the fact that maximizing the joint entropy of a set of signals results in minimizing their mutual information, increasing independence in the outputs. This does not guarantee independence since the objective goal of the method is to maximize entropy using gradient ascent in a nonconvex problem. The algorithm uses only a linear function to determine the learning rule, and it could only separate superGaussian sources.

iii. Extended Infomax is a variant of Infomax that enables the incorporation of sources from diverse distributions. Extended Infomax employs two learning rules to enhance the separation between subGaussian and superGaussian sources [33].

iv. Picard (Preconditioned ICA for Real Data) [34] also minimizes the mutual information between sources using the same gradient calculation but also explores the curvature of the objective function to speed up convergence. It uses L-BFGS [35] in the role of the optimization method, but it adds curvature information by approximating the Hessian of the objective function, leading to faster convergence.

v. Picard-O is a variant of Picard that further enforces decorrelation of the sources and is used in conjunction with the extended version of the algorithm that can separate both super and subGaussian ICs [36].

vi. The SOBI algorithm [37] jointly diagonalizes the time-lagged covariance matrices of the extracted sources, i.e., removes the source correlation of recent and past data samples. It assumes that the sources are either deterministic ergodic sequences or stationary multivariate processes. It also assumes mutually uncorrelated sources and additive, stationary, temporally white noise, independent of the source signals. We use 100 lags for calculating the covariance matrices.

vii. Hyvarinen's Fixed Point algorithm (FastICA) [38] assumes nonGaussian sources and iteratively adjusts the unmixing matrices while maximizing the sources' negentropy, which is equivalent to minimizing their mutual information in this case. A family of contrast (objective) functions can be used with this goal, and there is also a family of fixed-point algorithms to optimize them. Each contrast function is more adequate for different distributions of source signals, but some are good for general use, with the biggest decision being if the function is adequate for superGaussian or subGaussian sources. The only assumptions are that the number of sources is equal to the number of observed variables and that the source kurtosis is nonzero (nonGaussianity), depending on the contrast function. We use the parallel method of IC estimation and $G(x) = log(cosh(x))$ as the contrast function.

viii. An online algorithm has been proposed called online recursive ICA (ORICA) [27], in which a whitening matrix and an unmixing matrix are both adjusted in real-time. This approach has the advantage of being an adaptive solution, which in theory can at least partly deal with EEG nonstationarity. ORICA is derived from the same natural gradient learning rule as Infomax, which places it in the family of algorithms that minimize sources of mutual information. Hsu et al. [39] proposed a block update rule for the original algorithm, increasing processing speed with little accuracy decrease, and used it together with a recursive least squares whitening algorithm. For our training phase, we use the same values of the initial forgetting factor and the decay rate, but we use blocks of 8 samples, and during the testing phase, we assume that the system is being used at a much later time than the training, and the forgetting factor is kept constant at 0.001.

2.3   Feature extraction and selection

ERD is the phenomenon of activity attenuation in the *mu* (8–13 Hz) and *beta* (13–25 Hz) bands in the motor cortex (parietal lobe) during motor-imagery tasks. As it is related to the sensorimotor cortex, this effect can be of important use for detecting limb movement and imagination events, and a frequency analysis of the EEG can be used for discriminating the imagined task [2]. Eq. (11.3) shows Welch's method for power spectral density estimation (PSD), and we use it for inferring the power over these two bands [40]. We used a $N_H = 1s$ window, $D = 250ms$ of window overlap and the Hamming window for $H(s)$.

$$\widehat{S}(\omega) = \frac{1}{KN_H U} \sum_{k=1}^{K} | \sum_{n=1}^{N_H} H(n)x(n+kD)e^{-j\omega n}| \qquad (11.3)$$

We extract the PSD for both frequency bands for each channel or extracted source, so the feature vector contains two times more features than the number of channels. The mean and standard deviation of the training set are calculated, and a z-score transformation is applied to each feature in both training and test sets. Then, features are selected using a wrapper [41] that also maximizes kappa in a 4-fold cross-validation scheme.

We select a subset of the features for further training the classifiers. The wrapper [41] is an algorithm that incrementally constructs a set of features that achieves the highest performance score. The general algorithm is as follows:

1. Initially $S = \{\}$ is the selected feature set, and $F = \{f_1, f_2, ..., f_M\}$ is the complete feature set.
2. For each feature f_i in F, calculate a classifier's cross-validation score using features $S \cup \{f_i\}$.
3. Add the feature f_{best} with the highest score to S and remove it from F.
4. If $S \cup \{f_{best}\}$. Repeat (2) and (3) until the S score stops improving.

We use logistic regression as the classifier guiding the wrapper for its simplicity and also according to the premise that ERDs are attenuations in the motor cortex activity, so the weights for the corresponding bands and sources should capture the importance of the associated features. The feature subset is evaluated using Cohen's kappa as a score metric in four-fold cross-validation.

2.4   Classification

For prediction purposes, a classifier is a function that takes a mathematical object and outputs a class. In a BCI case, the mathematical object is the feature vector that we extracted from each window of the EEG or source signals, and the class is the MI task that this vector represents. Many different classifier algorithms have been used in BCI implementations, as reviewed in Ref. [14], and state-of-the-art methods are mostly based on deep learning techniques [42]. Apart from linear discriminant analysis (LDA), which is of recurrent use in more classical BCI systems [25], we chose a set of classifiers for being previously reported in BCI research and also for being readily available in machine learning computing toolkits.

For this work, we mainly differentiate the classifiers as being linear or nonlinear. Linear classifiers work by finding the best hyperplanes that separate the feature vectors.

For instance, a hyperplane in two dimensions is a line, a hyperplane in three dimensions is a plane, and so on. This hyperplane is called the decision boundary of the classifier, where data points lying on one side of the hyperplane are classified as one class and on the other side as the other class. For classifying more than two classes, two or more hyperplanes are needed. Nonlinear classifiers, on the other hand, also use decision boundaries that separate different classes, but they can engender a broader class of mappings. The boundaries are defined by each method and may be arbitrarily complex depending on the method, and this complexity may be necessary if, in the feature space, the data is not linearly separated.

A set of seven machine learning classifiers is employed to discriminate the feature vectors of each window, which are:

i. Support vector machine (SVM): In this family of algorithms, the feature vectors are implicitly projected onto a latent space where the data points can be linearly classified [43], as shown in Eq. (11.4), where ϕ is the latent transformation. The goal is to maximize the distance between data points and the decision boundary of the ensuing linear classifier, which leads to the solution in Eq. (11.5), where a_n is a scalar associated with the respective support vector z_n, with the corresponding class t_n, and v is the characteristic vector. The support vectors are the closest points to the decision boundary, but not all data points are support vectors, i.e., for some points $a_n = 0$. Here, $k(v, z_n) = \phi(v)\phi(z_n)$ and k is called the kernel function, which is an analytical solution to the dot product between the two feature vectors in the projected space. The goal of this function is to calculate the inner product of both feature vectors in the latent space without the need to map the features explicitly, which allows nonlinear decision boundaries in feature space. We explore the use of four kernel functions, leading to four scenarios: radial-basis function (RBF), polynomial (Poly), linear, and sigmoidal kernels (Sig.).

$$y(v) = w^T \phi(v) + b \qquad (11.4)$$

$$y(v) = \sum_{n=1}^{N} a_n t_n k(v, z_n) + b \qquad (11.5)$$

ii. Logistic regression is an algorithm in which each variable contributes linearly to a score, which is converted to a probability using the logistic function. The weights of the scorer can be constrained by adding an L1 or L2 penalty to the objective function during fitting [43]. The probability of the feature vector v belonging class c_k is given by Eq. (11.6), where w_j are weight vectors that can be interpreted as the importance of each feature to the probability score of the respective class. It is also possible to constrain the solution of the w_j weight vectors by using regularization, which is the use of penalty terms that are functions of those weights. Examples of regularizations are L1 regularization and L2 regularization.

$$p(c_k|v) = \frac{\exp\left(w_k^T v\right)}{\sum_j \exp\left(w_j^T v\right)} \tag{11.6}$$

iii. Multilayer perceptron classifier (MLP): Can be seen as a stack of multiple logistic regressions, each with multiple outputs that feed into the next one. It is also a class of neural networks (with fully connected layers) with the potential to approximate any function [44] that is constituted of a set of L layers, where layer l has N_l nodes (also called neurons). Each layer can have any number of nodes, except for the last layer, which has k nodes, one for each possible class. The output $z_{i,l}$, of the i-th node of the l-th layer is calculated using Eq. (11.7), where h is called the activation function, $w_{j,i,l}$, is the weight of the j-th node of the $l-1$ layer to the i-th node of the l-th layer, and $b_{i,l}$ is a bias term for the i-th node of the l-th layer. The inputs of the first layer $z_{i,0}$ are the values of the feature vector, and the following layers use the outputs of the preceding layer, so the inference is made from the first layer to the last layer. This is known as the feedforward process. The output of the model is a probability score calculated using the output of the last nodes, as in Eq. (11.8). MLPs are commonly adjusted using gradient descent, using the cross-entropy between the model outputs and the targets as the objective function. The sigmoid, hyperbolic tangent, and rectified linear unit are common choices for h.

$$z_{i,l} = h\left(\sum_{j=1}^{N_l} w_{j,l}z_{j,l-1} + b_{i,l}\right) \tag{11.7}$$
$$z_{i,0} = v_i$$

$$p(v) = \frac{\exp\left(z_{k,L}\right)}{\sum_j \exp\left(z_{j,L}\right)} \tag{11.8}$$

iv. Gaussian Naive Bayes: Naive Bayes is a family of algorithms that uses the Bayes rule, shown in Eq. (11.9), for classifying a feature vector. In this equation $v = (v_1, v_2, ..., v_d)$ is the feature vector and c is its class. It assumes that each feature is independent, so $p(v|c) = \prod_{i=1}^{d} p(v_i|c)$. The Gaussian variant assumes that the likelihood of each feature is Gaussian, as in Eq. (11.10), where μ_k and σ_k are estimated using maximum likelihood. This conditional probability is used with the Bayes rule to determine the most probable class [45].

$$p(v) = \frac{p(c)p(c|v)}{p(v)} \tag{11.9}$$

$$p(v_i|c_k) = \frac{1}{\sqrt{2\pi\sigma_k^2}}\exp\left(\frac{-(v_i - \mu_k)^2}{2\sigma_k^2}\right) \tag{11.10}$$

v. LDA: Works by using a linear mapping $v' = wv$ that maximizes between-class variance and minimizes within-class variance. This means that in the projected space, which always has fewer dimensions than the number of classes, data points from the same class will be clustered separately from other classes [46]. After the projection, the conditional distribution $p(v'|c_k)$ is modeled as a multivariate Gaussian distribution, and the predicted class c is the c_k that maximizes the posterior probability $p(c = c_k|v')$, using the Bayes rule.

vi. Random Forest [47]: This classifier is a set of individual decision trees. Decision trees are hierarchical classifiers in which decisions are taken at each node until a leaf is reached, determining the predicted class. At each node, the value of one feature from the feature vector is compared to a threshold, and being smaller or greater determines which subsequent branch the algorithm continues. In the conventional decision tree, the maximally discriminative feature is used at each node, but in the random forest, the individual classifiers are constructed using random features and a random sample of the dataset. This procedure may seem to harm the decision process, but it helps with generalization to unseen samples.

vii. Extremely randomized trees [48]: The algorithm also uses a set of decision trees, and the difference is in their construction. During learning, random thresholds are applied at each node (instead of the optimal), and the best among them is chosen. This reduces model variance but slightly increases bias.

2.5 EEGNet

EEGNet is a compact convolutional neural network for EEG signal classification for different BCI paradigms. It comprises signal processing, feature extraction, and classification procedures in a single network and has been applied for both within-subject and cross-subject applications [21]. The network is composed of a temporal filter, followed by a spatial filter, and then a separable convolution layer that serves the feature extraction purpose. Even though EEGNet learns spatial filters, we theorize that it could benefit from the ICA since some of the extracted sources would be more discriminative right from the start of the training epochs. With this in mind, we perform source extraction before the actual preprocessing.

We use the EEGNet-4,2 configuration, where there are 4 temporal filters for each input signal and 2 point-wise convolutions for each temporal filter. As in the original work, we perform the same preprocessing of the signal described in Schirrmeister et al. after the source extraction. The signals are resampled to 128 Hz, then we apply a third-order Butterworth bandpass filter between 4 and 38 Hz, and finally, we apply electrode-wise exponential moving standardization, shown in Eq. (11.11), where $x_i(t)$ is i-th electrode value at the time t and $x_i'(t)$ is its standardized value, and $\mu_i(t)$ and $\sigma_i^2(t)$ are exponential moving averages (EMA) of the electrodes' mean and variance at the time t, respectively. The EMA is controlled by α, which was set to 0.001. For $t \leq 1000$ $\mu_i(t)$ is set to the average of the first 1000 values, and $\sigma_i^2(t)$ is set to the variance of the first 1000 values.

$$x_i'(t) = \frac{x_i(t) - \mu_i(t)}{\sqrt{\sigma_i^2(t)}}$$

$$\mu_i(t) = \alpha.x(t) + (1-\alpha).\mu_i(t-1) \tag{11.11}$$

$$\sigma_i^2(t) = \alpha.(x(t) - \mu_i(t))^2 + (1-\alpha).\sigma_i^2(t-1)$$

3. Results

3.1   ICA and classifier comparison

In this section, we present the results from the two experiments, comparing how different ICA methods affect the evaluation metrics, including the case where raw signals are used. We evaluate the performance of the BCI in terms of Cohen's kappa, which is a measure of agreement between predicted and expected values and is proportional to accuracy, as shown in Eq. (11.12). In the equation, p_0 is the accuracy (or concordance) and p_e is the expected accuracy if predictions are made at random (if the number of samples from each class is the same, this equals the inverse of the number of classes). Note that when $p_0 = p_e$ kappa is zero, and when $p_0 = 1$, kappa is 1. For $p_0 < p_e$, i.e., the observed accuracy is worse than a random guess, kappa assumes negative values.

$$\kappa = 1 - \frac{1-p_0}{1-p_e} \tag{11.12}$$

In the first experiment, many ICA algorithms are used for the source extraction for each subject of each dataset, individually. Then, each classifier is trained and tested individually to evaluate if there are significant differences in the expected performance of the BCI. The first experiment is conducted using the BCI Competition IV dataset 2a (dataset D1) and the OpenBMI dataset (dataset D2). In the second experiment, we evaluate whether extracting sources before feeding the signals into the network improves the prediction's kappa or speeds up training convergence. In both experiments, if either the ICA or classifier is not deterministic (is affected by random initialization or has other sources of randomness), we run the experiment 10 times to calculate the mean, median, and standard deviation values of the kappa values.

Fig. 11.3 shows the full pipeline for the first experiment. First, the dataset is selected. The first session is used for fitting the ICA model and finding the unmixing matrix, but in one case no ICA is used, represented by the "None" box. Then, the sources' PSD is extracted, yielding two features per source, and a subset of features is selected using the wrapper method. One of the classifiers is selected, and in the first step, it uses k-fold cross-validation for hyperparameter tuning, and then it is trained using only the first session data. Then we extract sources from the second session trials using the calculated unmixing matrix, extract the PSD, select the same feature subset, and predict the task using the trained model.

The pipeline for the second experiment is shown in Fig. 11.4. For The first sessions from dataset D1 are selected, one ICA method is used for calculating an unmixing matrix, and then sources are extracted. EEGNet is trained using the sources as raw signals. Then, the second session is selected, and the unmixing matrix is used for extracting sources. The fitted EEGNet model is then applied to those sources of data for predicting MI tasks. ORICA 0 and ORICA 1 refer to the methods assuming 0 and 1 as subGaussian sources, respectively.

We will present an extensive analysis of ICA, classifiers, and subject performance. For the first experiment, the best combination between ICA and classifier for each subject and the whole dataset is shown, and then we analyze each one independently. For dataset D2, we only analyze aggregated results over all subjects due to the dataset size and find the ICA and classifier with the highest accuracy. For each ICA, we also find the classifier that better pairs with it, and we also do it for each classifier. For the EEGNet experiments, we used only dataset D1.

The yellow blocks in both figures indicate ICA methods or classifiers that have random initialization or learning algorithms. At each experiment, the corresponding pipeline is run 10 times for the random algorithms, selecting one of each ICA algorithm and each classifier independently for each subject.

Section 4.1 presents the results and discussions for dataset D1, and Section 4.2 is for dataset D2.

3.2   BCI Competition IV 2a

In this section, we apply the methodology to dataset D1 to investigate our specific goals. We first evaluate the best ICA and classifier combinations per subject. Then, we evaluate each source separation method and each classifier independently, averaging the results for each algorithm. We also find the ICA method that yields the highest kappa for each classifier and the classifier that yields the highest kappa for each ICA. Finally, we aggregate the results over all subjects and find the methods that maximize accuracy at both steps.

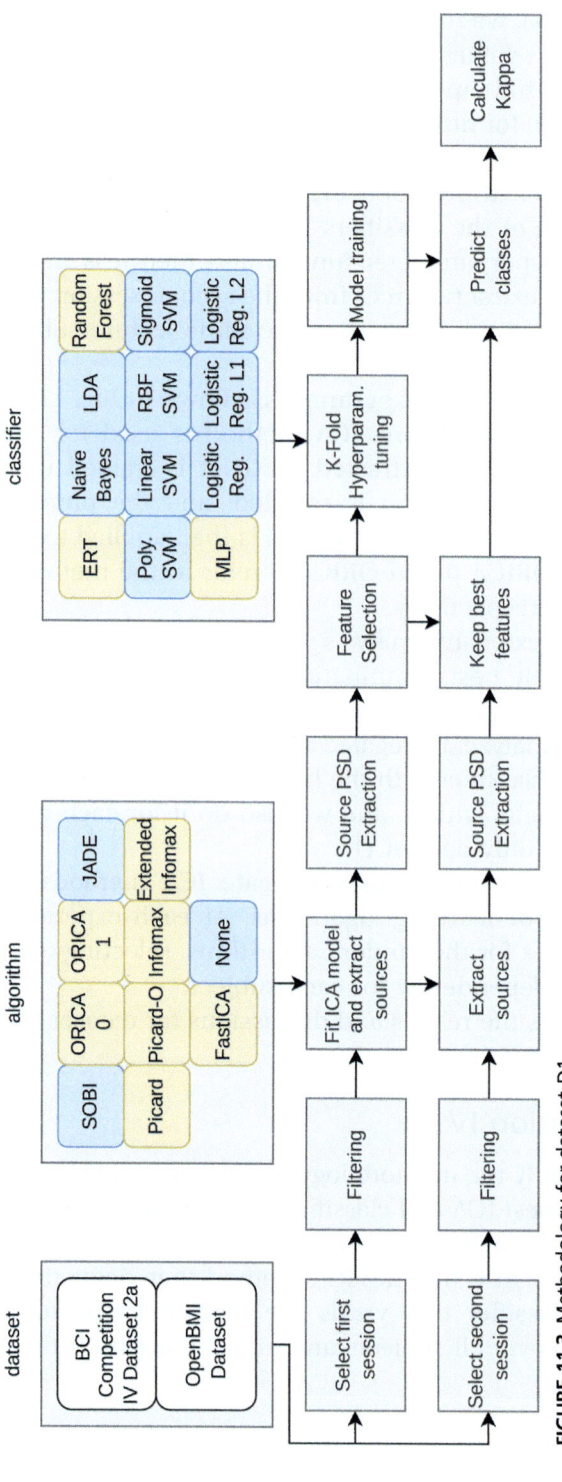

FIGURE 11.3 Methodology for dataset D1.

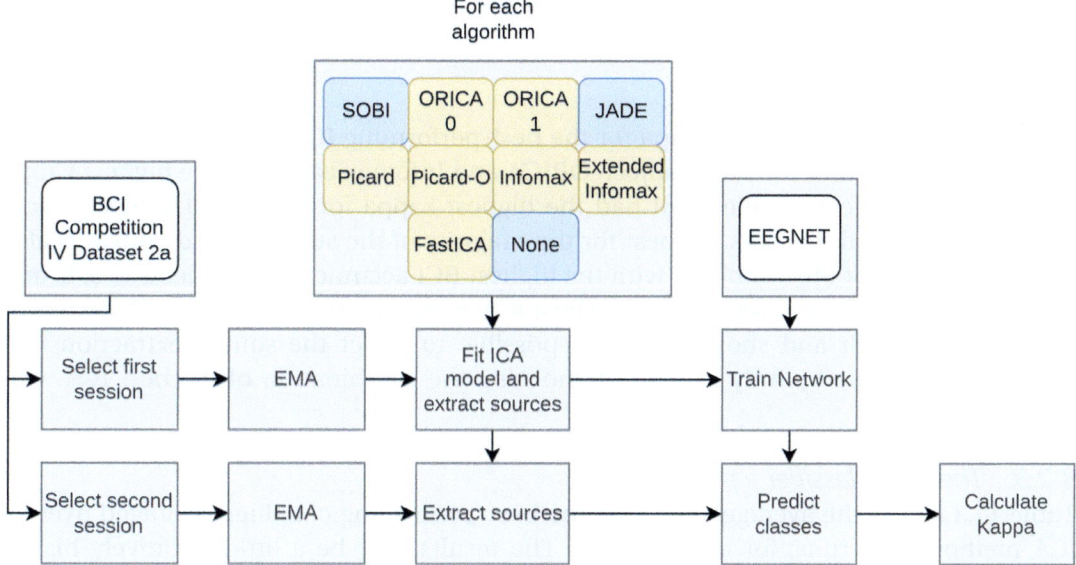

FIGURE 11.4 Methodology for dataset D2.

3.2.1   The best combination per subject

Table 11.1 shows the best possible combination of classifier and ICA for each subject (average and std. over the 10 runs for each combination). We see that ORICA and SOBI had the highest kappa for 3 out of 9 subjects. Subject 3 had the highest kappa among subjects, and both linear and nonlinear classifiers were tied for the highest accuracy. Results with zero standard deviation mean that both the classifier and ICA method are

Table 11.1   The best combination of ICA and classifier for each subject.

Subject	Classifier	ICA	Kappa Mean	Kappa Std.
1	Log Reg. (L1)	SOBI	0.574	0.000
2	Random forest	Infomax	0.247	0.035
3	Log. Reg	ORICA (1)	0.676	0.000
	Log. Reg. (L1)	ORICA (1)	0.676	0.000
	SVM (RBF)	ORICA (1)	0.676	0.000
4	SVM (Lin.)	Picard-O	0.284	0.013
5	Naïve Bayes	ORICA (1)	0.102	0.000
6	Log. Reg. (L1)	SOBI	0.278	0.000
7	Log. Reg. (L1)	ORICA (0)	0.611	0.000
8	SVM (RBF)	Picard	0.626	0.014
9	SVM (Lin.)	SOBI	0.662	0.000
	SVM (RBF)	SOBI	0.662	0.000

232 Signal Processing Strategies

deterministic (initial conditions do not affect the results), and subjects with more than one line in the table indicate ties.

3.2.2 The best ICA - per subject

Table 11.2 shows the average Kappa for the best-performing ICA method, averaged over all classifiers and runs, for each subject. ORICA and Infomax each had the highest kappa for 3 out of 9 subjects, while SOBI had the highest kappa for 2 out of 9 subjects each. There is no method that was the best for the majority of the subjects, and ORICA is also the best for 3 out of the 5 subjects with the highest BCI accuracy. For subjects 4, 5, 8, and 9, the best combination does not use the best ICA averaged on all classifiers. This was not the expected result and shows that it is possible to select the source extraction and classification method jointly instead of choosing the combination of the best ICA with the best classifier.

3.2.3 The best classifier - per subject

Table 11.3 shows the average kappa for the best-performing classifier, averaged over all ICA methods and runs, for each subject. The results may be a little negatively biased because the baseline source extraction is taken into consideration here. Even though LDA did not show up in the best ICA and classifier combination for subjects 4 to 8, it has the greater expected kappa among all ICA algorithms. Also, linear models are shown as the best option for all but one subject. As of Table 11.2, the best classifier is not necessarily the one chosen in the best combination (Table 11.1), and this is again not expected, but it shows again that a good selection of both at the same time can yield better results.

Table 11.2 Average kappa for the best ICA for each subject.

Subject	ICA	Kappa	
		Mean	Std.
1	SOBI	0.521	0.031
2	Infomax	0.185	0.033
3	ORICA (1)	0.643	0.035
4	Picard	0.259	0.024
5	Infomax	0.071	0.013
6	SOBI	0.231	0.054
7	ORICA (0)	0.527	0.078
8	Infomax	0.584	0.024
9	ORICA (1)	0.613	0.038

Table 11.3 Average kappa for the best classifier for each subject.

Subject	Classifier	Kappa	
		Mean	Std.
1	Log. Reg. (L2)	0.474	0.054
2	Random forest	0.183	0.084
3	Log. Reg.	0.561	0.085
	Log. Reg. (L2)	0.561	0.082
4	LDA	0.245	0.065
5	LDA	0.072	0.032
6	LDA	0.144	0.052
7	LDA	0.469	0.104
8	LDA	0.535	0.061
9	Log. Reg. (L2)	0.518	0.114

3.2.4   The best classifier per ICA - average over the dataset

Table 11.4 shows the classifier with the highest kappa, on average, for each ICA method. It is shown that ORICA(1) was the only method that matched with a nonlinear classifier, but with no statistical difference to the logistic regression variations ($p < 0.05$). Also, when not using ICA (none), even though LDA had the highest average kappa, logistic regression was not significantly different from it ($p < 0.05$). Since the statistics are calculated among all subjects, those values can be interpreted as the expected results independently of the subject. The median values vary greatly for each choice of source extraction method, and even though not using ICA shows as the worst alternative, JADE also performed badly.

Table 11.4 The best classifier for each ICA algorithm.

ICA	Classifier	Kappa		
		Median	Mean	Std.
Picard	Log. Reg. (L2)	0.498	0.380	0.199
FastICA	Log. Reg. (L2)	0.464	0.379	0.209
Picard-O	Log. Reg. (L1)	0.480	0.373	0.219
ORICA (1)	Random forest	0.438	0.373	0.231
Infomax	Log. Reg. (L2)	0.438	0.356	0.197
Ext. Infomax	Log. Reg. (L1)	0.439	0.355	0.211
SOBI	Log. Reg. (L1)	0.278	0.344	0.229
ORICA (0)	Log. Reg. (L1)	0.472	0.344	0.260
JADE	Log. Reg.	0.278	0.291	0.147
None	LDA	0.259	0.248	0.177

Table 11.5 The best ICA for each ICA classifier.

Classifier	ICA	Kappa		
		Median	Mean	Std.
Log. Reg. (L2)	Picard	0.498	0.380	0.199
Log. Reg.	FastICA	0.468	0.378	0.207
Log. Reg. (L1)	FastICA	0.486	0.377	0.207
LDA	Picard	0.477	0.373	0.187
Random forest	ORICA (1)	0.438	0.373	0.231
SVM (Lin.)	FastICA	0.452	0.371	0.203
SVM (Sig.)	FastICA	0.460	0.369	0.205
SVM (RBF)	Picard	0.435	0.368	0.190
MLP	FastICA	0.448	0.368	0.198
Extra trees	ORICA (1)	0.439	0.348	0.235
Naïve Bayes	FastICA	0.345	0.324	0.200
SVM (Poly)	Picard-O	0.377	0.319	0.194

3.2.5   The best ICA per classifier - average over the dataset

Table 11.5 shows the ICA with the highest kappa, on average, for each classifier. FastICA and Picard (including Picard-O) stand out as the best ICA methods for 10 out of 12 classifiers, and ORICA for 2 out of 12 classifiers. Most algorithms show a mean value considerably lower than their median value, indicating a long left tail in the kappa value distribution. As shown in Table 11.3, these statistics can be interpreted as the expected results independently of the subject. Note that even though FastICA and Picard did not show up in Tables 11.1 and 11.2, when we consider each classifier independently, their average over the subjects is higher, indicating that even with their results not being optimal for individual subjects, we can expect them to be populationally better. Also, we see a decrease of up to 29% in median kappa when we change the classifier, even when we jointly select the best ICA method.

3.2.6   The best ICA and the best classifier - average over the dataset

Figs. 11.5 and 11.6 show the average kappa for each ICA method and classifier, respectively. The results are ordered from lowest to highest mean kappa and are averaged over all runs, subjects, and classifiers, or ICA methods. Here, the red asterisk indicates that the one-sided Wilcoxon's signed-rank test between each bar and the highest (rightmost) bar had a P-value lower than 0.05, meaning that its value is significantly lower than the highest bar. It can be seen that Picard, FastICA, and ORICA (1) are tied for the best-performing ICA, and even though ORICA (1) is on average smaller than the highest kappa, there is no statistical significance. The logistic regression classifier had the highest average kappa among classifiers, but it is statistically tied with LDA in this experiment. SVM with a polynomial kernel had the lowest average kappa.

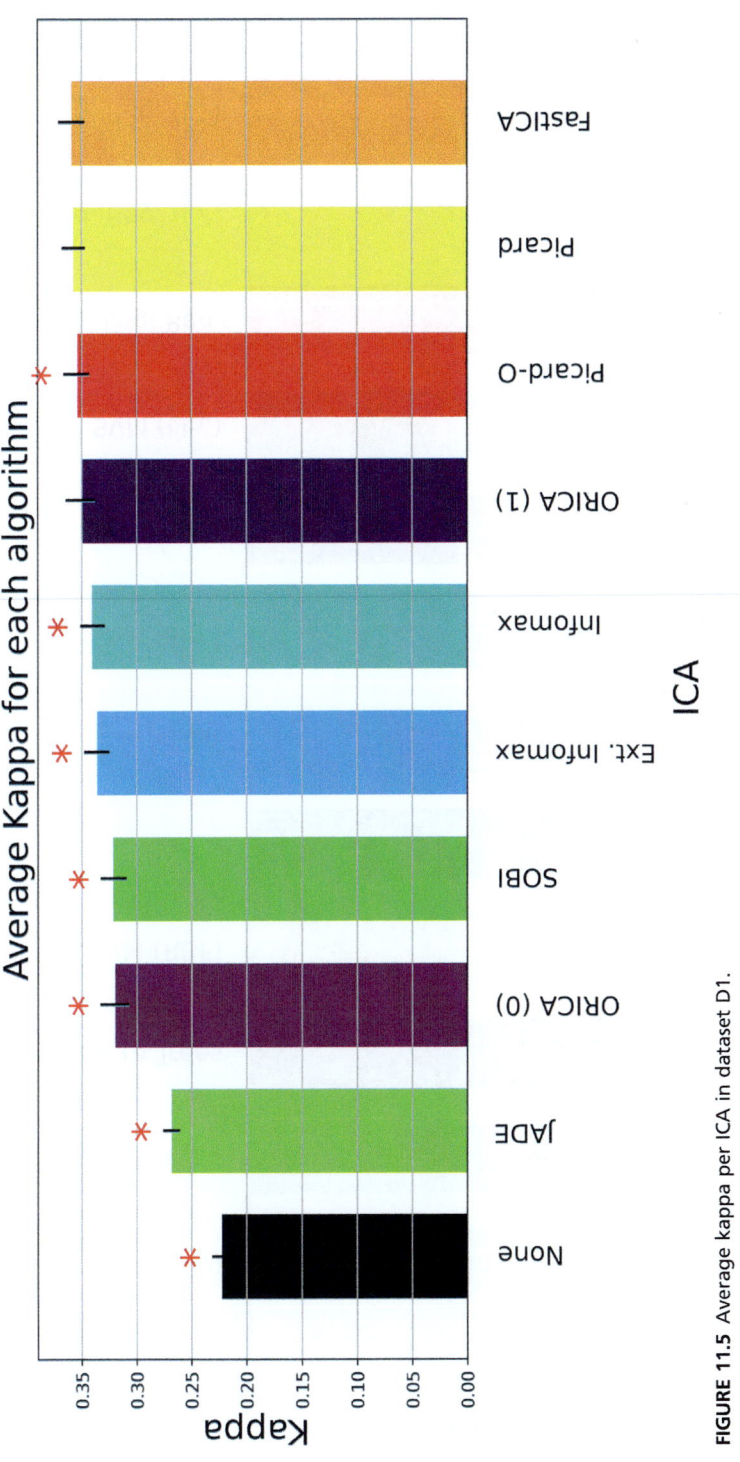

FIGURE 11.5 Average kappa per ICA in dataset D1.

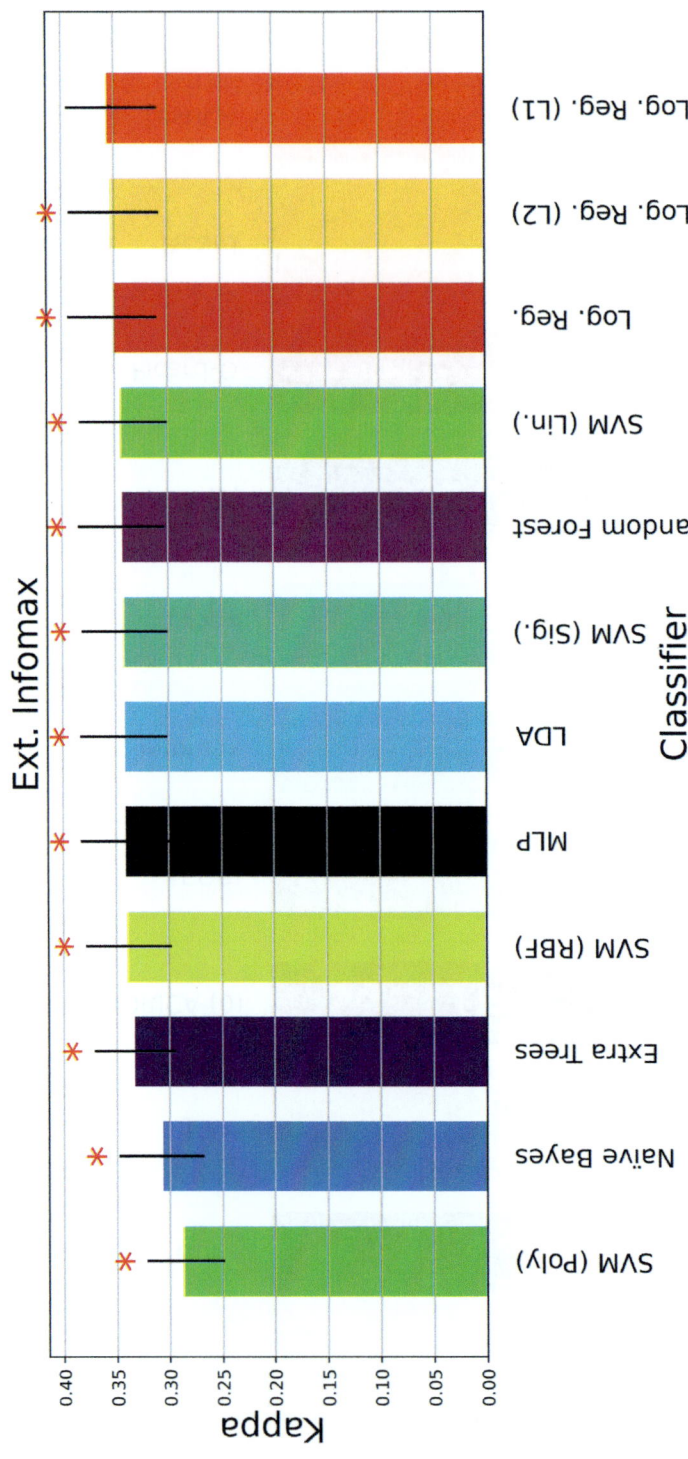

FIGURE 11.6 Average kappa per classifier in dataset D1.

Fig. 11.6 also explains why, in Table 11.3, LDA appears as the best classifier for many subjects while not being selected as the highest kappa for individual classifiers in Table 11.4. We see that there is no significant difference between it and the logistic regression, which had the highest average kappa value.

In Uribe et al. [49] JADE has shown great performance for a 2-class MI-BCI system, with better classification accuracy than SOBI and Infomax, contrasting with the results here, where JADE has been consistently a low-accuracy ICA method. This could be due to differences in dataset characteristics or ICA implementations.

Even though LDA is commonly used as a classifier in BCI systems, we can note that the logistic regression model and its variations (regularized versions) are equally great in terms of classification accuracy, especially in the case of using PSD features. It is also possible to conclude that, when using those features, it is always better to try extracting sources from the EEG signals before the feature extraction step. Out of the many classifiers in our experiments, there was no significant gain when using nonlinear approaches, such as logistic regression, LDA, and linear SVM, which were among the top average kappa values.

3.3   OpenBMI

In this section, we apply the methodology to dataset D2. Since this dataset has six times more subjects than D1, we only investigate results aggregated over the dataset. The goal of the experiment is to first find the classifier that has the highest accuracy in combination with each ICA, and also the ICA method that has the highest accuracy for each classifier. Finally, we find the ICA and classifier methods that have the highest average kappa over the whole dataset.

3.3.1   The best classifier per ICA - average over the dataset

Table 11.6 shows the ICA with the highest kappa, on average, for each classifier. ICA methods from Infomax to SOBI had the highest average kappa, with no statistical difference between them ($p > 0.05$) on one-sided Wilcoxon's signed-rank test, and then "None" ORICA (0), ORICA (1), and JADE are statistically different from this best-performing group. Even though FastICA has the highest median among them, it is also not statistically different from Infomax, Picard, or Picard-O on Mood's median test. When comparing dataset D1 results, we see that ORICA no longer shows as one of the ICA methods with the highest kappa. The average kappa values are considerably smaller, and this is due to the OpenBMI being a more diverse dataset with fewer subjects that can control a BCI. The median kappa is lower than the mean for all ICA methods, showing that, differently from dataset D1, the metric distribution has a longer right tail.

3.3.2   The best ICA per classifier - average over the dataset

Table 11.7 shows the ICA with the highest kappa, on average, for each classifier. It is noticeable that aggregating by the classifier, Infomax, and its extended version perform

Table 11.6　The best classifier for each ICA method.

ICA	Classifier	Kappa		
		Median	Mean	Std.
Infomax	Log. Reg.	0.108	0.238	0.295
Picard	Log. Reg. (L1)	0.106	0.238	0.292
Picard-O	Log. Reg. (L1)	0.106	0.238	0.292
Ext. Infomax	Log. Reg. (L1)	0.091	0.238	0.292
FastICA	Log. Reg. (L1)	0.136	0.237	0.294
SOBI	Random forest	0.091	0.231	0.294
None	Log. Reg. (L1)	0.085	0.209	0.278
ORICA (0)	SVM (Sig.)	0.040	0.095	0.224
ORICA (1)	Random forest	0.036	0.077	0.195
JADE	MLP	0.013	0.064	0.149

Table 11.7　The best ICA for each classifier.

Classifier	ICA	Kappa		
		Median	Mean	Std.
Log. Reg.	Infomax	0.108	0.238	0.295
Log. Reg. (L1)	Infomax	0.104	0.238	0.291
Random forest	Infomax	0.135	0.238	0.291
SVM (Lin.)	Ext. Infomax	0.088	0.236	0.288
MLP	Infomax	0.088	0.235	0.289
Log. Reg. (L2)	Ext. Infomax	0.094	0.234	0.289
Extra trees	Infomax	0.096	0.227	0.290
SVM (Sig.)	Ext. Infomax	0.099	0.226	0.284
SVM (RBF)	Ext. Infomax	0.075	0.224	0.283
LDA	Ext. Infomax	0.081	0.220	0.270
SVM (Poly)	Infomax	0.070	0.209	0.272
Naïve Bayes	FastICA	0.061	0.189	0.266

better than the other ICA algorithms. Also, the average kappa values are much more homogeneous, so we can conclude that different classifiers have less influence on the final metric than the different ICA possibilities. The random forest classifier with Infomax ICA combination had a higher median kappa than others, but with no statistical difference when compared to the first and second best combinations of the table (using Mood's median test, with a 95% confidence level).

3.3.3　The best ICA and best classifier - average over the dataset
The average kappa values for each ICA algorithm and classifier are presented in Figs. 11.7 and 11.8, respectively. The red asterisk indicates statistically significant differences

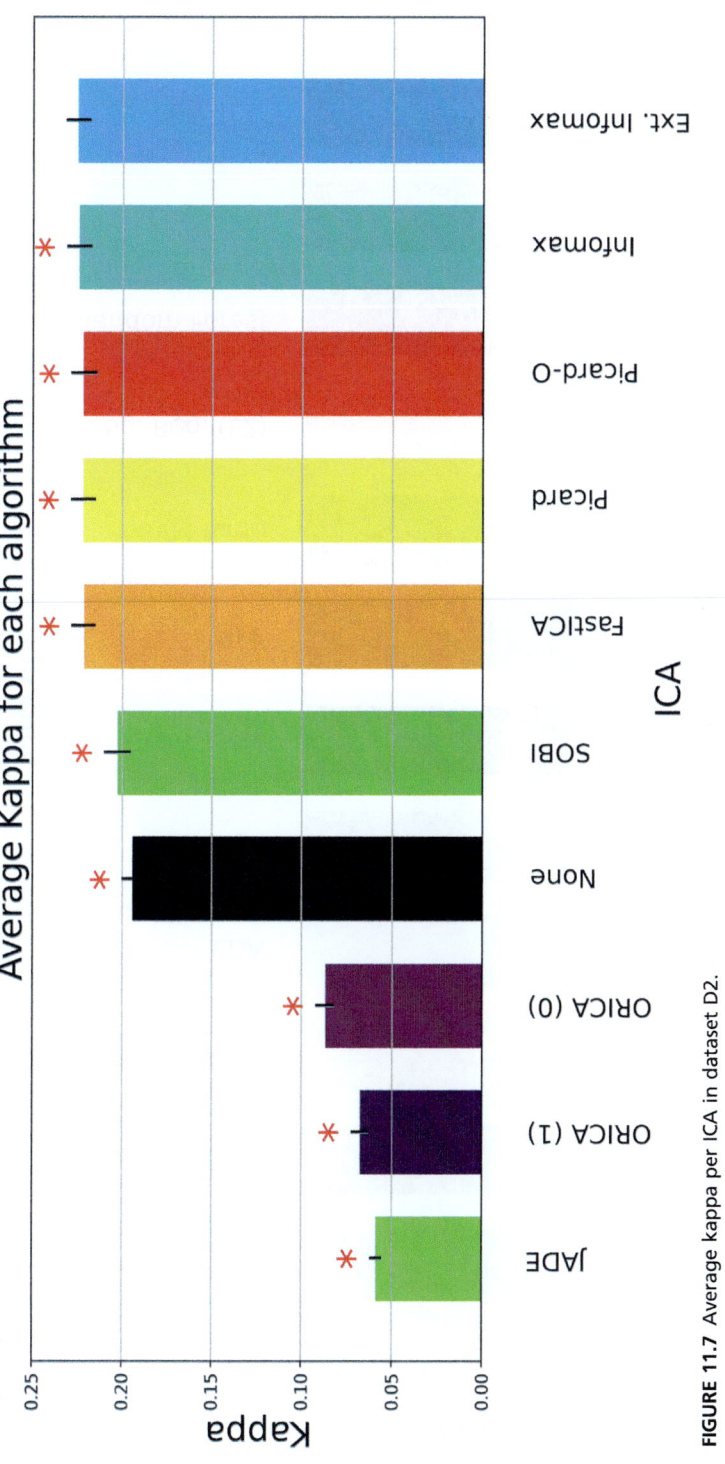

FIGURE 11.7 Average kappa per ICA in dataset D2.

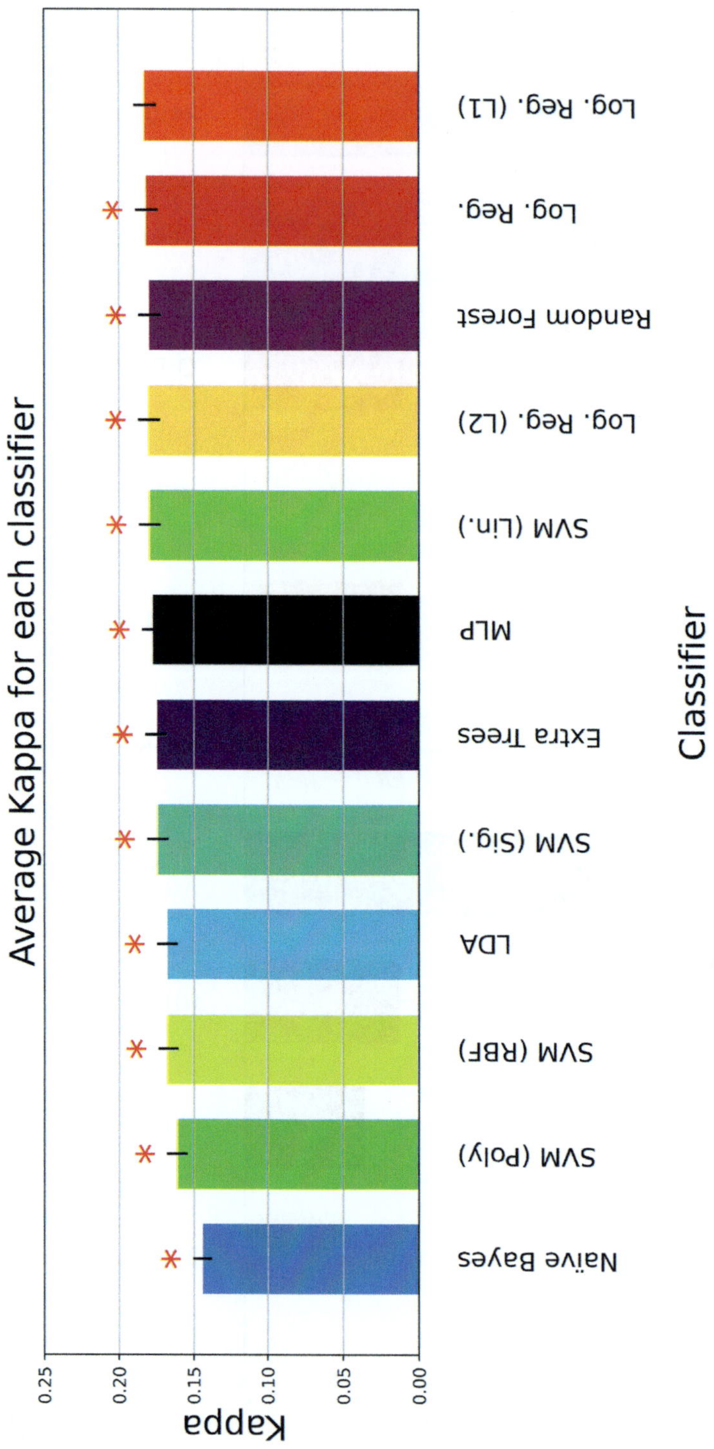

FIGURE 11.8 Average kappa per classifier in dataset D2.

compared to the rightmost bar (one-sided Wilcoxon's signed-rank test with $p > 0.05$). On average, Extended Infomax was the ICA with the highest kappa, followed by Infomax and Picard-O, while JADE demonstrated the lowest kappa value among the ICA algorithms. Logistic regression with L1 regularization was the classifier with the highest average kappa, followed by the nonregularized version and random forest. Naive Bayes had the lowest performance among the classifiers.

We counted the number of subjects each classifier had the highest kappa among all classifiers: Random forest, for 14 subjects; linear SVM, for 8 subjects; polynomial kernel SVM, for 6 subjects; Naïve Bayes, for 5 subjects; logistic regression (L1), for 4 subjects; extra trees, LDA, nonregularized logistic regression, and RBF-kernel SVM, 3 subjects each; and the remaining classifiers were selected as best for 2 or less subjects. When averaging over all ICA algorithms, only 9 subjects had accuracy greater than 75%; logistic regression (L1) was better for 3 of them. SVM variations, MLP, and nonregularized logistic regression did not show as the best classifiers for them, and the other classifiers each showed as the best for one subject.

We also counted for how many subjects each ICA algorithm had the highest kappa among all algorithms: No ICA was better for 15 out of 54 subjects, followed by SOBI (11 subjects), ORICA (both variations sum to 8 subjects), extended infomax and Picard (5 subjects each), FastICA and Infomax (4 subjects each), and JADE (2 subjects). When averaging all ICA classifiers, only 12 subjects had an accuracy greater than 75%: SOBI was the best ICA for 5 subjects; Infomax, Picard, and No ICA were better for 2 subjects each; Ext. Infomax was better for 1 subject.

These results contrast a little with the ones in dataset D1. Here, for many subjects, it was better not to use ICA, which did not happen before. LDA, which was the best classifier for 5 out of 9 subjects, now gave place to random forest and the linear SVM, while the latter did not even appear as the best for any subject in D1. Furthermore, even though not using ICA and random forest show up as the best preprocessing and classifiers, these are results averaged over many BCI-illiterate subjects, and as results show, for subjects with higher BCI control, there are better ICA and classifier options.

Finally, the ORICA algorithm had a degraded performance compared to the first dataset, on average, being worse than not using ICA at all, and this could indicate that either the method is not suitable for the average subject or that a better preprocessing method is needed before applying ORICA.

In Brunner et al. [50] and Naeem et al. [15], SOBI, FastICA, and Infomax were compared regarding classification accuracy in a 4-class MI task. In the testing phase, Infomax had the best accuracy, followed by FastICA and SOBI, in this order, and this is validated by our results in D2. JADE rarely outperformed other methods, similar to the findings of [26], where Infomax also had the top-2 classification accuracy among ICA methods.

4. EEGNet

EEGNet is a network that was designed for learning temporal filters, spatial filters, feature extraction, and classification in just one structure. This joint learning is generally better since all steps are optimized to minimize the same loss function in a supervised manner. It was designed to deal with EEG signals in the electrode space instead of the source space, which we were projecting in the last section.

In this section, we experiment with feeding signals in source space to this network and evaluate kappa and training convergence speed. We assume that some sources extracted using ICA have a lot more information about the classification task we are training the BCI system to identify, so the network learning phase should be easier because we expect that at least the temporal filters will not need to be as complex as if we were using raw signals. The experiments were carried out using only the BCI competition dataset due to the dataset size of D2.

4.1   The best ICA - per subject

Table 11.8 shows the average kappa for the best-performing ICA algorithm for each subject. The results are the average and standard deviation over 10 runs (different network initializations).

Fig. 11.9 extends the results from Table 11.8, showing the average kappa value along with the 95% confidence interval for each ICA algorithm and each subject. The red asterisk indicates that the respective ICA algorithm kappa value is significantly lower than the best ICA for the respective subject (one-sided Wilcoxon's signed-rank test with $p < 0.05$). For subjects 3, 6, and 9, using no ICA is almost always significantly better than

Table 11.8   The best ICA algorithms for each subject in the BCI Competition dataset.

Subject	ICA	Kappa	
		Mean	Std.
1	SOBI	0.687	0.042
2	ORICA (0)	0.268	0.038
3	None	0.779	0.049
4	Picard	0.375	0.041
5	ORICA (0)	0.307	0.103
6	None	0.279	0.041
7	None	0.540	0.055
8	SOBI	0.685	0.047
9	None	0.726	0.042

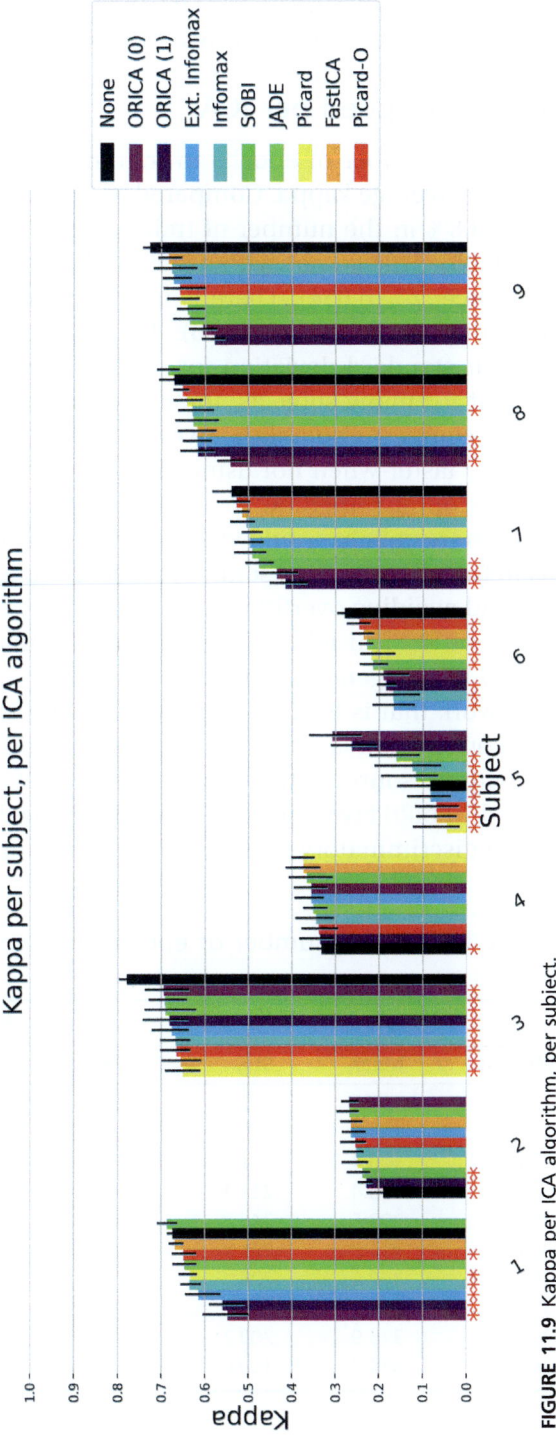

FIGURE 11.9 Kappa per ICA algorithm, per subject.

using any ICA; for subjects 1, 7, and 8, using ICA is statistically indifferent from using any ICA; and for subjects 2, 4, and 5, using ICA improved the average kappa.

4.2　Training convergence speed per ICA

Table 11.9 shows the average number of epochs to training convergence for each ICA algorithm and also shows the average kappa. Compared to SOBI, the use of no ICA leads to an average increase of 28.8% in the number of training epochs (one-sided Wilcoxon signed rank test with $p < 0.05$). This comes with a 3.8% reduction in average kappa but with no statistical significance ($p = 0.054$). There is no significant reduction in median kappa either (Mood's median test with $p < 0.05$). Fig. 11.10 illustrates the average number of epochs and their standard deviation for each subject and the ICA method, and it is possible to see that for many subjects, the difference between the numbers of epochs when not extracting sources is notable. However, interestingly, for 3 subjects, the convergence was on average faster when not using ICA.

Since we use an early stopping method for assessing training convergence, it is possible that training with more epochs would yield better kappa, but there is effectively no way of assessing this in this experimental setup because training for too many epochs could lead to overfitting, and avoiding overfitting is precisely why early stopping is used. As the kappa on the validation set is used for this, it is possible that using ICA, although making training faster, also harms generalization.

EEGNet is a neural network that is for EEG signal processing and classification, as it has temporal and spatial convolution layers. It was designed to deal with EEG signals in the electrode space, i.e., with no source extraction. Using ICA before passing the signals through EEGNet theoretically could favor spatial filtering learning, but in some cases, the source space does not have discriminatory signals, and the learning would be hampered.

Table 11.9　Average kappa and number of epochs until convergence for each ICA method.

ICA	Epochs			Kappa		
	Median	Mean	Std.	Median	Mean	Std.
None	492.5	523.1	262.5	0.512	0.475	0.252
SOBI	312.5	406.1	270.5	0.502	0.457	0.229
Picard	335.0	406.4	251.5	0.512	0.441	0.225
Ext. Infomax	322.5	398.4	232.1	0.484	0.438	0.225
FastICA	290.0	393.5	246.2	0.509	0.454	0.222
Infomax	305.0	391.2	221.7	0.493	0.444	0.224
Picard-O	295.0	389.8	226.4	0.512	0.451	0.225
JADE	312.5	385.3	229.7	0.495	0.455	0.207
ORICA (0)	300.0	375.9	203.2	0.435	0.439	0.178
ORICA (1)	245.0	334.3	190.6	0.456	0.429	0.188

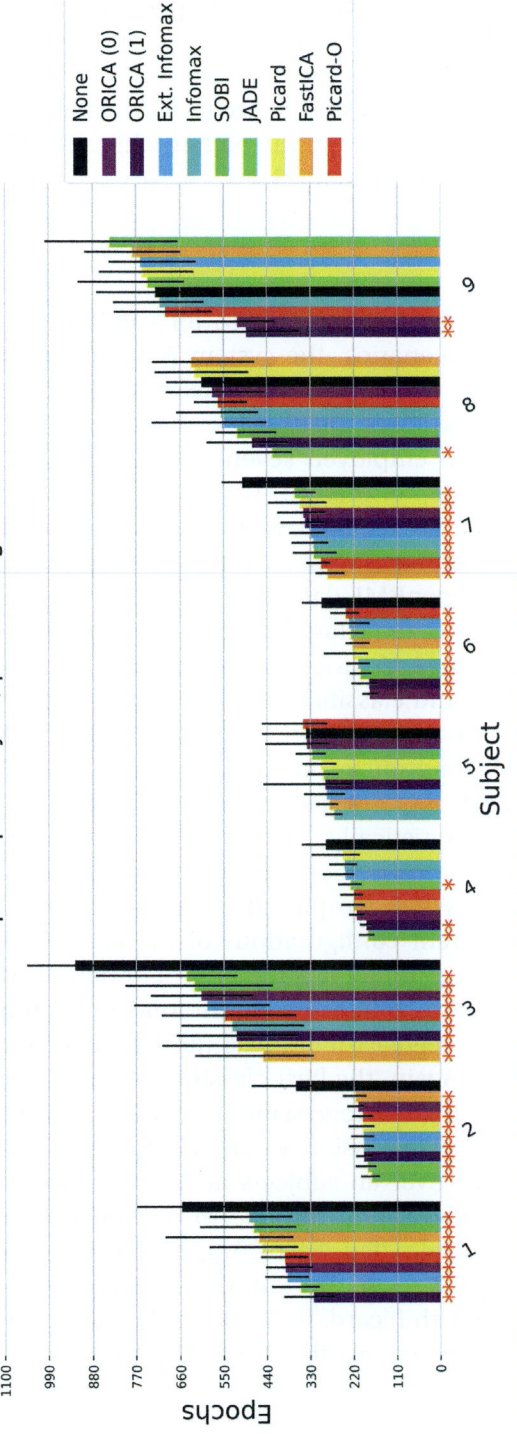

FIGURE 11.10 Average number of epochs and standard deviation for each subject and each ICA.

We see that in our experiments, the convergence speed was greatly improved when using ICA, and the decrease in kappa is not significant for some methods. This could mean that a little accuracy can be traded by a great amount of convergence speed, and in some cases, as is seen for subjects 2, 4, and 5, using ICA could improve the performance. Finally, we note that the average kappa is higher when not using ICA, and the result matches the accuracy reported by Ref. [21].

5. Conclusion

In this work, we designed three test scenarios for validating a MI-based BCI system's accuracy. In the first two scenarios, the system pipeline consisted of EEG recording, filtering, source estimation, feature extraction, feature selection, and classification. The source estimation step was based on the use of six different ICA methods in eight configurations, and we also employed seven classification algorithms in 12 different configurations for the last step of the BCI. We calibrated the system for each subject in two datasets and evaluated its accuracy between sessions. Each scenario made use of this same system, but in the first one we used the BCI Competition IV Dataset 2a, and in the second we used the OpenBMI dataset. In the third scenario, we substituted the feature extraction, selection, and classification steps for the EEGNet, a deep neural network for MI classification, and varied the ICA used at the source extraction step. The goals were to find the ICA and classifier methods that contributed the most to the system's accuracy, the combinations of both that maximized the kappa, and evaluate the EEGNet-based BCI used alongside a source extraction method.

For the first dataset, D1, FastICA had the highest average accuracy over all subjects and classifiers but was statistically similar to Picard and ORICA, assuming one subGaussian source. Logistic regression with L2 regularization appeared to be the best classifier for this method. When we picked the best classifier of choice for each ICA method, we saw that different configurations of the logistic regression almost always outperformed other classifiers for all ICA methods. For dataset D2, FastICA had only the fifth-highest average accuracy, with extended Infomax being the top one method. Here, logistic regression with L1 regularization was the best choice of classifier for the BCI system that used this ICA. Again, the best classifier of choice for each ICA method was different configurations of logistic regression, in particular using L1 regularization.

In both datasets, logistic regression variants achieved the top four highest kappa metrics when we averaged over all subjects and ICA techniques. For dataset D1, the configuration with no regularization resulted in the most accurate BCI, but the three configurations were statistically tied. FastICA appeared as the best ICA choice for this classifier, but when using L2 regularization, the logistic regression got a higher kappa when used in conjunction with Picard. In the second dataset, the results were somewhat different, as Infomax appeared as the best ICA choice for L1-regularized logistic regression.

In the third experimental scenario, we saw that four out of the nine subjects using ICA decreased the system's accuracy, showing that EEGNet is better fitted for raw EEG signals. Out of all ICA methods, the SOBI algorithm was the one that penalized the metric the least, but the accuracy loss was not significant. On the other hand, using SOBI leads to a decrease of 22.4% in the necessary number of training epochs, which could be further decreased by 36.1% by using ORICA with one subGaussian source, but with the disadvantage of also reducing kappa by 9.6%. This result is interesting because it shows that there was no correlation between the ICA being good in a BCI system with hand-crafted features and it being good for a deep learning-based system.

In both datasets, the logistic regression configurations outperformed the other classifiers. But for each dataset, different ICA methods lead to the highest accuracy. It can be argued that dataset D2 better represents the general population, given its size and the original goal of the researchers when constructing it. This would privilege the choice of Infomax as a more secure source extraction method when designing a BCI. Since the BCI's accuracy is highly dependent on the features that one uses, it is important to consider that those results may differ for different EEG feature extraction methods and different paradigms. For instance, due to the neurophysiological nature of the MI task, it makes sense that a linear classifier such as logistic regression can discriminate between different tasks.

5.1 Future work

For the next steps, we would be interested in the theoretical and practical reasons why each ICA method worked the best in each case, since there was no single method that stood out in the majority of subjects. Understanding and identifying the environmental, technical, and physiological variables at the time of data collection could give insight on how to choose the method with the most probability of really improving the BCI system. As important as those external factors are the assumptions that each ICA method uses about its input signals that may or may not hold in each situation. The use of an adaptive solution, as the environment and even the subject's physiology change, could also improve the long-term functionality of the system. The development of new methods that take all the peculiarities of the MI paradigm could also give us a more fitting method for more diverse BCI application contexts.

Since using ICA with EEGNet proved to speed the training speed but not the accuracy, the next step would be to understand how the ICA objective function interacts with the general optimization problem that the network solves, and for this one could try to jointly optimize both problems at the same time. ICA could theoretically substitute or at least be used to initialize the spatial filter layer of EEGNet, leading to fewer parameters to learn or even faster convergence. Since ICA is a nonsupervised method, nonlabeled data could be used in conjunction with transfer learning techniques for augmented model performance in more challenging scenarios, such as subject-to-subject transfer.

Acknowledgments

R. Attux thanks CNPq for the financial support (process 308811/2019-4).

References

[1] Wolpaw J, Birbaumer N, Mcfarland D, Pfurtscheller G, Vaughan T. Brain–computer interfaces for communication and control. Clin Neurophysiol 2002;113(6):767–91. ISSN 1388-2457.

[2] Pfurtscheller G, Neuper C. Motor imagery and direct brain-computer communication. Proc IEEE 2001;89(7):1123–34.

[3] Mulder T. Motor imagery and action observation: cognitive tools for rehabilitation. J Neural Trans June 2007;114(10):1265–78. Springer Science and Business Media LLC.

[4] Singh A, Hussain A, Lal S, Guesgen H. A comprehensive review on critical issues and possible solutions of motor imagery based electroencephalography brain-computer interface. Sensors March 2021;21(6):2173. MDPI AG.

[5] Haufe S, Tomioka R, Dickhaus T, Sannelli C, Blankertz B, Nolte G, Muller K-R. Localization of class-related mu-rhythm desynchronization in motor imagery based brain-computer interface sessions. Annu Int Conf IEEE Eng Med Biol Soc 2010;2010:5137–40. https://doi.org/10.1109/IEMBS.2010.5626177. PMID: 21095811.

[6] Pfurtscheller G, Neuper C, Flotzinger D, Pregenzer M. EEG-based discrimination between imagination of right and left hand movement. Elsevier BV Electroencephalogr Clin Neurophysiol 1997; 103(6):642–51. dez.

[7] Hyvarinen A, Karhunen J, Oja E. Independent component analysis. John Wiley & Sons, Inc.; 2001.

[8] Chen W, Du C, Zhang Y, Wu X. Combine ICA and ensemble learning methods for motor imagery EEG classification. In: Advances in natural computation, fuzzy systems and knowledge discovery. Springer International Publishing; 2021. p. 1376–84.

[9] Varsehi H, Firoozabadi S. An EEG channel selection method for motor imagery based brain-computer interface and neurofeedback using granger causality. Neural Net January 2021;133:193–206. Elsevier BV.

[10] Ruan J, Wu X, Zhou B, Guo X, Lv Z. An automatic channel selection approach for ICA-based motor imagery brain computer interface. J Med Syst November 2018;42(12). Springer Science and Business Media LLC.

[11] Gong X, Chen S, Ban Y, Wang M. Feature processing of multi-classification motor imagery EEG based on improved ICA and SVM. In: 2021 2nd international conference on intelligent computing and human-computer interaction (ICHCI). Los alamitos, Ca, USA. IEEE Computer Society; 2021. p. 318–21.

[12] Qin L, Ding L, He B. Motor imagery classification by means of source analysis for brain-computer interface applications. IOP Publishing J Neural Eng 2004;1(3):135–41. set.

[13] Kachenoura A, Albera L, Senhadji L, Comon P. Ica: a potential tool for bci systems. IEEE Signal Proc Mag 2008;25(1):57–68.

[14] Bashashati A, Fatourechi M, Ward RK, Birch GE. A survey of signal processing algorithms in brain–computer interfaces based on electrical brain signals. J Neural Eng March 2007;4(2):R32.

[15] Naeem M, Brunner C, Leeb R, Graimann B, Pfurtscheller G. Seperability of four-class motor imagery data using independent components analysis. J Neural Eng jun. 2006;3(3):208–16. IOP Publishing.

[16] Sadiq MT, Yu X, Yuan Z. Exploiting dimensionality reduction and neural network techniques for the development of expert brain–computer interfaces. Expert Syst Appl 2021;164:114031. ISSN 0957-4174.

[17] Winkler I, Haufe S, Tangermann M. Automatic classification of artifactual ICA-components for artifact removal in EEG signals. Behav Brain Func 2011;7(1):30. Springer Science and Business Media LLC.

[18] Wang Y, Jung T-P. Improving brain–computer interfaces using independent component analysis. In: Towards practical brain-computer interfaces: bridging the gap from research to real-world applications. Berlin, Heidelberg: Springer Berlin Heidelberg; 2013. p. 67–83.

[19] Hsu S, Pion-Tonachini L, Jung T-P, Cauwenberghs G. Tracking non-stationary EEG sources using adaptive online recursive independent component analysis. In: 2015 37th annual international conference of the IEEE engineering in medicine and biology society (EMBC). IEEE; 2015.

[20] Schirrmeister RT, Springenberg JT, Fiederer LDJ, Glasstetter M, Eggensperger K, Tangermann M, Hutter F, Burgard W, Ball T. Deep learning with convolutional neural networks for EEG decoding and visualization. Human Brain Mapp 2017;38(11):5391–420. Wiley.

[21] Lawhern VJ, Solon AJ, Waytowich NR, Gordon SM, Hung CP, Lance BJ. EEGNet: a compact convolutional neural network for EEG-based brain–computer interfaces. J Neural Eng July 2018;15(5):056013. IOP Publishing.

[22] Santamaría-Vázquez E, Martínez-Cagigal V, Vaquerizo-Villar F, Hornero R. EEG-inception: a novel deep convolutional neural network for assistive ERP-based brain-computer interfaces. IEEE Trans Neural Syst Rehabilit Eng 2020;28(12):2773–82.

[23] Brunner C, Leeb R, Muller-Putz G, Schlogl A, Pfurtscheller G. BCI Competition IV dataset IIa. 2004. at: https://www.bbci.de/competition/iv/desc_2a.pdf.

[24] Lee M-H, Kwon O-Y, Kim Y-J, Kim H-K, Lee Y-E, Williamson J, Fazli S, Lee S-W. EEG dataset and OpenBMI toolbox for three BCI paradigms: an investigation into BCI illiteracy. GigaScience 2019; 8(5). ISSN 2047-217X. Giz002.

[25] Saibene A, Caglioni M, Corchs S, Gasparini F. EEG-based BCIs on motor imagery paradigm using wearable technologies: a systematic review. Sensors March 2023;23(5):2798. MDPI AG.

[26] Wu X, Zhou B, Lv Z, Zhang C. To explore the potentials of independent component analysis in brain-computer interface of motor imagery. IEEE J Biomed Health Inform March 2020;24(3): 775–87. Institute of Electrical and Electronics Engineers (IEEE).

[27] Lin X, Wang L, Ohtsuki T. Online recursive ICA algorithm used for motor imagery EEG signal. In: 2020 42nd annual international conference of the IEEE engineering in medicine &amp biology society (EMBC). IEEE; 2020.

[28] Zhou B, Wu X, Zhang L, Lv Z, Guo X. Robust spatial filters on three-class motor imagery EEG data using independent component analysis. J Biosci Med Sci Res Pub Inc 2014;02(02):43–9.

[29] Gouy-Pailler C, Congedo M, Brunner C, Jutten C, Pfurtscheller G. Nonstationary brain source separation for multiclass motor imagery. Institute of Electrical and Electronics Engineers (IEEE) IEEE Trans Biomed Eng 2010;57(2):469–78. fev.

[30] Delorme A, Palmer J, Onton J, Oostenveld R, Makeig S. Independent EEG sources are dipolar. PLoS One Febuary 15, 2012;7(2):e30135. Public Library of Science (PLoS).

[31] Cardoso J-F, Souloumiac A. Blind beamforming for non-Gaussian signals. IEE Proc F Radar Signal Process 1993;140(6):362. Institution of Engineering and Technology (IET).

[32] Bell A, Sejnowski J. An information-maximization approach to blind separation and blind deconvolution. Neural Computa MIT Press J November 1995;7(6):1129–59.

[33] Lee T-W, Girolami M, Sejnowski J. Independent component analysis using an extended infomax algorithm for mixed subgaussian and supergaussian sources. Neural Comput 1999;11(2):417–41. MIT Press - Journals.

[34] Ablin P, Cardoso J-F, Gramfort A. Faster independent component analysis by preconditioning with hessian approximations. IEEE Trans Signal Process ago. 2018;66(15):4040–9. Institute of Electrical and Electronics Engineers (IEEE).

[35] Liu DC, Nocedal J. On the limited memory BFGS method for large scale optimization. Mat Program 1989;45(1–3):503–28. Springer Science and Business Media LLC.

[36] Ablin P, Cardoso J-F, Gramfort A. Faster ICA under orthogonal constraint. arXiv; 2017.

[37] Belouchrani A, Abed-Meraim K, Cardoso J-F, Moulines E. A blind source separation technique using second-order statistics. IEEE Trans Signal Process 1997;45(2):434–44.

[38] Hyvarinen A. Fast and robust fixed-point algorithms for independent component analysis. Institute of Electrical and Electronics Engineers (IEEE) IEEE Trans Neural Net 1999;10(3):626–34. maio.

[39] Hsu S, Mullen T, Jung T, Cauwenberghs G. Online recursive independent component analysis for real-time source separation of high-density EEG. In: 2014 36th annual international conference of the IEEE engineering in medicine and biology society. IEEE; 2014.

[40] Welch P. The use of fast Fourier transform for the estimation of power spectra: a method based on time averaging over short, modified periodograms. IEEE Trans Audio Electroacoust 1967;15(2): 70–3.

[41] Kohavi R, John H. Wrappers for feature subset selection. Elsevier BV Artif Intell 1997;97(1–2): 273–324. dez.

[42] Altaheri H, Muhammad G, Alsulaiman M, Amin SU, Altuwaijri GA, Abdul W, Bencherif MA, Faisal M. Deep learning techniques for classification of electroencephalogram (EEG) motor imagery (MI) signals: a review. Neural Comput Appl August 25, 2021;35(20):14681–722. Springer Science and Business Media LLC.

[43] Bishop CM. Pattern recognition and machine learning, vol 4; 2006 [s.n.].

[44] Hornik K, Stinchcombe M, White H. Multilayer feedforward networks are universal approximators. Neural Net 1989;2(5):359–66. ISSN 0893-6080.

[45] Zhang H. The optimality of naive Bayes. 2004.

[46] Theodoridis S, Koutroumbas K. Pattern recognition. 3. San Diego, CA: Academic Press; 2006.

[47] Breiman L. Random forests. Mach Learn 2001;45(1):5–32. Springer Science and Business Media LLC.

[48] Geurts P, Ernst D, Wehenkel L. Extremely randomized trees. Mach Learn March 2006;63(1):3–42. Springer Science and Business Media LLC.

[49] Uribe LS, Costa T, Carvalho S, Soriano D, Suyama R, Attux R. Two-class motor imagery bci based on the combined use of ica and feature selection. 2016.

[50] Brunner C, Naeem M, Leeb R, Graimann B, Pfurtscheller G. Spatial filtering and selection of optimized components in four class motor imagery EEG data using independent components analysis. Pattern Recog Lett 2007;28(8):957–64. ISSN 0167-8655.

12

Advancing neural engineering: Hierarchical control strategies with human-centered focus for hand prosthetics

Tanaya Das[1,2] and Dhruba Jyoti Sut[3]

[1]INDEPENDENT RESEARCHER, DIGBOI, ASSAM, INDIA; [2]SCHOOL OF BIOMEDICAL ENGINEERING, THE UNIVERSITY OF SYDNEY, SYDNEY, NSW, AUSTRALIA; [3]DEPARTMENT OF MECHNAICAL ENGINEERING, SRM INSTITUTE OF SCIENCE AND TECHNOLOGY, CHENNAI, TAMIL NADU, INDIA

1. Introduction

Amid the growing incidence of upper limb loss cases, assistive devices, notably prosthetic hands, have emerged as a paramount source of hope and empowerment. The challenges presented by the absence of an upper limb, particularly the hand, within contemporary society are profoundly impactful. The hand plays a pivotal role in our daily lives, encompassing functions that range from grasping objects to the navigation of our environment through the utilization of our intricate tactile system. Without the presence of such an essential body part, the ability to perform routine tasks is significantly compromised for the majority of individuals affected by this condition. It is noteworthy that prosthetic hands have indeed proven to be a viable solution for the substitution of an amputated or missing hand.

Over the past decade, significant strides have been taken in the development of prosthetic hands, leading to a surge in their popularity and igniting a greater curiosity for further research in this domain. For a prosthetic hand to be truly effective, it must embody two desirable features: first, an intuitive and natural control mechanism; and second, a sufficient level of sensory feedback mechanism. To meet these needs, various types of prosthetic hands, spanning from body-powered prosthetic hands to externally powered prosthetic hands and even bionic hands, have been researched and developed. Each category of prosthetic hand incorporates a distinct control strategy and sensory mechanism, aimed at maximizing its functionality. Presently, the convergence of advanced engineering and medical innovation has enabled these assistive devices to

Signal Processing Strategies. https://doi.org/10.1016/B978-0-323-95437-2.00007-0

closely replicate the intricate dexterity and sensory capabilities of natural hands, thereby empowering users to reclaim their independence and enhance their overall quality of life.

Neural engineering is one such prime instance of this convergence, and since its advent, the field of hand prosthetics has been able to realize an anthropomorphic hand. Through direct interfacing with the nervous system, neural engineering enables bidirectional communication between the user's intentions (UIs) and the prosthetic device. This communication involves not only the extraction of neural signals that represent movement commands but also the delivery of sensory feedback to the user, closing the loop of natural interaction. Such a control strategy mirrors the hierarchical nature of motor control in humans and offers a more intuitive and versatile interaction between the user and prosthetic hand. This chapter aims to provide a review focusing on the hierarchical human-centered control strategies utilized in hand prosthesis in the context of neural engineering.

Approaches inspired by human hierarchical motor control hold great potential among various other control strategies, as they prioritize the natural interaction between the user and the prosthesis. Such strategies strive to create an intimate connection between the human brain, the prosthetic limb, and the external environment, with the ultimate goal of enhancing motor function, providing a sense of embodiment, and facilitating seamless control over the device. Conventional prosthetic control strategies that rely solely on mechanical mechanisms and muscle signals have only been able to achieve basic functionality for the prosthetic hand. In contrast to this, hierarchical control strategies that depend on neural signals alone or in combination with other body signals have demonstrated an enhanced ability to achieve humanlike hand control. Consequently, this leads to the exploration of state-of-the-art technologies such as brain-computer interfaces (BCIs), neural interfaces, and artificial intelligence (AI), which lie behind the development of such human-centered control strategies in the domain of neural engineering.

BCIs, in particular, are crucial technological elements in enabling human-centric control of prosthetic hands. They establish a direct communication pathway between the human brain and prosthetic hand device, allowing users to control the movements of their artificial limbs directly through their brain signals [1]. The integration of a BCI with a prosthetic hand significantly amplifies its dexterity and functionality, enabling more finer-tuned movements and precise grasping capabilities, thereby imparting a sense of embodiment to the user. Typically, a BCI system first acquires the user's motor intention from their neural activity, decodes this recorded information, and subsequently directs the prosthetic hand to perform the intended actions.

The recording of the user's neural data for hand prosthesis control, consisting of hand movement information, is achieved through either noninvasive or invasive methods, while the decoding of movement information largely involves AI algorithms. This chapter endeavors to delve into the role of BCI in controlling prosthetic hands, with a specific focus on noninvasive neural signal acquisition technologies, neural interfaces, and AI-driven control algorithms. In addition, it seeks to analyze their advantages, limitations, and real-world applications.

Apart from neural signal-based control of hand prosthetics, sensory feedback from the prosthetic limb constitutes an indispensable element in the hierarchical control and regulation of these assistive devices. By providing real-time feedback on the state of the prosthesis, this sensorimotor integration allows for feeling a sensation of touch [2], pain [3], and movement [4], akin to the experience of a human hand. Additionally, it also facilitates the execution of precise and synchronized movements by the artificial hand. This chapter thus explores the sensory feedback control mechanisms in prosthetic hands, placing an emphasis on the human experience and interaction. Furthermore, it highlights the human-centered designs applied to hand prosthetics. This synergy between hierarchical control strategies and human-centered, focused designs has propelled the field of neural engineering for hand prosthetics forward, resulting in innovations that effectively address the needs and preferences of users.

The remainder of the chapter addresses the ongoing challenges and limitations in the development and adoption of these hand prostheses and presents an overview of the future directions that these approaches are taking in shaping the landscape of prosthetic technologies.

1.1   A neurological background on the motor control in the human body

The human brain serves as the main source of control upon whose commands all motor movements are initiated. This control of movements from the brain is primarily hierarchical in nature, as described by the widely accepted theory of *hierarchical motor control* [5]. When commands from the central nervous system (CNS) in the form of motor signals are sent to the motor muscles, they initiate voluntary limb movements through a structured pathway that is both anatomically and functionally hierarchical [6]. Though structured, the pathway involves a complex system of neural circuits and brain regions that work together to produce purposeful and coordinated movements.

Considering the control hierarchy, there are several levels through which the motor signal flows before reaching the limb for the execution of movements. At the top of this hierarchy is the primary motor cortex within the cerebral cortex, where the generation of these motor signals originates. In the motor cortex, these signals represent the intention to move and carry information about specific muscles and limbs involved in the action. From the cerebral cortex, the signals then proceed to subcortical structures like the basal ganglia and cerebellum, and eventually to the spinal cord—the final level of this hierarchy. In the basal ganglia, these signals are involved in modulating and refining movements, while in the cerebellum, they maintain balance and coordination of movements after receiving sensory feedback from various sensory systems and the somatosensory cortex. Upon reaching the spinal cord, they integrate with local circuitry and reflex pathways, facilitating rapid and automatic responses to sudden stimuli

FIGURE 12.1 An illustration depicting the hierarchical motor control of the hand in the human body. The *yellow arrows* indicate the flow of signals from the primary cortex to the initiation of hand movement.

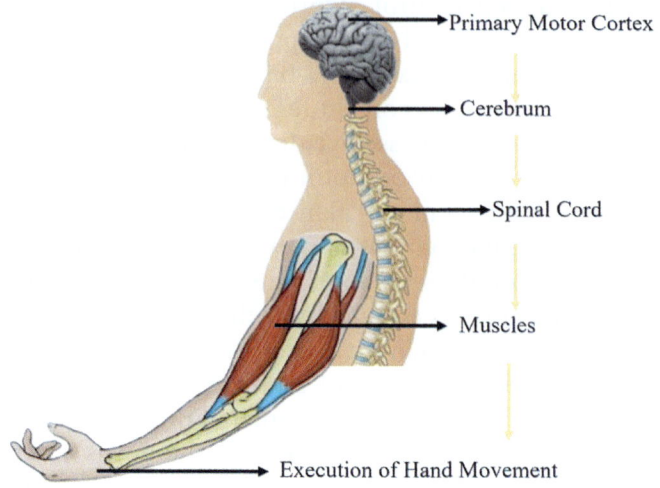

Primary Motor Cortex

Cerebrum

Spinal Cord

Muscles

Execution of Hand Movement

received from the limbs [7]. It is worth noting that the spinal cord also possesses semiautonomy from the cortex and can generate simple motor patterns.

From the spinal cord, these signals travel through peripheral nerves and eventually reach the muscles, inducing muscle contractions and finally resulting in a limb movement. Likewise, in a hand, finger movements, grasps, and wrist movements are controlled through this hierarchical mechanism. Fig. 12.1 illustrates the hierarchy of signal flow from the primary cortex to the spinal cord at the initiation of a hand movement.

Although the generation of movement intention occurs in the primary cortex, different regions within it are responsible for controlling the upper and lower limbs [8]. Specific muscle commands for initiating particular finger movements or grasping actions are transmitted from the cortex to the spinal cord. Motor neurons in the spinal cord then activate the muscles responsible for the specific task, and a task is thus performed. Throughout this process of signal flow from the CNS to the musculoskeletal system, sensory feedback is constantly integrated, allowing for adjustments to the movement.

To replicate the hierarchical control seen in natural hand movements within a prosthetic hand, the control mechanism of the prosthetic hand is organized into hierarchical levels, with each level being assigned a specific task. The signals generated in the motor cortex, mentioned earlier, are essentially electrical impulses resulting from neural activity in the brain. In order to control a hand prosthesis, this neural activity is decoded and utilized in the development of a control algorithm for the prosthesis controller. The algorithm then classifies the intended limb actions and subsequently transmits commands to the prosthesis, enabling it to execute the correct actions. An example of a simple hierarchical system is depicted in Fig. 12.2. The figure shows a two-stage hierarchical hand prosthesis controller where a high-level controller receives task context

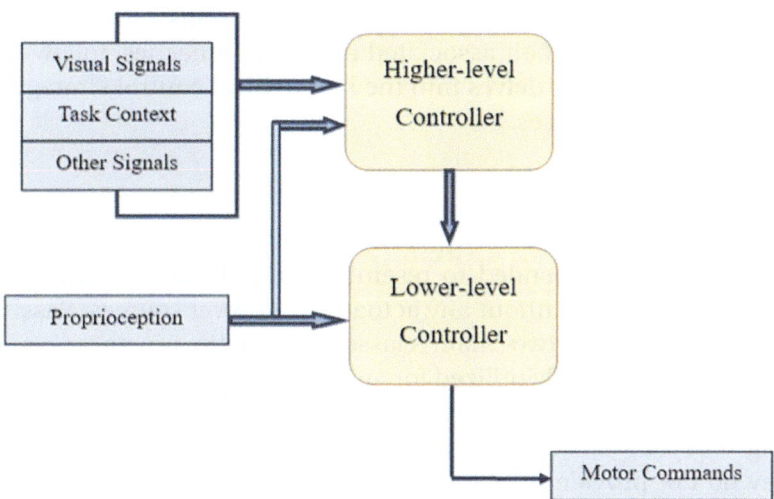

FIGURE 12.2 A simple hierarchical control mechanism with two levels of hierarchy for hand prosthesis control. *Image is modified from Ref. [6] and used under Creative Commons Attribution 4.0 International license http:// creativecommons.org/licenses/by/4.0/*

signals along with some other signals and sends them to a low-level controller. Upon receiving input from the high-level controller and proprioceptive sensors, the low-level controller generates control signals and commands the prosthetic hand's effector to perform the intended actions.

It is evident from the existence of such a complex hierarchical motor pathway within our system that precise and coordinated movements are possible to be performed by our hands. Currently, research in AI and prosthetic limbs has witnessed a shift toward designing limb control models based on the hierarchical motor control structure of the human body [6]. The need for this has been driven by the fact that it will be able to solve complex problems in the field of prosthesis efficiently. Moreover, such models will provide greater flexibility and a comprehensive skill set, overcoming obstacles while performing hand actions like a human hand.

2. Hierarchical control strategies in hand prosthesis

Control strategies for hand prostheses are critical components that allow people who have undergone upper limb amputations to regain functional and intuitive control of their prosthetic hands. Control approaches that are particularly hierarchical in nature streamline prosthesis control by orchestrating simple motor units to execute complex actions, enhancing both functionality and user integration. Basic movements are controlled at lower levels, while intricate tasks are managed at higher levels, optimizing overall performance. This merging of natural hierarchy with prosthetic demands optimizes the prosthetic hand's performance across a wide spectrum of tasks, offering a holistic and versatile user experience. This section presents an overview of different types of hand prostheses and their associated control approaches, followed by the need for hierarchical control. It then delves into the hierarchical control strategies reported in the literature on hand prostheses.

2.1 Types of hand prosthesis

2.1.1 Passive hand prosthesis

These prosthetic types are intended to resemble natural hands and are designed in a simplistic way. They operate without any actuators or power sources. Passive prostheses are further categorized into two main classes: Cosmetic prostheses and functional prostheses. The former is mostly utilized for aesthetic reasons and by individuals who do not require extensive hand functionality. The latter class is designed to assist users in accomplishing everyday basic bimanual tasks like holding and moving objects [9]. However, they do not provide the same dexterity level as other types of prostheses. In terms of control mechanisms, passive prostheses lack a controller and, hence, without any sensory feedback, provide only very basic functionality with a simple user interface.

2.1.2 Active hand prosthesis

These prosthetic types are designed to offer control that mimics a natural hand and are generally complex in their architecture. They are powered devices and can be divided into different classes too based on power source: Self-powered prostheses and externally powered prostheses [10]. Self-powered prostheses, also known as body-powered prostheses, are the ones that operate by harnessing the user's own biomechanical energy or other ambient energy produced during the user's limb movement. These prostheses are typically actuated through cables and mechanical linkages.

On the contrary, the externally powered ones are operated based on an external power source and actuators. The externally powered prostheses are further subdivided into three categories: myoelectric prosthetic hands, hybrid prosthetic hands, and bionic hands. While the self-powered prostheses provide simple grasping movements, the externally powered prostheses provide hand movements and grasp actions similar to those of a human hand. However, the level of dexterity is not the same for all externally powered prostheses and depends on their category. The following points present a brief overview of the major categories of these types of prostheses.

2.1.2.1 Myoelectric prosthetic hand

Myoelectric prosthetic hands use muscle signals generated by the user's residual muscles to control the movements of their end effector. These signals are commonly detected by muscle sensors or electrodes placed on an area of the user's skin surface, located remotely from the site of the lost limb [11]. This detection technique is referred to as electromyography (EMG), so the signals also called surface EMG signals. Additionally, they are referred to as intramuscular EMG signals when detected using implantable electrodes placed beneath the skin surface. The EMG signal is the main controlling element here, offering the users of these prostheses a greater degree of dexterity and precision compared to self-powered prostheses. Myoelectric prosthetic hands enable users to perform everyday tasks such as grasping objects, typing, and picking up small items.

2.1.2.2 Hybrid prosthetic hand

Hybrid prosthetic hands are designed by combining components from both self-powered and myoelectric systems. In these prosthetics, certain portions of the prosthetic hand are controlled by mechanical components such as joints or cables, and some by EMG signals. The incorporation of this combination enables users to execute a wide range of activities.

2.1.2.3 Bionic hand

Bionic hands replicate the natural movements of a human hand as closely as possible and are designed to provide users with an intuitively wide range of functionalities [12]. These prosthetic hands are driven using various control mechanisms. The majority of them use EMG as control signals in alliance with advanced sensor technology to mimic the muscular control of a physical hand, while the more advanced ones use neural signals integrated with AI and sensory feedback loops. This enables the prosthetic hand's users to experience feelings of touch, pain, temperature, and proprioception, thereby enhancing their interaction with the environment. When control is primarily based on neural signals, these hands are referred to as neuroprosthetic hands. To use neural signals as control signals in the controller of these devices, they are acquired from the brain through neural interfaces, or BCIs, and decoded using AI. The acquisition process is usually carried out using either noninvasive techniques like EEG or invasive neural implants. However, the chapter's focus will be on noninvasive techniques, especially EEG and noninvasive BCI systems. For instance, the Open Bionics Hero Arm [13] emerges as a striking exemplar, boasting a customizable and instinctive system. This system empowers users to map distinct movements to control various gripping patterns, culminating in a personalized interaction that feels natural and tailored. On the other hand, the i-LIMB [14] by Touch Bionic introduces a seamless transition between diverse grip patterns and gestures driven by muscle signals. This innovation substantially augments usability and user control using advanced sensors. Equally noteworthy is the BeBionic hand [15], which employs myoelectric sensors to facilitate control over individual finger movements and grip patterns. This emulation of natural hand motions enhances dexterity and usability. Furthermore, 'ENRICH', a prosthetic hand developed by Kakoty et al. [16], bestows users with a natural sense of control by mimicking the human hand responses during grasp operations enabled by one-channel EMG. The current market has two highly advanced dexterous bionic hands. The first, developed by BrainRobotics [17], uses AI to provide amputees with precise control over each finger, enabling them to perform a wide range of gestures and grips. The other is the Esper hand [18] by Esper Bionics, which combines AI and cloud-based technology, incorporating inputs from 24 advanced noninvasive sensors to perform nearly all tasks that a human hand can accomplish. Both of these hands are controlled through thought.

2.1.2.4   Neuroprosthetic hand

Neuroprosthetic hands focus on restoring motor control by directly interfacing with the brain's motor cortex or peripheral nerves through invasive and noninvasive implants. Neuroprosthetic hands that interface invasively employ microelectrode arrays or electrodes implanted surgically in the brain to record neural signals, whereas neuroprosthetic hands that interface noninvasively employ EEG, magnetoencephalography (MEG), functional magnetic resonance imaging, electrocorticography (ECoG), and near-infrared spectroscopy. Among all, EEG is the most widely used due to its noninvasive nature and portability of its signal acquisition equipment for clinical applications. In addition to these methods, BCI systems also utilize either invasive or noninvasive techniques to control these prosthetic devices. Although invasive methods have shown significantly higher accuracy compared to noninvasive methods, they come with disadvantages such as surgical risks, long-term complications, limited accessibility, and ethical considerations [19].

Eventually, the decoded neural information is used to control the prosthetic hand's movements. To raise the level of embodiment in such hands, artificial sensors are embedded in them to detect tactile sensations like pressure, temperature, and texture. The sensory feedback from these sensors is conveyed to the user's brain through electrical stimulation of sensory areas, creating a sense of embodiment and improving the interaction with the environment. Neuroprosthetic hands thus allow users to perform various grasp patterns and manipulate objects with very high precision and dexterity, restoring in them the sense of touch and proprioception.

Recent neuroprosthetic hands are primarily laboratory-developed prototypes and are still in the research phase before they can become commercial products. For example, Fukuma et al. [20] pioneered the development of a noninvasive neuroprosthetic hand controlled by MEG with two degrees of freedom (2-DoF) for paralyzed individuals. This prosthetic hand can determine both the movement timing and movement type from brain signals in the user's motor cortex. Furthermore, it can adapt to individual patients based on their neural information and perform grasping or opening hand movements. In contrast, other neuroprosthetic hands have relied on invasive implants to control their movements. Nguyen et al. [21] developed a neuroprosthetic hand that integrated AI with an implantable nerve chip to decode movement intentions from the peripheral nerve in the user's arm. This AI integration enables the prosthetic hand to interpret neural signals from the chip and perform finger movements solely through thought. To get more information about peripheral nerve interfaces [22], refer to. Another innovative study [23] built a neuroprosthetic hand for individuals with spinal cord injuries (SCIs) based on an implanted functional electrical stimulation system combined with an intracortical BCI to execute reach and grasp movements. These hands exhibit high dexterity and precision performance, indicating great potential for the future of prosthetic hands.

Previously developed neuroprosthetic hands employed EEG to control the prosthetic hand's movements. The Modular Prosthetic Limb (MPL) developed by the Johns Hopkins University Applied Physics Lab [24] has a humanlike appearance, strength, and dexterity with sophisticated tactile and position-sensing capabilities. It uses biosignals, including the EEG, to intuitively control individual fingers and perform fine motor tasks. Another noteworthy development is the EEG-based BCI-controlled neuroprosthetic hand [25], designed by researchers at the University of Houston. This hand allows the user to execute grasp actions and hold objects by decoding the UIs alone.

2.2   Need for hierarchical control approach in hand prosthesis

Over the period of development of hand prostheses, spanning from basic ones to advanced neuroprosthetic hands, the primary goal has consistently been to render these prosthetic limbs as human-centric as possible. The application of a human-centric approach stands as a paramount requirement for achieving dexterity within prosthetic hands. Integral to the realization of this objective is the incorporation of a hierarchical pathway.

When designing a prosthetic hand, an important consideration lies in ensuring its user-friendliness. A user-friendly prosthesis seamlessly harmonizes with the user's distinct needs and expectations. This harmonization, in turn, fosters the establishment of better dexterity, anthropomorphic qualities, and overall performance output [26]. The significance of hierarchical control becomes pronounced within this context, as it enables users to have reliable and better control over their prosthetic devices.

Nevertheless, the fundamental need for hierarchical control in hand prostheses arises from the desire to restore lost or impaired hand function and improve the quality of life for individuals with upper limb amputations or motor disabilities. Arranging the control in a hierarchical manner, from high-level intentions to low-level motor commands, has demonstrated the prosthetic hand's ability to autonomously control a number of degrees with minimal supervision by the user [27]. This increased range of motion and control capabilities empowers users to engage in complex tasks that were previously challenging or impossible with simple conventional prosthetic devices. Moreover, integration of the hierarchical pathway has leveraged laboratory-developed prosthetic hands to be used in clinical trials, which was previously a problem due to a lack of sufficient user interfaces [28]. Mirroring a human hand's versatility and adaptability to adjust grip strength, finger movements, and wrist movement, a hierarchical control system lets the prosthetic dynamically adapt to various tasks and environments in real-time. Apart from this, the presence of sensory feedback, such as a vibrotactile system in the hierarchy, endows the prosthesis with the possibility to activate different prehension patterns that can be useful in everyday life.

The use of hierarchical control that is human-centered further reduces the user's cognitive burden. By leveraging the brain's natural motor planning and execution mechanisms, users can operate the prosthesis more effortlessly, allowing them to focus on the task at hand rather than the mechanics of controlling the device. Finally, incorporating this human-centric edge into the control overall enhances user satisfaction and acceptance of the prosthetic hand, strengthening their trust to use the device in daily activities.

Consequently, there is no doubt that the implementation of such a control approach will revolutionize the way individuals with upper limb amputations or motor disabilities interact with and benefit from hand prosthetic technology, fostering ongoing research and advancements in hand prostheses with improved capabilities.

2.3   Hierarchy explored in control approaches for hand prosthesis

A control method in any limb prosthetic device is designed with the goal of accomplishing intended limb actions without any flaw. The earliest attempts at controlling hand prostheses date back to the 1960s, involving hierarchical controls inspired by the phases of human prehension [29]. In hand prosthesis, control strategies that are hierarchical in structure replicate the concept of a human body's neuromusculoskeletal hierarchical organization to bridge the gap between the complexity of natural hand movements and the capabilities of the prosthetic device.

At the highest tier of this hierarchical organization, control primarily revolves around planning movements and coordinating the fingers to perform complex tasks like picking up an object, holding it, and placing it down. This level considers the UI for intended usage of the object. The UI and goals are decoded from neural or myoelectric signals. The UI is thereafter converted into motor commands using advanced bio-signal processing techniques. These signal processing techniques are explored in Section 2.4. After the highest level, the mid-level control focuses on translating the motor commands into movements and joint angles of the prosthetic hand to perform the intended action. This control ensures accurate execution of gestures, grasps, and movements based on the UI. At the lowest level, the control mechanism integrates sensory feedback, which can be either somatotypically matched (sensory signals acquired from the user's residual limb provide real-time feedback on the prosthetic hand's interaction with the environment), modality matched (feedback is provided from the prosthesis's sensors like tactile sensors to the prosthesis to mimic natural sensations of touch and pressure), or sensory substituted (feedback is provided from the prosthesis's sensors like vibrotactile, auditory, or visual cues into the user's perception about the environment) [30]. Additionally, some hierarchical control mechanisms incorporate adaptability to enable the prostheses to adjust control parameters, sensitivity, and preferences based on changing circumstances or individual needs. Thus, structuring the control mechanism into levels not only enhances the prosthetic hand's functionality but also facilitates a more seamless integration into users' daily lives.

2.3.1 Myoelectric control

The myoelectric control strategy has been used in prostheses for decades. EMG signals, generated by the UI and captured from the user's muscle surface, act as the main commanding element to control the hand prosthesis in this strategy. The extraction of such a user-centric signal is noninvasive and hassle free, making it widely used in prosthesis control. Traditionally, depending on the nature of myoelectric control, it has been divided into sequential control and simultaneous control. The sequential control is further classified into switch on-off control, proportional control, direct control, finite-state machine (FSM) control, pattern recognition-based control, posture control, and regression control [31]. The main limitations faced in these control schemes are their inability to accurately decode human intention and the absence of a sensory feedback system, making them less intuitive and natural. Nevertheless, recent myoelectric control strategies, such as feedback control and shared control have been able to alleviate these limitations to a certain extent and have demonstrated greater dexterity than their previous counterparts. Regardless of the control scheme, a hierarchical architecture has often been reported as being incorporated in these strategies to obtain an anthropomorphic prosthetic hand.

A hierarchical grip control, as proposed by Kyberd et al. [32] was able to actuate a 2-DoF myoelectric signal-driven prosthetic hand, demonstrating that more than one DoF can be controlled when control is organized in levels. Followed by this, Cipriani et al. [26] introduced a hierarchical control with vibrotactile sensory feedback into the CyberHand [33,34], achieving a more interactive autonomous control than other control architectures. Their work implemented an FSM control strategy with shared control between the high-level controller and the low-level controller of the prosthetic hand. The low-level controller maintained grasp stability, while the high-level controller selected the grasp configuration and force level to be exerted by the prosthesis as interpreted by the user. Additionally, sensory feedback was provided to both levels of control to ensure better execution of movements. Apart from this, Cipriani et al. [35] also adopted a hierarchical architecture for a bio-inspired multifingered underactuated prosthesis, where they divided the control into high-level and low-level components. The former component was responsible for decoding intention signals and grasp selection, whereas the latter component was responsible for preshaping and force control. Emulating the CyberHand, Dosen et al. [36] integrated a new control method termed the cognitive vision system into a hierarchical control structure to mimic human biological control. In this, the user initiates the system and manages hand orientation. A high-level controller then takes action by choosing the appropriate grasp type and size, while the low-level (embedded hand) controller executes the chosen grasp through closed-loop position or force control.

Furthermore, the hierarchical architecture proposed by Quinayás et al. [37] encompassed a human-machine interface, haptic perception, high-level control, mid-level control, and low-level command stages. The architecture aimed at replicating brain areas and employed sensor feedback to control various hand movements. Even though the complete integration of a hierarchy level in hand prosthesis control is a recent addition, it was introduced much before 1970. The Southampton Adaptive Manipulation Scheme, or SAMS [38] which evolved from the study in 1960 [39], introduced a three-level hierarchy of control systems: Reflex systems (automatic actions), intermediate systems (object decision and force control), and supervisor systems (user commands). A validation study showed the SAMS strategy to outperform conventional myoelectric prostheses in various tasks. Another strategy [40] focused on reflex control, involving geometric primitives for object representation, preshaping, standard configurations for grasping, and separate target approach and shape adaptation phases. The approach aimed to mimic reflex arcs of the human hand, activating movement patterns based on sensory input. Most recently, Cha et al. [41] developed a novel EMG-based control for providing tactile feedback on information grasped by a robotic prosthetic hand, creating a comprehensive closed-loop system.

The studies reported employed various hierarchical control structures inspired by the human biological system. They utilized force, touch, and slippage information to achieve stable object manipulation in prosthetic hands.

2.3.2 Brain-computer interfaces and neural control

A BCI, or brain machine interface technology serves as a bridge, facilitating bidirectional communication between the brain and the prosthetic hand. It detects the neural activity, interprets the UIs, and translates these intentions into motor commands for prosthetic control. Neural control in hand prosthetics is often integrated with BCI to aid individuals with motor and sensory impairments resulting from limb amputations, strokes, SCIs, and other cognitive disorders. As such, the residual limb muscles of these individuals are no longer able to generate myoelectric signals, and they may experience a lack of sensation. However, the motor cortex continues to generate motor signals whenever the need to perform a movement arises. A neuroprosthetic hand takes advantage of this fact to achieve superior dexterity and anthropomorphic control, as motor intentions obtained directly from the source of generation contain an abundance of information required to execute any limb action. Achieving fine-grained control in this aspect involves a hierarchical approach.

Furthermore, neuroprosthetic hands controlled by BCIs were able to mimic the neuromuscular motor control of our bodies and achieve high-level precision in controlling the prosthetic hand. Recent works in hand prosthesis control mostly utilize EEG-based BCI systems due to their ability to control complex tasks flexibly and robustly. This way of controlling also lets the cognitive load on the user be minimized while still reserving the power in the user to intervene. Hierarchical and adaptive control was developed using an EEG-based BCI to drive a robotic arm where the high-level commands invoked learned skills, relieving the user of lower-level control [42]. This hierarchical BCI sets itself apart from traditional BCI systems by its ability to learn new behaviors from user demonstrations. The learning process occurs within the robotic component of the BCl system and is then abstracted into the hierarchical menu system.

Advanced neuroprosthetic hands have also incorporated myoelectric control to enhance their functionality and performance. In one study [43], a myoelectric control-based soft neuroprosthetic hand successfully controlled six degrees of freedom under pneumatic actuation, providing tactile feedback to simulate natural touch sensation. Another investigation [44] detailed a hierarchical control framework designed for a robotic arm. Within this arm's framework, a planner devises behaviors for the flexible arm based on the given task, while a high-level controller establishes objectives for the low-level motion controller. This allows the soft arm to be guided toward a predetermined goal, facilitating the execution of the desired behavior. Furthermore, the MPL hand [24] also employed a hierarchical architecture with different levels of control, allowing the user to retain the power to intervene if necessary. Despite significant progress in this field, there still remain challenges to developing a bidirectional hand neuroprosthetic that enables complete artificial sensory-motor restoration [45].

2.3.3   Collaborative and shared autonomy control

The ideas of collaborative control and shared autonomy have revolutionized how prosthetic devices interact with users, enhancing functionality and restoring natural control to those who've lost hand function. Collaborative control merges UIs with the prosthetic's capabilities, enabling smoother, more accurate movements. This approach allows users to perform intricate actions. Shared autonomy takes this further, giving the prosthetic decision-making abilities to optimize task completion based on user intent. This reduces the cognitive load on users, particularly for complex tasks, creating a seamless interface. It is particularly advantageous as the prosthetic intelligently adjusts its actions without requiring constant manual input. Both collaborative control and shared autonomy lead to quicker task completion and less user effort, which is crucial in robotics and AI for prosthetic hand control. To move toward a hierarchical strategy, these controls are thus beneficial. While the user's input remains pivotal, shared autonomy empowers the prosthetic to adapt its movements in real-time, accommodating unforeseen environmental changes or the UIs.

A study orchestrated by researchers from the University Medical Center Göttingen delved into implementing and comparing three shared control modes (sequential, simultaneous, and continuous) in terms of their efficacy and their impact on users' cognitive and physical demands. In the sequential approach, the volitional input temporarily suspends autonomous control [46]. The fusion of robotic or prosthetic arms with BCI systems has recently embraced the concept of shared autonomy. This innovative approach involves an ongoing partnership between the BCI user's neural signals and intelligent external sensors in controlling the robotic effector, all while maintaining a certain level of awareness of the surrounding environment and the intended task's outcome [47]. It is important to emphasize that the intent behind shared autonomy is not to create a fully autonomous system but rather to alleviate the user from the more demanding aspects of the task. ECoG recordings have also been used in shared-control BCI applications [48]. The system ingeniously combined computer vision to identify objects within the robotic arm's workspace, eye tracking to select targets, and ECoG signals to initiate the task. The findings unveiled that all three semiautonomous modes expedited task completion and necessitated lesser user-directed control inputs than the manual baseline (purely volitional control). A depth sensor placed on the back of the prosthetic hand was used to create a collaborative control system by Castro et al. [49]. This development made it easier for users to interact with the prosthetic hand in real-time. Starke et al. [50] introduced a semiautonomous control strategy based on visual object recognition using a single EMG channel and a multimodal sensor system. Zhuang et al. [51] developed a collaborative control system utilizing hidden touch sensors in a prosthetic hand through tactile feedback. This approach showed improved user endurance and stable grasping. In a related breakthrough, Federico et al. [52] pioneered a "hand-eye learning" approach, relying on data from a wrist-mounted camera to control hand preshaping and grasp aperture adjustment. Their research encompassed various grasping scenarios and provided comprehensive control over the Hannes prosthetic hand. Similarly, another study [53] demonstrated that employing a shared control with a robotic arm led to the accomplishment of complex tasks of reach and grasp using only two-class motor-imagery-based BCIs.

2.4   Signal processing techniques in hand prosthesis

The decoding of myoelectric or neural signals for advanced prosthetic control requires advanced signal processing schemes. For a prosthetic hand to be truly humanlike , it must rely on high-quality movement decoding and signal processing. Machine learning (ML), artificial neural networks, AI, and adaptive algorithms are the current advanced bio-signal processing techniques used in prosthesis to achieve high-level dexterous control. They are pivotal in customizing the prosthesis's functionality to cater to each user. These intelligent algorithms consider factors such as the user's fatigue levels, alterations in limb position, and the specific environment in which the prosthesis is being used. Hierarchical, human-centered control systems include such processing techniques to develop a dynamic and responsive prosthesis. Integration of self-learning capabilities into hierarchical control frameworks allows for the development of generic control strategies that can adapt to changing conditions while maintaining robustness and stability. For instance, Gentile et al. [54] focused on novel modeling and adaptive control design approaches for hierarchical human-machine cooperation in hand prosthetics. The research aimed to improve the cooperation between human operators and artificial agents in highly advanced prosthetic hands. The approach demonstrated improved safety by reducing human error, which is the most common cause of accidents. Likewise, in the study [55], Pascal et al. proposed the development of prosthetic hands with semiautonomous grasping abilities that lead to more intuitive control by the user. The prominent ML and AI techniques employed in prosthetic hands include:

Pattern recognition and ML algorithms: One of the critical challenges in prosthetic hand control is interpreting the UIs to perform specific movements. ML algorithms, such as support vector machines, random forests, and neural networks, are being employed to recognize motor patterns in EMG signals. The MPL and BeBionic hands mostly use pattern recognition to associate specific patterns of muscle activity with different hand movements or functions.

Reinforcement learning (RL): The RL technique involves an agent who learns to make decisions through trial and error in an environment. In prosthetic hands, RL is used to optimize the control strategy over time. The agent's (prosthetic hand) learns from the user's actions and the environment's feedback to improve its grasp and manipulation capabilities.

Computer vision and image processing: Computer vision techniques are vital in prosthetic hands equipped with cameras and sensors. By capturing real-world scenes, the prosthetic hand can better understand the user's environment and assist in object manipulation and interaction. Starke et al. [50] implemented a vision-based approach in their prosthetic hand to grasp objects semiautonomously, which reduced the user workload.

Natural language processing (NLP): NLP enables natural communication between users and their prosthetic devices. Voice commands or text inputs can be processed using NLP algorithms, allowing users to control their prosthetic hands verbally or through written instructions.

Sensor fusion: Sensor fusion integrates data from various sensors, such as accelerometers, gyroscopes, force sensors, and tactile sensors. ML techniques can process this combined sensor data to improve the prosthetic hand's accuracy, stability, and responsiveness during grasping and manipulation tasks.

Deep learning and convolutional neural networks (CNN): Deep learning techniques, especially CNN, have shown promising results in enhancing prosthetic hand vision systems. These networks can accurately identify objects and hand gestures, enabling the prosthetic hand to respond more precisely to UIs. Over the past several years, there has been a shift in the approach of researchers, who have begun to utilize deep learning techniques [56–58], thereby moving away from the traditional focus on crafting features and instead emphasizing the process of learning features. Côté-Allard et al. [59] developed an ADANN algorithm for obtaining myoelectric control in gesture recognition, transitioning from traditional hand-crafted EMG signal features to deep learning methods. The ADANN algorithm improved intersubject classification accuracy, showcasing a 19.40% average enhancement over standard training. Hae-June Park et al. [60] introduced a deep learning-based algorithm to control a robotic prosthetic hand using 2D images and 3D point clouds. The algorithm involved object recognition, grasping pose selection, and the determination of grasping time. It achieved 89% accuracy in grasping the intended object, though challenges such as object localization inaccuracies and occlusion were encountered. The algorithm showed promise in enhancing prosthetic hand usability for users. Atzori et al. [57] explored the application of CNN for real-time myoelectric control of prosthetic hands, enabling the prosthetic hand to recognize hand movements accurately and instantaneously.

A deep learning architecture represents a multitiered method for learning representations, wherein input embeddings aid in detection or classification. In the context of sEMG-based gesture recognition, deep learning has demonstrated competitiveness with the current state-of-the-art methods. Several studies have used EEG-based deep learning classification models to decode motor intentions. Alazrai et al. [61] decoded motor imagery (MI) tasks using CNNs to learn features from EEG recordings and recognize 11 MI tasks. A recent study proposed a genetic algorithm-optimized long-short-term memory model to classify arm movements [62]. The model was then integrated to control a 3-DOF prosthetic arm. Ramos-Murguialday et al. [63] used EEG-based BCIs to control a robotic hand for chronic stroke patients. The EEG signals recorded from the scalp were translated into hand movements, showing potential for rehabilitation applications. Another study by Das et al. [64] used a hierarchical approach and a deep learning method fusing EEG and EMG to achieve simultaneous finger movement control for a prosthetic hand. A CNN model with common spatial patterns was also used to identify palm extension and hand grasp motion using EEG signals [65]. The studies showed significant accuracy and recognition ability in identifying hand movements and grasp types from EEG than their EMG-based control counterparts.

3. Sensory feedback in hierarchical control

Over the past decade, substantial advancements have been made in the technology and functionality of upper limb prostheses, particularly in control and sensory feedback. Since our hands are vital for daily functioning, including touch sensitivity, it becomes imperative for an upper limb prosthesis not only to facilitate grasping and manipulation but also to replace some of the lost sensory components. Integrating a sensory mechanism into prosthetic hands can better emulate the functions of our natural limbs and significantly improve the lives of those who rely on them. Several studies have incorporated sensory feedback to simulate sensations of movement, touch, and pain. Fig. 12.3 depicts the general block diagram of sensory feedback in hand prosthetics.

Natural and reliable sensory feedback is essential for creating a lifelike and fully functional prosthetic limb that could enhance the user's daily activities and quality of life. A concept referred to as the "sense of agency" is highly desired in prosthesis use, as it enables the feeling of movement through motor commands. A way to assess this sense is by asking the person if they feel in control of the prosthesis movement [66,67]. Present-

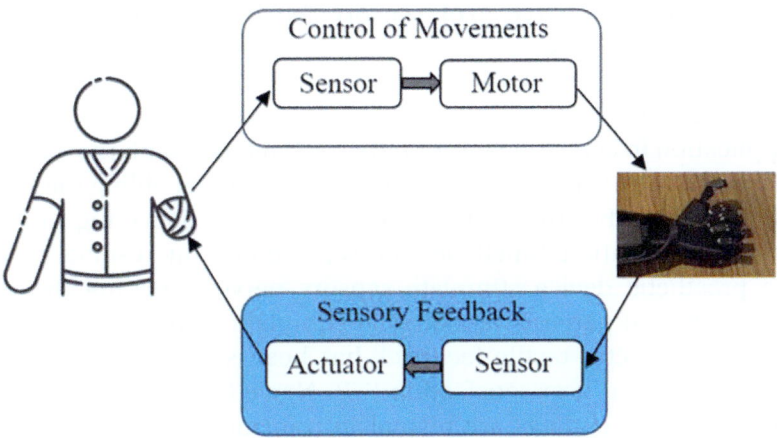

FIGURE 12.3 A block diagram illustrating the sensory feedback in prosthesis control.

day prostheses offer users a sense of agency regarding the device. Still, the absence of sensory feedback remains a significant factor limiting the feeling of body ownership of the prosthesis [68]. In addition to comfort, function, appearance, and durability, sensory feedback is essential to a prosthesis [69,70]. While grasping performance may already be satisfactory, feedback becomes particularly beneficial for complex tasks and situations where visual feedback is limited. Moreover, sensory feedback enhances the subjective experience of embodiment, making the user feel more connected to the prosthesis [71]. Furthermore, some studies have observed that including sensory feedback in a prosthetic hand may reduce phantom limb pain (PLP) experienced by amputees [72−74]. The desire for sensory feedback among prosthesis users, coupled with its potential benefits in performance and reducing PLP, highlights its importance for improving overall functionality and user satisfaction with hand prostheses. Sensory feedback is essential in hierarchical control to enable natural interaction between the user and the prosthesis, experience enhanced sensory perception, know proprioception and spatial awareness of the prosthesis and ensure safety.

3.1   Types of sensory feedback modalities

While developing advanced sensory feedback systems for prosthetic hands is challenging, significant progress has been made in recent years. Researchers are exploring various technologies, including advanced sensors, neural interfaces, and haptic feedback systems, to replicate the complexity and sensitivity of natural human touch. Various methods have been investigated to establish sensory feedback communication between prosthetics and humans through a prosthetic and human interface. Innovative approaches such as vibration, rotational stress, electrical stimulation, and temperature feedback [75] used to transmit information to the hand prosthesis via the skin surface have shown an immersive sensory experience for users with hand loss. This section presents an overview of the types of sensory feedback modalities that have been embedded in hand prosthetics.

3.1.1 Electro-tactile feedback

Beyond its application in activating motor nerves to facilitate limb movements, electrical stimulation has also been utilized to trigger sensory nerves, enabling the transmission of tactile information [76]. Electro-tactile feedback involves delivering tactile sensations through electrical stimulation. Small electrodes placed on the residual limb or the skin overlying the prosthetic device stimulate sensory nerves, creating the perception of touch. The intensity and patterns of electrical stimulation can be controlled to convey different sensations. Electro-tactile feedback has been studied and shown to improve grasping performance and user satisfaction [77]. Nerve fibers that innervate cutaneous receptors and free nerve endings are pivotal in transmitting diverse sensations to the brain's somatosensory cortex. The technique, Transcutaneous Electrical Nerve Stimulation (TENS), harnesses electrical pulses from surface electrodes on the skin, triggering a volley of sensory afferents. These electrodes' spatial configuration can encode sensory input intensity, such as gripping forces, providing multiple modalities of sensation [78]. Chai et al. [79] demonstrated that TENS can evoke distinct feelings classified as light touch, pressure, pressure & buzz, pressure & vibration, pressure & numbness, and tingling & pain. However, the quality of sensation produced by TENS in amputee subjects may depend on factors such as the skin condition at the amputated stump and the size of the surface electrodes used [80]. Each subject may exhibit varying thresholds and modulation ranges for these parameters. To enhance sensory feedback, Shin et al. [81] introduced a stimulation grid with 16 electrodes strategically placed along the upper arm's medial side beneath the short head of the biceps brachii. This innovative setup aimed to encode precise stimulation locations, enriching the user's tactile perception. Interestingly, a remarkable phenomenon called evoked tactile sensation has been observed in the stump skin of numerous transradial amputees [82–84]. Amputees can experience the vivid sensation of their lost fingers and even digits by stimulating specific skin areas in the stump, either mechanically or electrically. This finding opens up new possibilities for enhancing the prosthetic experience and improving the sense of embodiment for individuals with upper limb amputations.

3.1.2 Vibrotactile feedback

Vibrotactile feedback utilizes mechanical vibrations to provide tactile sensations to the user. Vibrators or actuators embedded in the prosthetic hand or socket generate vibrations that users can feel on their skin. These vibrations can be modulated to convey information about grip force, object properties, or contact with the environment. Vibrotactile feedback has demonstrated potential for enhancing object recognition and grip force control [85]. The vibrating feedback system comprises two essential components: pressure sensors and vibrating motors. Pressure sensors are crucial in detecting the pressure exerted on the prosthetic hand. This pressure information is then relayed to a microcontroller, which adjusts a motor's vibration intensity. The system strategically varies the vibration's frequency and amplitude to enable subjects to distinguish the applied pressure. In a previous study, researchers mapped the force applied by the prosthetic hand onto the vibration frequency, creating a dynamic range typically spanning from 10 to 500 Hz [86].

In various research experiments [85,87], vibration devices were affixed to an arm cuff, strategically stimulating the subjects' skin. To optimize performance, subjects underwent training to associate specific vibrations with stimuli applied to the prosthetic hand, such as grasping force. However, the stability of vibrotactile feedback can be influenced by factors like the stimulation site's location, the movement of the prosthetic hand, and sensory adaptation. Despite indications that vibrotactile feedback aids subjects in enhancing their prosthetic hand control and dexterity, further investigations are necessary to assess its long-term effectiveness. As researchers continue to delve into this field, understanding the dynamics of vibrotactile feedback and its potential benefits for prolonged use will pave the way for more advanced and user-friendly prosthetic technologies.

3.1.3   Visual feedback
Visual feedback involves using cameras or sensors on the prosthesis to relay information about the hand's position and orientation to the user. The user then observes this information through displays or augmented reality interfaces. Visual feedback helps users align the prosthetic hand with the target object and understand its spatial relationship with the environment. Visual feedback has been shown to improve hand-manipulative skills. Visual attention is commonly necessary for lifting and manipulating delicate objects, even when using the natural hand [88]. Erik et al. [89] took an innovative approach by integrating force feedback into the prosthetic hand's control algorithm and visually relaying this information to the user through a bicolor LED experimentally mounted to the thumb. Visual feedback proved beneficial for both experienced and inexperienced prosthetic hand operators, effectively reducing the likelihood of dropping or breaking objects during manipulation tasks. This groundbreaking study opens up new avenues for enhancing the functionality and safety of prosthetic technologies, ultimately improving the user experience for individuals with limb loss.

Each feedback modality has its benefits and limitations, and researchers are continuously exploring ways to combine multiple modalities to create a comprehensive sensory experience for prosthesis users. By integrating effective sensory feedback systems, prosthetic devices can offer users a more intuitive and immersive interaction with their artificial limb and the surrounding environment.

3.2   Sensor technologies for sensory feedback

Numerous endeavors have been made to offer upper-limb amputees realistic and meaningful sensory feedback. As a complex and multidimensional sensation, touch works in synergy with muscle movements, facilitating intricate manipulation tasks and tactile perceptions. However, providing this sensory feedback poses significant challenges. It requires capturing comprehensive touch information through sensors and effectively closing the loop by delivering that information back to the user, as depicted in Fig. 12.4. Achieving this seamless integration of sensory input and motor control remains crucial to developing advanced prosthetic technologies.

The integration of sensory feedback is a crucial element in the creation of a lifelike prosthesis [90]. Firstly (A), tactile sensations can be perceived in the phantom hand of amputees, allowing them to experience a sense of touch despite the absence of a physical limb. To achieve this, (B) sensors placed on the prosthesis play a vital role in

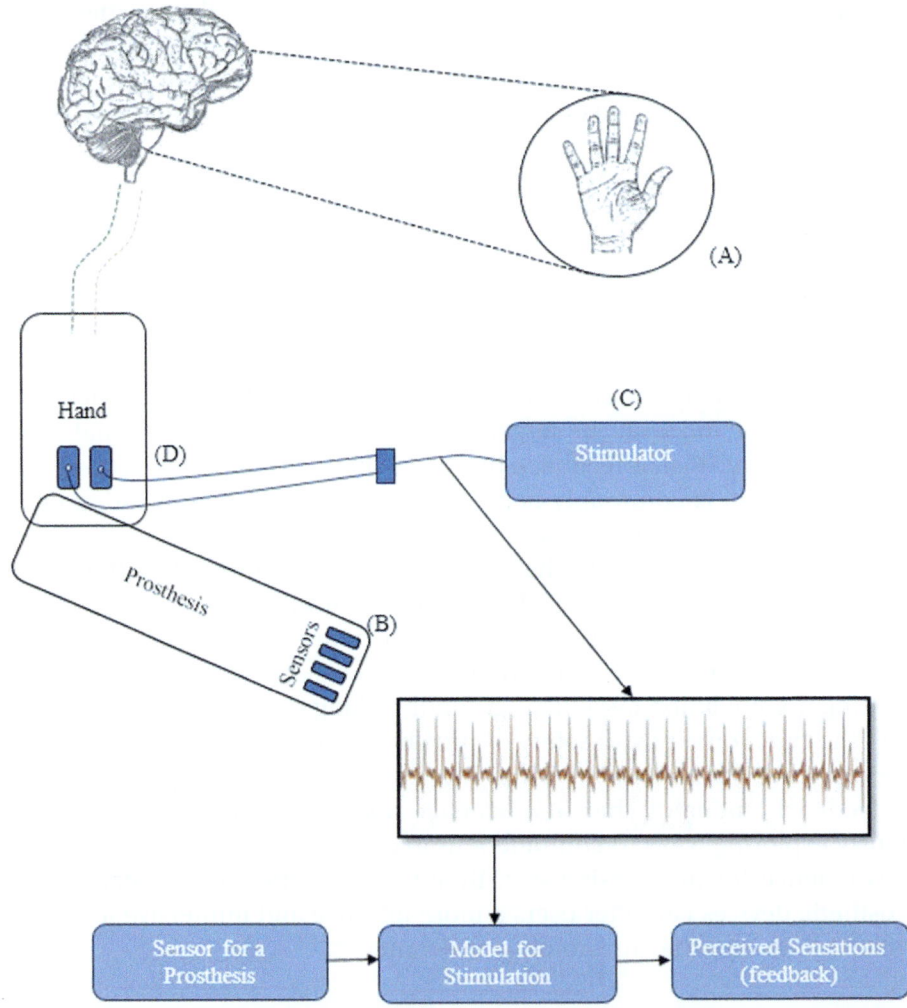

FIGURE 12.4 An example depicting the sensory feedback process in an user equipped with a hand prosthesis. (A) Phantom hand of amputees can experience sense of touch (B) Touch information from the sensors placed on prosthesis are captured, and (C) Stimulator returns this information back to the amputee. Models for stimulation translates the sensor information into sensory feedback as electrical stimulation (D) Tactile perceptions to the phantom hand are provided by stimulating the reinnervated sensory nerves.

capturing touch information. Subsequently (C), this sensory data is returned to the user through a stimulator. The stimulator employs sophisticated models to transform the sensor information into potentially natural and relevant sensory feedback conveyed through electrical stimulation (D). By stimulating reinnervated sensory nerves, the prosthesis can provide tactile perceptions of the phantom hand, contributing to a more immersive and realistic experience for the user. This ongoing research and development in sensory feedback technology holds great promise for creating prosthetic devices that offer a more intuitive and lifelike interaction with the environment, ultimately improving the quality of life for individuals with limb loss.

Indeed, sensor technologies play a crucial role in providing sensory feedback in prosthetic devices. There are various types of sensors that are used to detect and measure different sensory information, allowing users to perceive sensations similar to those of natural touch. Some of the most commonly used sensor technologies are highlighted in this section.

3.2.1   EMG sensors

EMG sensors are placed on the surface of the residual limb or within the prosthesis to detect muscle-generated electrical signals to control various hand movements like opening, closing, and rotating. Obtaining cortical and peripheral signals to control prosthetic hands often involves delicate and invasive neural interfaces. As a noninvasive alternative, surface EMG signals have proven to be highly practical and are widely employed in commercial prostheses today. The advantage of EMG lies in its ease of acquisition and minimal setup time. While invasive approaches involving implanted electrodes for intramuscular activity measurement are available, they are less commonly utilized.

However, decoding movements based on physiological signals comes with its own set of challenges. ML methods typically depend on incoming signals resembling those acquired during training, and variations in signals over time due to factors like fatigue or positional changes can negatively impact signal stability [91]. Additionally, loading effects on the prosthesis can degrade the classification performance [92]. A comprehensive discussion on EMG and its role in controlling prosthetic hands can be found in Ref. [93].

3.2.2   EEG sensors

EEG sensors can be used to detect brain signals related to the UI to move or control the prosthesis. These brain signals can be decoded to trigger specific movements or provide feedback about the prosthesis's position and orientation. EEG signals acquired noninvasively from the scalp have been explored for controlling multiple DoFs of a hand [94]. However, using EEG for prosthesis control is generally considered due to complications with the electrode-scalp interface during head movement and the low-pass filtering effect on signals.

A more invasive approach involving recording neural signals directly from the brain might be necessary for improved signal quality to achieve a higher signal-to-noise ratio.

One promising method involves capturing action potentials directly from the cortex using microelectrode arrays. These particular action potentials have successfully decoded reaching and grasping movements in a prosthetic arm [95]. Another alternative approach involves recording action potentials from the peripheral nervous system for prosthesis control [96]. This utterly different method offers an alternative path for achieving robust and precise control of prosthetic devices.

3.2.3   Force and pressure sensors

To enhance the gripping capabilities of prosthetic hands, force and pressure sensors are integrated into the fingers and palm. These sensors detect the amount of force or pressure applied when grasping objects, enabling the prosthetic hand to adjust its grip strength accordingly. This feature is particularly useful when handling delicate objects that require a gentle touch or when dealing with heavier items that necessitate a stronger grip. By providing users with real-time feedback on the pressure exerted, the risk of damaging objects or losing their grasp is significantly reduced.

3.2.4   Tactile sensors

Tactile sensors are instrumental in bridging the gap between the user and the environment. These sensors are embedded in the fingertips and palm of the prosthetic hand, allowing it to "feel" the surfaces and textures of objects it comes into contact with. The information gathered by tactile sensors is relayed back to the user as sensory feedback, enabling them to perceive the shape, texture, and hardness of the objects they are holding. This sensory input is crucial for fine-tuning manipulation tasks and provides a more natural and immersive experience when interacting with the environment.

Tactile sensors can be categorized based on their sensing principles and energy conversion methods, resulting in various types such as resistive, capacitive, inductive, optical, magnetic, piezoelectric, piezoresistive, and triboelectric sensors. A resistive tactile sensor operates on the principle that the resistance of its surface changes with the amount of applied load. Recent advancements have introduced microstructure layers into tactile sensing devices, enhancing their ability to encode tactile sensations. These microstructured sensors aim to mimic human skin, providing high sensitivity to low pressure, improved response time, and stability. In one study by Zhu [97], microstructured tactile sensors were designed to imitate human skin, offering enhanced sensitivity to subtle mechanical stimuli and quick response times. Another research effort by Park et al. [98] involved attaching electronic skin to human skin, enabling the interpretation of air flow, vibrations, and other mechanical cues. Inspired by the interlocked microstructures found in human skin, their work aimed to replicate similar functionality. Ji et al. [99] implemented microstructured elastomers in a flexible capacitive sensing array to develop a sensing mechanism for robotic skin. This design allowed for improved sensitivity and adaptability, simulating humanlike tactile perception. These advancements in microstructured tactile sensors hold promise for various applications, including robotics, prosthetics, and human-machine interfaces, as they bring us closer to achieving more humanlike and responsive touch interactions in various technological systems.

3.2.5   Position and inertial sensors
Position and inertial sensors, such as accelerometers and gyroscopes, are utilized to monitor the orientation and movement of the prosthetic hand in space. By continuously tracking the hand's position and motion, the prosthesis can adjust its grip and orientation to accommodate various activities and tasks. For instance, if the user wants to pick up an object from a table or reach for a specific item, the position and inertial sensors provide the necessary information to execute the desired movement accurately and efficiently.

3.2.6   Computer vision and camera sensors
Some advanced prosthetic hands may incorporate cameras and computer vision technology. These sensors enable the prosthesis to recognize and interpret visual cues from the user's environment. For example, the camera can identify different objects or gestures, allowing the prosthetic hand to perform context-aware actions based on the UIs or surroundings. This integration of computer vision enhances the prosthetic hand's adaptability and responsiveness, enhancing the user's overall experience.

3.2.7   Vibration sensors
Vibration sensors are utilized to deliver vibrotactile feedback to the user. They can be embedded in the prosthetic hand or socket to convey tactile sensations through mechanical vibrations.

The combination of these sensors and advanced signal processing techniques has revolutionized prosthetic hand technology, enabling users to regain functionality, independence, and a closer sense of integration with their environment. Ongoing research and development in sensor technology continue to push the boundaries, promising even more sophisticated and versatile prosthetic hands in the future. Ultimately, the incorporation of sensors in prosthetic hands has significantly contributed to enhancing the lives of individuals with limb loss, offering them newfound opportunities and possibilities in their daily activities and interactions with the world.

4. Challenges and future directions

Hierarchical control of hand prosthetics presents both exciting opportunities and significant challenges. While progress has been made to achieve natural and intuitive control through the use of bionic and BCI-based systems, there often remains a struggle to achieve these goals. Ethical aspects surrounding the integration of invasive or noninvasive technology to acquire neural activity in the human body often pose hurdles and impede progress. User-efficient usability of hand prostheses and satisfaction play a significant role but still require significant improvements due to the design and high cost of the prostheses. To overcome these challenges, research should focus on developing signal processing algorithms that can accurately decode motor signals. Efficient adaptive control and training strategies need to be developed that are user-centric and easily customizable. Finally, ongoing efforts should focus on enhancing the reliability of control systems through redundancy, fault tolerance, and robust sensor fusion techniques. More focus should be directed toward 3D-printed advanced prosthetic hands to reduce costs and enable customization, while also focusing on conducting rigorous evaluations before launching them on the market.

5. Conclusion

Hierarchical control strategies offer a promising paradigm for advancing the capabilities of neuroprosthetic hands. By aligning the control mechanisms with the hierarchical nature of human motor control, these strategies empower users with enhanced functionality, versatility, and ease of interaction. As research and development in this field continue to evolve, the integration of hierarchical control approaches with human-centered design principles holds the potential to redefine the landscape of hand prosthetics, enabling individuals with upper limb impairments to regain a remarkable degree of dexterity and independence in their daily lives. Integration of ML and AI techniques in such control strategies has empowered prosthetic users with greater control and natural interactions. Moreover, by utilizing neural interfaces, closed-loop systems, and various feedback mechanisms, researchers are advancing the field to enable intuitive, precise, and natural control of prosthetic hands. As research in this field progresses, we can expect even more groundbreaking developments, further blurring the boundaries between human and machine interactions.

Acknowledgment

We thank Prof. Dr. Nayan M. Kakoty, Professor, IEEE Sr. Member, Department of ECE, School of Engineering, Tezpur University, Tezpur, India for his valuable suggestion and comments.

References

[1] Vilela M, Hochberg LR. Applications of brain-computer interfaces to the control of robotic and prosthetic arms. Handb Clin Neurol 2020;168:87–99.

[2] Raspopovic S, Capogrosso M, Petrini FM, Bonizzato M, Rigosa J, Di Pino G, Micera S. Restoring natural sensory feedback in real-time bidirectional hand prostheses. Sci Transl Med 2014;6(222). 222ra19-222ra19.

[3] Osborn LE, Dragomir A, Betthauser JL, Hunt CL, Nguyen HH, Kaliki RR, Thakor NV. Prosthesis with neuromorphic multilayered e-dermis perceives touch and pain. Sci Robot 2018;3(19):eaat3818.

[4] Marasco PD, Hebert JS, Sensinger JW, Shell CE, Schofield JS, Thumser ZC, Orzell BM. Illusory movement perception improves motor control for prosthetic hands. Sci Transl Med 2018;10(432): eaao6990.

[5] Cano-De-La-Cuerda R, Molero-Sánchez A, Carratalá-Tejada M, Alguacil-Diego IM, Molina-Rueda F, Miangolarra-Page JC, Torricelli D. Theories and control models and motor learning: clinical applications in neurorehabilitation. Neurologia 2015;30(1):32−41.

[6] Merel J, Botvinick M, Wayne G. Hierarchical motor control in mammals and machines. Nat Commun 2019;10:5489. https://doi.org/10.1038/s41467-019-13239-6.

[7] Markin SN, Klishko AN, Shevtsova NA, Lemay MA, Prilutsky BI, Rybak IA. Afferent control of locomotor CPG: insights from a simple neuromechanical model. Ann N Y Acad Sci 2010;1198(1): 21−34.

[8] Kawato M, Furukawa K, Suzuki R. A hierarchical neural-network model for control and learning of voluntary movement. Biol Cybern 1987;57:169−85.

[9] Maat B, Smit G, Plettenburg D, Breedveld P. Passive prosthetic hands and tools: a literature review. Prosthet Orthot Int 2018;42(1):66−74.

[10] Dunai L, Novak M, García Espert C. Human hand anatomy-based prosthetic hand. Sensors 2020; 21(1):137.

[11] Geethanjali P. Myoelectric control of prosthetic hands: state-of-the-art review. Med Dev Evid Res 2016:247−55.

[12] Moradi A, Rafiei H, Daliri M, Akbarzadeh-T MR, Akbarzadeh A, Naddaf-Sh AM, Naddaf-Sh S. Clinical implementation of a bionic hand controlled with kineticomyographic signals. Sci Rep 2022; 12(1):14805.

[13] The hero arm overview is a prosthetic arm made by open bionics. Open Bionics. https://openbionics.com/en/hero-arm-overview/.

[14] Academy O. i-Limb Quantum. 2019. https://www.ossur.com/de-de/prothetik/arm/i-limb-quantum?tab=specification.

[15] bebionic Hand EQD. The most lifelike prosthetic hand. https://www.ottobock.com/en-ex/product/8E70.

[16] Kakoty NM, Gohain L, Saikia JB, Kalita AJ, Borah S. Real-time EMG based prosthetic hand controller realizing neuromuscular constraint. Int J Intell Robot Appl 2022;6(3):530−42.

[17] Home - BrainRobotics. BrainRobotics. March 13, 2023. https://brainrobotics.com/.

[18] Esper Bionics. Esper Bionics. https://esperbionics.com/.

[19] Waldert S. Invasive vs. non-invasive neuronal signals for brain-machine interfaces: will one prevail? Front Neurosci 2016;10:295.

[20] Fukuma R, Yanagisawa T, Saitoh Y, Hosomi K, Kishima H, Shimizu T, Yoshimine T. Real-time control of a neuroprosthetic hand by magnetoencephalographic signals from paralysed patients. Sci Rep 2016;6(1):21781.

[21] Nguyen AT, Drealan MW, Luu DK, Jiang M, Xu J, Cheng J, Yang Z. A portable, self-contained neuroprosthetic hand with deep learning-based finger control. J Neural Eng 2021;18(5):056051.

[22] Yildiz KA, Shin AY, Kaufman KR. Interfaces with the peripheral nervous system for the control of a neuroprosthetic limb: a review. J NeuroEng Rehabil 2020;17:1−19.

[23] Ajiboye AB, Willett FR, Young DR, Memberg WD, Murphy BA, Miller JP, Kirsch RF. Restoration of reaching and grasping movements through brain-controlled muscle stimulation in a person with tetraplegia: a proof-of-concept demonstration. The Lancet 2017;389(10081):1821−30.

[24] Bridges M, Beaty J, Tenore F, Para M, Mashner M, Aggarwal V, Thakor N. Revolutionizing prosthetics 2009: dexterous control of an upper-limb neuroprosthesis. Johns Hopkins APL Tech Dig 2010;28(3):210−1.

[25] Kever BJ. Researchers build brain-machine interface to control prosthetic hand. University of Houston; August 7, 2017. https://www.uh.edu/news-events/stories/2015/March/0331BionicHand.php.

[26] Cipriani C, Zaccone F, Micera S, Carrozza MC. On the shared control of an EMG-controlled prosthetic hand: analysis of user–prosthesis interaction. IEEE Trans Robot 2008;24:170–84.

[27] Fu Q, Santello M. Improving fine control of grasping force during hand-object interactions for a soft synergy-inspired myoelectric prosthetic hand. Front Neurorobot 2018;11:71. https://doi.org/10.3389/fnbot.2017.00071.

[28] Cheu LR, Casals A, Cuxart A, Parra A. Towards the definition of a functionality index for the quantitative evaluation of hand-prosthesis. In: 2005 IEEE/RSJ international conference on intelligent robots and systems. IEEE; August 2005. p. 541–6.

[29] Salisbury LL, Colman AB. A mechanical hand with automatic proportional control of prehension. Med Biol Eng 1967;5:505–11.

[30] Schofield JS, Evans KR, Carey JP, Hebert JS. Applications of sensory feedback in motorized upper extremity prosthesis: a review. Expert Rev Med Dev 2014;11:499–511. https://doi.org/10.1586/17434440.2014.929496.

[31] Ciancio AL, Cordella F, Hoffmann KP, Schneider A, Guglielmelli E, Zollo L. Current achievements and future directions of hand prostheses controlled via peripheral nervous system. The Hand: perception. Cognit Action 2017:75–95.

[32] Kyberd PJ, Holland OE, Chappell PH, Smith S, Tregidgo R, Bagwell PJ, Snaith M. MARCUS: a two degree of freedom hand prosthesis with hierarchical grip control. IEEE Trans Rehabil Eng 1995;3(1):70–6.

[33] Carrozza MC, Dario P, Vecchi F, Roccella S, Zecca M, Sebastiani F. The CyberHand: on the design of a cybernetic prosthetic hand intended to be interfaced to the peripheral nervous system. In: Proceedings 2003 IEEE/RSJ international conference on intelligent robots and systems (IROS 2003)(cat. No. 03CH37453), vol 3. IEEE; October 2003. p. 2642–7.

[34] Carrozza MC, Cappiello G, Micera S, Edin BB, Beccai L, Cipriani C. Design of a cybernetic hand for perception and action. Biol Cybern 2006;95:629–44.

[35] Cipriani C, Zaccone F, Stellin G, Beccai L, Cappiello G, Carrozza MC, Dario P. Closed-loop controller for a bio-inspired multi-fingered underactuated prosthesis. In: Proceedings 2006 IEEE international conference on robotics and automation, 2006. ICRA 2006. IEEE; May 2006. p. 2111–6.

[36] Dosen S, Cipriani C, Kostić M, Controzzi M, Carrozza MC, Popović DB. Cognitive vision system for control of dexterous prosthetic hands: experimental evaluation. J NeuroEng Rehabil August 23, 2010;7:42. https://doi.org/10.1186/1743-0003-7-42. PMID: 20731834; PMCID: PMC2940869.

[37] Quinayás C, Ruiz A, Torres L, Gaviria C. Hierarchical-architecture oriented to multi-task planning for prosthetic hands controlling. In: Biomedical applications based on natural and artificial computing: international work-conference on the interplay between natural and artificial computation, IWINAC 2017, corunna, Spain, June 19-23, 2017, proceedings, Part II. Springer International Publishing; 2017. p. 157–66.

[38] Codd RD, Nightingale JM, Todd RW. An adaptive multi-functional hand prosthesis. J physiol 1973;232(2):55P–6P.

[39] Kyberd PJ, Chappell PH. The Southampton Hand: an intelligent myoelectric prosthesis. J Rehabil Res Dev 1994;31(4):326.

[40] Tomovic R, Bekey G, Karplus W. A strategy for grasp synthesis with multifingered robot hands. In: Proceedings. 1987 IEEE international conference on robotics and automation, vol 4. IEEE; March 1987. p. 83–9.

[41] Cha H, An S, Choi S, Yang S, Park S, Park S. Study on intention recognition and sensory feedback: control of robotic prosthetic hand through EMG classification and proprioceptive feedback using rule-based haptic device. IEEE Trans Haptics 2022;15(3):560–71.

[42] Chung M, Cheung W, Scherer R, Rao RP. Towards hierarchical BCIs for robotic control. In: 2011 5th international IEEE/EMBS conference on neural engineering. IEEE; April 2011. p. 330–3.

[43] Gu G, Zhang N, Xu H, Lin S, Yu Y, Chai G, Zhao X. A soft neuroprosthetic hand providing simultaneous myoelectric control and tactile feedback. Nat Biomed Eng 2023;7(4):589−98.

[44] Jiang H, Wang Z, Jin Y, Chen X, Li P, Gan Y, Chen X. Hierarchical control of soft manipulators towards unstructured interactions. Int J Robot Res 2021;40(1):411−34.

[45] Iberite F, Mendez V, Mazzoni A, Shokur S, Micera S. Biomimetic bidirectional hand neuroprostheses for restoring somatosensory and motor functions. In: Somatosensory feedback for neuroprosthetics. Academic Press; 2021. p. 321−45.

[46] Mouchoux J, Bravo-Cabrera MA, Dosen S, Schilling AF, Markovic M. Impact of shared control modalities on performance and usability of semi-autonomous prostheses. Front Neurorobot 2021; 15:768619.

[47] Tang J, Zhou Z. A shared-control based BCI system: for a robotic arm control. In: 2017 first international conference on electronics instrumentation & information systems (EIIS). IEEE; June 2017. p. 1−5.

[48] McMullen DP, Hotson G, Katyal KD, Wester BA, Fifer MS, McGee TG, Crone NE. Demonstration of a semi-autonomous hybrid brain−machine interface using human intracranial EEG, eye tracking, and computer vision to control a robotic upper limb prosthetic. IEEE Trans Neural Syst Rehabil Eng 2013;22(4):784−96.

[49] Castro MN, Dosen S. Continuous semi-autonomous prosthesis control using a depth sensor on the hand. Front Neurorobot 2022;16:814973.

[50] Starke J, Weiner P, Crell M, Asfour T. Semi-autonomous control of prosthetic hands based on multimodal sensing, human grasp demonstration and user intention. Robot Autonom Syst 2022; 154:104123.

[51] Zhuang KZ, Sommer N, Mendez V, Aryan S, Formento E, D'Anna E, Micera S. Shared human−robot proportional control of a dexterous myoelectric prosthesis. Nat Mach Intell 2019;1(9):400−11.

[52] Vasile F, Maiettini E, Pasquale G, Florio A, Boccardo N, Natale L. Grasp pre-shape selection by synthetic training: eye-in-hand shared control on the Hannes prosthesis. In: 2022 IEEE/RSJ international conference on intelligent robots and systems (IROS). IEEE; October 2022. p. 13112−9.

[53] Xu Y, Ding C, Shu X, Gui K, Bezsudnova Y, Sheng X, Zhang D. Shared control of a robotic arm using non-invasive brain−computer interface and computer vision guidance. Robot Autonom Syst 2019; 115:121−9.

[54] Gentile C, Cordella F, Zollo L. Hierarchical human-inspired control strategies for prosthetic hands. Sensors 2022;22(7):2521.

[55] Weiner P, Starke J, Rader S, Hundhausen F, Asfour T. Designing prosthetic hands with embodied intelligence: the kit prosthetic hands. Front Neurorobot 2022;16:815716.

[56] Allard UC, Nougarou F, Fall CL, Giguère P, Gosselin C, Laviolette F, Gosselin B. A convolutional neural network for robotic arm guidance using sEMG based frequency-features. In: 2016 IEEE/RSJ international conference on intelligent robots and systems (IROS). IEEE; October 2016. p. 2464−70.

[57] Atzori M, Cognolato M, Müller H. Deep learning with convolutional neural networks applied to electromyography data: a resource for the classification of movements for prosthetic hands. Front. Neurorobot. 2016;10:9. https://doi.org/10.3389/fnbot.2016.00009.

[58] Phinyomark A, Scheme E. EMG pattern recognition in the era of big data and deep learning. Big Data Cognit Comput 2018;2(3):21.

[59] Côté-Allard U, Campbell E, Phinyomark A, Laviolette F, Gosselin B, Scheme E. Interpreting deep learning features for myoelectric control: a comparison with handcrafted features. Front Bioeng Biotechnol 2020;8:158.

[60] Park HJ, An BH, Joo SB, Kwon OW, Kim MY, Seo J. Grasping time and pose selection for robotic prosthetic hand control using deep learning based object detection. Int J Cont Autom Syst 2022; 20(10):3410−7.

[61] Alazrai R, Abuhijleh M, Alwanni H, Daoud MI. A deep learning framework for decoding motor imagery tasks of the same hand using EEG signals. IEEE Access 2019;7:109612–27.

[62] Kansal S, Garg D, Upadhyay A, Mittal S, Talwar GS. DL-AMPUT-EEG: design and development of the low-cost prosthesis for rehabilitation of upper limb amputees using deep-learning-based techniques. Eng Appl Artif Intell 2023;126:106990.

[63] Ramos-Murguialday A, Curado MR, Broetz D, Yilmaz Ö, Brasil FL, Liberati G, Birbaumer N. Brain-machine interface in chronic stroke: randomized trial long-term follow-up. Neurorehabil Neural Rep 2019;33(3):188–98.

[64] Das T, Gohain L, Kakoty NM, Malarvili MB, Widiyanti P, Kumar G. Hierarchical approach for fusion of electroencephalography and electromyography for predicting finger movements and kinematics using deep learning. Neurocomputing 2023;527:184–95.

[65] Jiang Y, Zhang X, Chen C, Lu Z, Wang Y. Deep learning based recognition of hand movement intention EEG in patients with spinal cord injury. In: 2020 10th institute of electrical and electronics engineers international conference on cyber technology in automation, control, and intelligent systems (CYBER). IEEE; October 2020. p. 343–8.

[66] Kalckert A, Ehrsson HH. Moving a rubber hand that feels like your own: a dissociation of ownership and agency. Front Human Neurosci 2012;6:40.

[67] Haggard P. Sense of agency in the human brain. Nat Rev Neurosci 2017;18(4):196–207.

[68] Wijk U, Carlsson I. Forearm amputees' views of prosthesis use and sensory feedback. J Hand Ther 2015;28(3):269–78.

[69] Biddiss E, Beaton D, Chau T. Consumer design priorities for upper limb prosthetics. Disabil Rehabil Assist Technol 2007;2(6):346–57.

[70] Cordella F, Ciancio AL, Sacchetti R, Davalli A, Cutti AG, Guglielmelli E, et al. Literature review on needs of upper limb prosthesis users. Front Neurosci 2016;10:209. https://doi.org/10.3389/fnins.2016.00209.

[71] Markovic M, Schweisfurth MA, Engels LF, Bentz T, Wüstefeld D, Farina D, Dosen S. The clinical relevance of advanced artificial feedback in the control of a multi-functional myoelectric prosthesis. J NeuroEng Rehabil 2018;15(1):1–15.

[72] Dietrich C, Walter-Walsh K, Preißler S, Hofmann GO, Witte OW, Miltner WH, Weiss T. Sensory feedback prosthesis reduces phantom limb pain: proof of a principle. Neurosci Lett 2012;507(2):97–100.

[73] Page DM, George JA, Kluger DT, Duncan C, Wendelken S, Davis T, Clark GA. Motor control and sensory feedback enhance prosthesis embodiment and reduce phantom pain after long-term hand amputation. Front Human Neurosci 2018;12:352.

[74] Paterson K, Lolignier S, Wood JN, McMahon SB, Bennett DL. Botulinum toxin-A treatment reduces human mechanical pain sensitivity and mechanotransduction. Ann Neurol 2014;75(4):591–6.

[75] Stephens-Fripp B, Alici G, Mutlu R. A review of non-invasive sensory feedback methods for transradial prosthetic hands. IEEE Access 2018;6:6878–99.

[76] Szeto AY, Saunders FA. Electrocutaneous stimulation for sensory communication in rehabilitation engineering. IEEE Trans Biomed Eng 1982;29(4):300–8.

[77] Antfolk C, D'alonzo M, Rosén B, Lundborg G, Sebelius F, Cipriani C. Sensory feedback in upper limb prosthetics. Exp Rev Med Dev 2013;10(1):45–54.

[78] Schweisfurth MA, Markovic M, Dosen S, Teich F, Graimann B, Farina D. Electrotactile EMG feedback improves the control of prosthesis grasping force. J Neural Eng 2016;13(5):056010.

[79] Chai G, Sui X, Li S, He L, Lan N. Characterization of evoked tactile sensation in forearm amputees with transcutaneous electrical nerve stimulation. J Neural Eng 2015;12(6):066002.

[80] Li P, Chai GH, Zhu KH, Lan N, Sui XH. Effects of electrode size and spacing on sensory modalities in the phantom thumb perception area for the forearm amputees. In: 2015 37th annual international conference of the IEEE engineering in medicine and biology society (EMBC). IEEE; August 2015. p. 3383–6.

[81] Shin H, Watkins Z, Huang HH, Zhu Y, Hu X. Evoked haptic sensations in the hand via non-invasive proximal nerve stimulation. J Neural Eng 2018;15(4):046005.

[82] Björkman A, Wijk U, Antfolk C, Björkman-Burtscher I, Rosén B. Sensory qualities of the phantom hand map in the residual forearm of amputees. J Rehabil Med 2016;48(4):70–365.

[83] Hunter JP, Katz J, Davis KD. Stability of phantom limb phenomena after upper limb amputation: a longitudinal study. Neuroscience 2008;156(4):939–49.

[84] Ramachandran VS. Behavioral and magnetoencephalographic correlates of plasticity in the adult human brain. Proc Natl Acad Sci 1993;90(22):10413–20.

[85] Clemente F, D'Alonzo M, Controzzi M, Edin BB, Cipriani C. Non-invasive, temporally discrete feedback of object contact and release improves grasp control of closed-loop myoelectric trans-radial prostheses. IEEE Trans Neural Syst Rehabil Eng 2015;24(12):1314–22.

[86] Kaczmarek KA, Webster JG, Bach-y-Rita P, Tompkins WJ. Electrotactile and vibrotactile displays for sensory substitution systems. IEEE Trans Biomed Eng 1991;38(1):1–16.

[87] Walker JM, Blank AA, Shewokis PA, O'Malley MK. Tactile feedback of object slip facilitates virtual object manipulation. IEEE Trans Haptics 2015;8(4):454–66.

[88] Johansson RS, Westling G, Bäckström A, Flanagan JR. Eye–hand coordination in object manipulation. J Neurosci 2001;21(17):6917–32.

[89] Engeberg ED, Meek S. Enhanced visual feedback for slip prevention with a prosthetic hand. Prosthet Orthot Int 2012;36(4):423–9.

[90] Weber DJ, Hao M, Urbin MA, Schoenewald C, Lan N. Sensory information feedback for neural prostheses. Biomed Inform Technol 2020:687–715.

[91] Fougner A, Scheme E, Chan AD, Englehart K, Stavdahl Ø. Resolving the limb position effect in myoelectric pattern recognition. IEEE Trans Neural Syst Rehabil Eng 2011;19(6):644–51.

[92] Cipriani C, Sassu R, Controzzi M, Carrozza MC. Influence of the weight actions of the hand prosthesis on the performance of pattern recognition based myoelectric control: preliminary study. In: 2011 annual international conference of the IEEE engineering in medicine and biology society. IEEE; August 2011. p. 1620–3.

[93] Scheme E, Englehart K. Electromyogram pattern recognition for control of powered upper-limb prostheses: state of the art and challenges for clinical use. J Rehabil Res Develop 2011;48(6).

[94] Bradberry TJ, Gentili RJ, Contreras-Vidal JL. Reconstructing three-dimensional hand movements from noninvasive electroencephalographic signals. J Neurosci 2010;30(9):3432–7.

[95] Hochberg LR, Bacher D, Jarosiewicz B, Masse NY, Simeral JD, Vogel J, Donoghue JP. Reach and grasp by people with tetraplegia using a neurally controlled robotic arm. Nature 2012;485(7398):372–5.

[96] Wendelken S, Page DM, Davis T, Wark HA, Kluger DT, Duncan C, Clark GA. Restoration of motor control and proprioceptive and cutaneous sensation in humans with prior upper-limb amputation via multiple Utah Slanted Electrode Arrays (USEAs) implanted in residual peripheral arm nerves. J NeuroEng Rehabil 2017;14:1–17.

[97] Zhu B. Skin-inspired flexible tactile sensing devices [Doctoral dissertation]. 2016.

[98] Park J, Lee Y, Hong J, Lee Y, Ha M, Jung Y, Ko H. Tactile-direction-sensitive and stretchable electronic skins based on human-skin-inspired interlocked microstructures. ACS Nano 2014;8(12): 12020–9.

[99] Ji Z, Zhu H, Liu H, Liu N, Chen T, Yang Z, Sun L. The design and characterization of a flexible tactile sensing array for robot skin. Sensors 2016;16(12):2001.

13

Advances in non-invasive EEG-based brain-computer interfaces: Signal acquisition, processing, emerging approaches, and applications

Shiu Kumar[1] and Alok Sharma[2,3]

[1]SCHOOL OF ELECTRICAL AND ELECTRONICS ENGINEERING, FIJI NATIONAL UNIVERSITY, SUVA, FIJI; [2]LABORATORY FOR MEDICAL SCIENCE MATHEMATICS, RIKEN CENTER FOR INTEGRATIVE MEDICAL SCIENCES, YOKOHAMA, JAPAN; [3]INSTITUTE FOR INTEGRATED AND INTELLIGENT SYSTEMS, GRIFFITH UNIVERSITY, BRISBANE, QLD, AUSTRALIA

1. Introduction

The human brain, with its intricate workings and vast capabilities, has long captivated researchers and scientists. Understanding how the brain functions and developing methods to directly interact with it has been a longstanding goal in the fields of neuroscience and neural engineering. In recent years, significant advancements in noninvasive electroencephalography (EEG)-based brain-computer interfaces (BCIs) have revolutionized these fields, offering exciting possibilities for direct communication and control between the human brain and external devices. The general framework of an EEG-based BCI system is shown in Fig. 13.1. The EEG signal is acquired from the scalp using EEG sensors and is passed to the processing device, where preprocessing, feature extraction, feature selection, etc. processes take place. The features are then used to train a classifier, which is later used to predict new unseen signals. The recognized signal is then used as a command to control external devices. Usually, an embedded system is used to accept these commands from the processor and generate appropriate signals to control the external devices.

This chapter presents a comprehensive exploration of recent progress made in noninvasive EEG-based BCIs, with a specific focus on signal acquisition, signal processing, emerging approaches, and applications. By measuring the brain's electrical

Signal Processing Strategies. https://doi.org/10.1016/B978-0-323-95437-2.00014-8

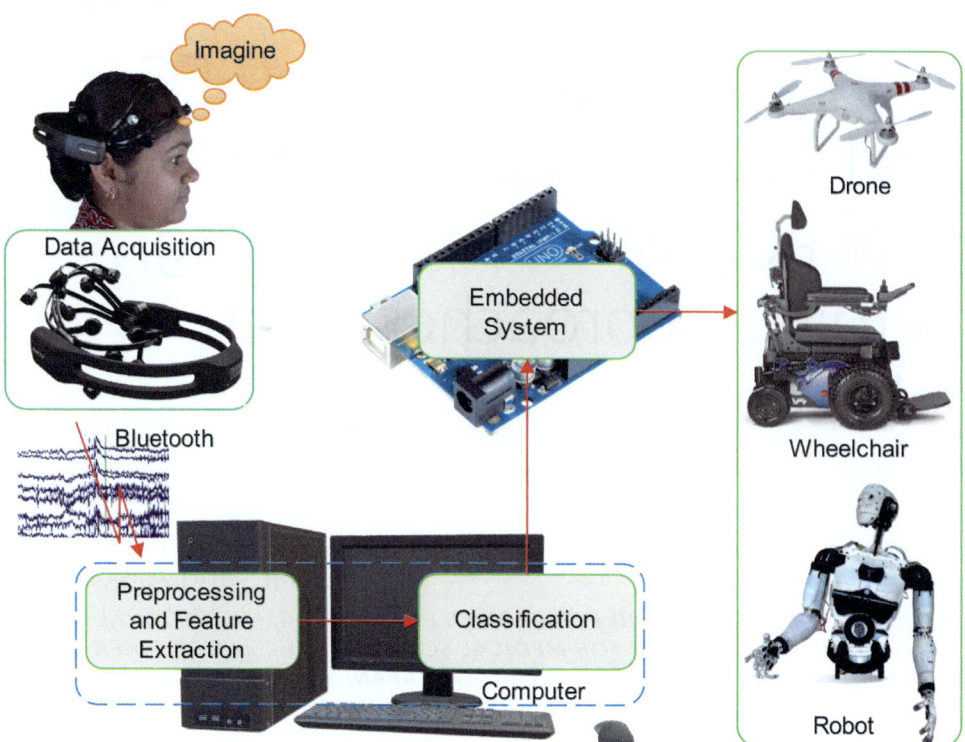

FIGURE 13.1 General framework of an EEG-based BCI system.

activity through electrodes placed on the scalp, EEG offers a noninvasive, portable, cost-effective, and easy-to-use approach with high temporal resolution. These advantages make EEG-based BCIs ideal for a wide range of applications, such as assistive technologies, neurorehabilitation, gaming, cognitive enhancement, and human-computer interaction.

Signal acquisition is a critical aspect of noninvasive EEG-based BCIs, as it directly influences the quality and reliability of the acquired brain wave data. Traditional EEG systems have utilized a limited number of electrodes, which restricts the spatial resolution and coverage of the recorded signals. However, recent advancements in sensor technology and signal processing algorithms have led to the development of high-density EEG systems, allowing for the acquisition of signals from a larger number of electrodes. These systems offer improved spatial resolution and a more comprehensive representation of brain activity, significantly enhancing the performance of EEG-based BCIs.

Emerging approaches in EEG-based BCIs, such as the integration of deep learning techniques like convolutional neural networks (CNNs) [1—3] and recurrent neural networks (RNNs) [4], have shown tremendous potential in automatically extracting relevant features from EEG data and achieving higher performance measures. Advanced signal processing techniques, including time-frequency analysis [5] and connectivity analysis, have also been employed to extract meaningful information from EEG signals, enabling more precise decoding of mental states or tasks and leading to context-aware control of BCIs.

Recent research has provided valuable insights into subject-specific frequency bands, wherein customizing the frequency bands of interest based on each user's EEG patterns significantly enhances the performance of EEG-based BCIs [6,7]. This personalized approach ensures better adaptability and robustness of the BCI system, resulting in more accurate and consistent control.

The subsequent sections of this chapter are structured as follows: Section 2 provides a detailed explanation of signal acquisition techniques, including advancements in electrode technologies and placement methods. Section 3 explores various signal processing methods, including preprocessing techniques for noise reduction and artifact removal, as well as feature extraction, connectivity analysis, and other signal processing approaches. Section 4 highlights emerging approaches and potential applications of EEG-based BCI systems. Finally, concluding remarks in the last section shed light on the challenges faced and future directions for EEG-based BCIs.

Through the exploration of these advancements, this chapter aims to provide readers with a comprehensive understanding of the state-of-the-art in noninvasive EEG-based BCIs. By elucidating the progress made in signal acquisition, signal processing, emerging approaches, and applications, we lay the groundwork for further advancements and innovations in this exciting field.

2. Signal acquisition techniques

Signal acquisition techniques are essential for capturing and recording brain activity in EEG-based BCIs. These techniques involve specialized equipment that detects and measures the electrical signals generated by the brain. Various methods and technologies have been developed to enhance signal acquisition and improve the overall accuracy and reliability of EEG measurements.

2.1 Traditional wet electrodes: Principles and limitations

Traditional wet electrodes are commonly used in EEG-based BCIs. They consist of metal discs or cups filled with conductive gel, which helps establish a low-impedance interface between the scalp and the electrode. This conductive gel enhances the electrical contact, ensuring accurate signal acquisition. However, wet electrodes have certain limitations.

Firstly, the use of conductive gel can be messy and time-consuming, requiring frequent reapplication to maintain good conductivity. Additionally, the gel may cause skin irritation or allergies in some individuals. The need for gel application and removal can also be inconvenient, especially in real-world applications that require frequent electrode placement and removal.

2.2   Advancements in dry electrodes

Advancements in dry-electrode technology have addressed many of the limitations associated with traditional wet electrodes. Dry electrodes do not require the use of conductive gel, making them more user-friendly and convenient. Instead, they utilize alternative methods to establish a reliable electrical connection with the scalp, such as dry contact materials, microneedles, or capacitive coupling.

Dry electrodes offer several advantages over wet electrodes. They are easier to apply and remove, reducing setup time and improving user comfort. Moreover, dry electrodes eliminate the need for messy gels, addressing issues related to skin irritation or allergies. These advantages make dry electrodes particularly suitable for long-term and practical BCI applications.

However, challenges exist with dry electrodes, including higher impedance compared to wet electrodes, which can result in reduced signal quality. Researchers are actively working on developing innovative designs and materials to overcome these challenges. Recent developments include the use of nanomaterials, flexible substrates, and active impedance reduction techniques to improve the signal quality and reliability of dry electrodes.

2.3   Introduction to alternative electrode technologies

Apart from wet and dry electrodes, alternative electrode technologies are being explored in the field of EEG-based BCIs. Active electrodes incorporate built-in amplification and noise reduction circuitry, allowing for signal amplification closer to the source and reducing susceptibility to external noise. Active electrodes offer improved signal quality and are particularly beneficial for applications requiring high signal fidelity.

Passive electrodes, on the other hand, do not have integrated amplification circuitry. They rely on external amplification to boost the signal. Passive electrodes are simpler in design and less expensive than active electrodes. They are commonly used in research settings and provide a cost-effective option for basic BCI experiments.

The choice between active and passive electrodes depends on specific application requirements, signal quality considerations, and cost constraints. Both technologies contribute to the advancement of EEG-based BCIs by offering alternative options for signal acquisition and processing.

2.4   Exploring electrode placement methods for improved signal acquisition

Electrode placement is a critical factor in EEG-based BCIs, as it influences signal quality, spatial resolution, and the ability to detect specific brain activities. The 10−20 system is a widely used electrode placement method that defines electrode positions based on specific anatomical landmarks on the scalp. This system ensures consistent and reproducible electrode positioning across individuals and facilitates comparison between studies.

In recent years, high-density EEG systems have gained popularity, allowing for more extensive coverage of the scalp and enhanced spatial resolution. High-density EEG systems utilize a larger number of electrodes, enabling more precise mapping of brain activity. They provide finer-grained spatial information, which is particularly useful for source localization and studying brain dynamics at a localized level.

Exploring electrode placement methods, including the 10–20 system and high-density EEG, has led to improved spatial resolution, better source localization, and increased sensitivity to subtle changes in brain activity. These methods have significantly contributed to the advancement of EEG-based BCIs, enabling more accurate decoding of brain signals and enhancing the overall performance of the interfaces.

3. Electroencephalography (EEG) and signal processing techniques

EEG is a noninvasive technique widely used to record and measure electrical activity in the brain. It provides valuable insights into the brain's functioning and is employed in various fields, including neuroscience, clinical diagnostics, and cognitive research. However, the raw EEG signals obtained from the scalp are often complex and noisy, posing challenges to extracting meaningful information.

3.1 Signal processing for enhancing EEG data

To extract valuable information from the vast amount of EEG data, signal processing techniques are employed. These techniques encompass computational methods and algorithms that aim to preprocess, analyze, and interpret EEG signals. By applying these techniques, researchers and clinicians can uncover patterns, extract features, and gain deeper insights into brain activity.

3.1.1 Noise reduction

One primary objective of EEG signal processing is noise reduction. EEG signals are susceptible to various interferences that can obscure meaningful brain activity. These include environmental electrical noise, muscle artifacts, and eye movements. Signal processing techniques such as filtering, artifact removal, and baseline correction are utilized to enhance the signal-to-noise ratio, thereby improving the quality and reliability of the recorded data.

3.1.2 Feature extraction

Feature extraction plays a crucial role in EEG signal processing, involving the identification of specific characteristics or patterns in the EEG signals. Techniques such as spectral analysis, time-frequency analysis, and statistical methods are employed to extract meaningful features, such as power spectral density, event-related potentials

(ERPs), or event-related desynchronization. These features can be utilized for tasks such as event detection, classification of brain states, or characterizing brain activity in response to stimuli.

3.1.3   Analyzing brain connectivity and network dynamics

Analyzing brain connectivity and network dynamics is crucial for understanding how different brain regions interact and communicate, providing insights into cognitive processes and neurological disorders. Techniques such as coherence analysis, graph theory, and connectivity metrics are employed to explore the complex network properties of the brain.

3.1.4   Coherence analysis

Coherence analysis measures the consistency or synchronization between two EEG signals from different brain locations. It quantifies the degree of functional connectivity between brain regions in specific frequency bands. Coherence analysis helps identify brain networks and assess the strength and dynamics of their interactions.

3.1.5   Graph theory

Graph theory provides a framework for analyzing the brain as a network of interconnected nodes, where each node represents a brain region or electrode and edges represent the functional or structural connections between them. Graph-based measures, including node degree, clustering coefficient, or betweenness centrality, can be calculated to characterize the topological properties and organization of brain networks derived from EEG signals.

3.1.6   Connectivity metrics

Various connectivity metrics, such as phase coherence, phase-amplitude coupling, or Granger causality, quantify the directional or functional connectivity between different brain regions. These metrics enable researchers to investigate how information flows between brain areas and identify key hubs or influential regions in the network.

In summary, EEG signal processing techniques are essential tools for extracting valuable information from EEG data. These techniques enable researchers and clinicians to reduce noise, extract meaningful features, and analyze brain connectivity, ultimately advancing our understanding of brain function, cognitive processes, and neurological disorders. By combining the power of EEG and signal processing, we can unravel the intricate workings of the human brain and pave the way for new discoveries and applications in neuroscience and healthcare.

3.2   Signal pre-processing methods for noise reduction and artifact removal

Signal preprocessing methods play a crucial role in EEG signal processing, as they are essential for enhancing the quality of the recorded data and extracting meaningful information by removing unwanted noise and artifacts. Here are some commonly used techniques:

(a) Filtering: Filtering is a fundamental preprocessing step that aims to remove unwanted noise and artifacts from the EEG signals, ensuring that the relevant brain activity is preserved. There are two main types of filters used in EEG signal processing:

  (i) Low-pass filter: This filter attenuates high-frequency noise and artifacts while allowing low-frequency brain signals to pass through. It helps remove high-frequency noise sources such as electrical interference and muscle artifacts.

  (ii) High-pass filter: This filter eliminates low-frequency noise and drifts while preserving the high-frequency components of the EEG signals. It removes baseline drift and slow artifacts caused by electrode impedance and movement.

  (iii) Notch filter: A notch filter is used to eliminate specific frequencies, such as power line interference (e.g., 50 Hz or 60 Hz), from the EEG signals. Power line interference is a common source of noise in EEG recordings, and a notch filter can effectively remove this unwanted signal.

  It is common to apply a combination of low-pass and high-pass filters to achieve a desired frequency range for the EEG signals.

(b) Spatial filtering: Spatial filtering methods, such as common average reference (CAR) or Laplacian filtering, are used to improve the spatial specificity of EEG signals. These techniques aim to enhance the desired brain activity while reducing artifacts and noise by considering the spatial distribution of the electrodes. The CAR technique calculates the average voltage across all EEG channels and subtracts it from each channel.

(c) Adaptive filtering: Adaptive filtering techniques, such as the recursive least squares algorithm or the Kalman filter, can be applied to estimate and remove specific artifacts from the EEG signals. These methods rely on adaptive modeling of the artifact sources and provide efficient artifact removal capabilities.

(d) Averaging and epoching: Averaging and epoching techniques are particularly useful when studying ERPs or specific brain responses to stimuli. By segmenting the EEG signals into smaller windows or epochs and averaging them over multiple repetitions, these techniques reduce random noise and enhance the signal-to-noise ratio.

(e) Independent component analysis (ICA): ICA is a powerful technique used to separate EEG signals into independent components. It can effectively identify and separate artifacts, such as eye blinks, eye movements, and muscle activity, from brain-related signals. By isolating these artifacts as separate components, they can be removed or further analyzed independently.

(f) Artifact subspace reconstruction (ASR): ASR is a data-driven method that automatically identifies and removes artifacts from EEG signals. It uses statistical techniques to model the data and identify portions of the signal that deviate significantly from the overall distribution. ASR can effectively remove a wide range of artifacts, including eye blinks, muscle artifacts, and sudden signal changes.

(g) Wavelet transform: The wavelet transform is a powerful time-frequency analysis technique that can reveal transient and nonstationary features in the EEG signals, making it suitable for identifying and removing artifacts with specific spectral and temporal characteristics.

(h) Artifact rejection: Manual or automated artifact rejection methods are used to identify and remove segments of EEG data that contain prominent artifacts, thus ensuring the integrity of subsequent analysis techniques.

These preprocessing methods are often used in combination to achieve optimal noise reduction and artifact removal in EEG signals. The choice of specific techniques depends on the characteristics of the noise and artifacts present in the signals, as well as the specific research or clinical goals. Proper preprocessing is crucial for obtaining reliable and meaningful results from EEG data, as it enhances the accuracy of subsequent analysis techniques such as feature extraction and classification. By effectively preparing the EEG data through preprocessing, researchers and clinicians can extract valuable insights into brain activity, cognitive processes, and neurological disorders, advancing our understanding of the human brain and its complex dynamics.

3.3   Feature extraction and feature selection techniques

EEG feature extraction techniques are methods used to derive meaningful and informative features from EEG signals. These features capture different aspects of brain activity and can be used for various applications, such as BCIs, cognitive state analysis, or neurological disorder detection. Here are some commonly used EEG feature extraction techniques:

(a) Power spectrum analysis: This technique involves calculating the power distribution across different frequency bands of EEG signals [8,9]. It provides information about the relative contribution of different frequency components, such as delta, theta, alpha, beta, and gamma waves, to the overall signal.

(b) Spectral entropy: Spectral entropy measures the complexity or randomness of the frequency content in EEG signals. It quantifies the distribution of power across different frequency bands and can be used to assess the brain's state or level of arousal.

(c) Time-frequency analysis: Time-frequency analysis methods capture the dynamic changes in the frequency content of EEG signals over time [10]. Techniques such as the short-time Fourier transform [11], the wavelet transform [12], or the Hilbert-Huang transform [13] provide insights into the time-varying spectral characteristics of the signal.

(d) ERPs: ERPs are EEG components that are time-locked to specific events or stimuli. They represent brain responses associated with cognitive processes. Features extracted from ERPs include peak amplitudes, latencies, or waveform morphology, providing information about cognitive processing and event-related brain activity.

(e) Connectivity measures: Connectivity measures quantify the functional interactions between different brain regions based on EEG signals. Techniques such as coherence, phase synchronization, or graph theory-based measures analyze the degree of synchronization or connectivity between EEG channels.

(f) Statistical features: Statistical features capture different statistical properties of the EEG signal, such as mean, variance, skewness, kurtosis, or higher-order moments. These features provide information about the distribution, shape, or asymmetry of the EEG data.

(g) Common spatial pattern (CSP): CSP is a powerful and widely used technique for extracting relevant features from EEG signals and has found widespread use in BCI applications. It aims to enhance the discriminative information in the EEG data by finding spatial filters that maximize the differences between the two.

(h) Fractal analysis: Fractal analysis investigates the self-similarity or complexity of EEG signals across different scales. Features such as fractal dimension [14] or Hurst exponent capture the long-range temporal correlations or scaling properties of the signal.

(i) Machine learning-based features: Machine learning techniques can be employed to extract more advanced features from EEG signals. These features can be learned automatically using algorithms like CNNs, RNNs, or support vector machines (SVMs), enabling the discovery of complex patterns and relationships within the data.

The selection of specific EEG feature extraction techniques depends on the research question, the characteristics of the EEG data, and the specific application or analysis objectives. Often, a combination of multiple techniques is used to capture a comprehensive set of features that represent different aspects of brain activity.

Feature selection techniques are methods used to identify and select a subset of relevant features from the extracted features. Feature selection is crucial to reduce dimensionality, improve computational efficiency, mitigate the curse of dimensionality, and enhance the interpretability and performance of the analysis. Here are some commonly used EEG feature selection techniques:

(a) Filter methods: Filter methods evaluate the relevance of individual features based on their statistical properties or other criteria. These methods typically rank features using metrics like mutual information, correlation analysis, t-tests, or analysis of variance. Features are then selected based on their ranking or a predefined threshold.

- Mutual information measures the amount of information shared between two variables. In the context of feature selection, mutual information is calculated between each feature and the target variable. Features with high mutual information scores are selected as they exhibit a strong relationship with the target [15].
- Correlation analysis measures the linear relationship between each feature and the target variable. Features with high correlation coefficients are considered more relevant and selected. Techniques such as Pearson correlation [16] or Spearman correlation can be used for this purpose.

(b) Wrapper methods: Wrapper methods assess feature subsets by training and evaluating a specific machine learning algorithm. They use a search strategy, such as forward selection, backward elimination, or genetic algorithms (GAs), to iteratively add or remove features and evaluate their impact on the model's performance. This process considers the interaction between features and the specific learning algorithm.

- Recursive feature elimination or selection (RFE or RFS): RFE is an iterative technique that starts with all features and progressively eliminates the least important ones. It uses a machine learning model to rank the features and remove the least ranked features in each iteration until a desired number of features remains [17]. RFS, on the other hand, begins with the full feature set and eliminates the least relevant feature at each step. Both methods aim to find an optimal subset of features by evaluating the impact of adding or removing features. RFE and RFS are commonly combined with cross-validation for robust feature selection.
- GAs: GAs [7] use an evolutionary approach to search for an optimal feature subset. They encode potential feature subsets as chromosomes and employ genetic operators such as selection, crossover, and mutation to evolve and improve the subsets based on a fitness function that measures their performance in a given task.

(c) Regularization methods: Regularization techniques like L1 (Lasso [18]) or L2 (Ridge) regularization impose penalties on the model coefficients during training, encouraging sparsity in the feature set. As a result, some coefficients become zero, effectively selecting the corresponding features. Regularization methods are often used with linear models, or SVMs.

(d) Principal component analysis (PCA): PCA is a dimensionality reduction technique that transforms the original EEG feature space into a new set of uncorrelated variables called principal components. These components capture the maximum variance in the data. By selecting a subset of the most significant principal components, the dimensionality of the feature space can be reduced while retaining as much information as possible [19]. However, the interpretability of the selected components may be challenging.

The choice of feature selection technique depends on the specific characteristics of the dataset, the machine learning algorithm used, and the goals of the analysis. It is often recommended to perform feature selection in conjunction with appropriate validation methods to ensure the selected features generalize well to unseen data. Validation techniques such as cross-validation or hold-out validation should be employed to ensure the selected feature subset's robustness and generalizability. Proper feature extraction and feature selection are essential steps in EEG signal processing, as they significantly impact the performance and interpretability of subsequent analysis and classification tasks.

3.4   Classification algorithms for decoding brain signals

Decoding brain signals, such as EEG, into different categories or states is a fundamental task in BCIs and cognitive research. Machine learning algorithms play a vital role in this process, as they can learn patterns and relationships in the data to accurately classify brain signals based on their features. Here are some commonly used classification algorithms for decoding brain signals:

(a) SVM: SVM is a powerful supervised learning algorithm widely used in EEG-based classification tasks. It aims to find an optimal hyperplane that best separates different classes in the feature space. SVM can handle high-dimensional data and can handle cases where the classes are not linearly separable.

(b) Random forest: Random forest is an ensemble learning algorithm that combines multiple decision trees to make predictions. Each decision tree is trained on a random subset of features and samples, and the final prediction is made by aggregating the predictions of individual trees. Random Forest is known for its robustness against overfitting and its ability to handle high-dimensional data.

(c) k-nearest neighbors (k-NN): k-NN is a straightforward and effective algorithm for brain signal classification. It classifies a sample based on the classes of its k nearest neighbors in the feature space. The choice of k determines the influence of nearby samples on the classification. k-NN is nonparametric and does not assume any underlying distribution of the data, making it suitable for both multiclass and binary classification tasks.

(d) Artificial neural networks (ANN): ANN models, inspired by the biological neural networks in the brain, are widely used for brain signal classification. Feedforward neural networks and CNNs are commonly employed in EEG-based BCIs. CNNs are particularly effective for analyzing spatial patterns in brain signals, such as in EEG-based BCIs.

(e) Naive Bayes: Naive Bayes is a probabilistic classification algorithm based on Bayes' theorem. It assumes that the features are conditionally independent, given the class label. Naive Bayes is computationally efficient and can handle high-dimensional data. It has been used for various EEG classification tasks, such as emotion recognition or mental state classification.

(f) Logistic regression: Logistic regression is a simple and interpretable linear classification algorithm. It models the relationship between features and the probability of belonging to a specific class. Logistic regression is suitable for the binary or multiclass classification of brain signals.

(g) Hidden Markov models (HMM): HMM is a probabilistic model that captures the temporal dependencies in sequential data, making it well-suited for time-series brain signal classification, such as EEG data. HMMs assume that the underlying system is a Markov process with hidden states, and they consider the temporal dynamics and transitions between different brain states.

(h) Deep learning models: Deep learning models, such as deep neural networks, or RNNs, have gained popularity in recent years for decoding brain signals. These models can learn complex hierarchical representations from the data and capture temporal dependencies. They have been successfully used for various brain signal classification tasks, including EEG-based emotion recognition or motor imagery (MI)-based BCIs.

Selecting the appropriate classification algorithm is crucial, and it depends on factors such as the nature of the brain signal data, the complexity of the classification task, the interpretability requirements, and the availability of labeled training data. Experimenting with multiple algorithms and comparing their performance is often recommended to determine the most suitable approach for a specific application. Accurate and reliable classification is essential to unlocking the valuable information contained within brain signals and advancing our understanding of brain function and cognition.

4. Existing approaches, packages, datasets, and applications of EEG signal processing

The development of improved EEG signal processing methods has been driven by recent technological advancements and the growing awareness of EEG signals' potential for various applications, particularly in BCIs. In this subsection, we present some state-of-the-art techniques for EEG signal processing.

4.1 Common Spatial Pattern

CSP is a widely used technique for EEG signal processing, which leverages the noninvasive nature of EEG to record the brain's electrical activity. It offers valuable insights into brain function and finds applications in BCIs, neurofeedback, and the clinical diagnosis of neurological disorders. The main objectives of EEG signal processing are to extract relevant information from the acquired brain signals and enhance their discriminative features. CSP achieves this by spatially filtering EEG signals. Here's why CSP is used for EEG signal processing:

(a) Feature extraction: EEG signals are typically multichannel in nature, recorded from multiple electrodes placed on the scalp. CSP helps in identifying the spatial patterns of brain activity that are most relevant for distinguishing between different cognitive states or tasks. It extracts features from the EEG signals that maximize the differences between different brain states and minimize variations within the same state.

(b) Dimensionality reduction: CSP reduces the dimensionality of the EEG data by projecting the original multichannel EEG signals onto a lower-dimensional subspace that captures the most discriminative information. This simplifies subsequent classification or analysis tasks while retaining essential information.

(c) Discrimination of brain states: CSP excels at distinguishing between different brain states or conditions. It identifies the spatial filters that amplify the signal components related to the target brain activity and suppress the interference from other brain activity or noise sources. This enables accurate discrimination between different mental states, such as different cognitive tasks or the presence of specific brain abnormalities.

(d) Increased signal-to-noise ratio: EEG signals are often contaminated with various sources of noise, such as eye movements, muscle activity, and environmental artifacts. CSP improves the signal-to-noise ratio by enhancing the brain-related components while attenuating the noise components. This makes the subsequent analysis or interpretation of the EEG signals more reliable.

(e) Adaptability: CSP can be adapted to specific EEG datasets or applications. It can be customized to target specific frequency bands or brain regions based on the experimental requirements. This flexibility allows researchers and clinicians to focus on the brain activity of interest and enhances EEG signal interpretability.

The CSP algorithm decomposes multichannel EEG signals into a set of spatial filters. These filters are designed to maximize the variance of one class while minimizing the variance of the other class. By applying these filters to the EEG signals, the resulting transformed signals highlight differences between the classes. CSP has proven particularly effective in BCI applications, where it is used to distinguish between different mental states or tasks based on EEG data. It has been successfully applied in various domains, including MI tasks, where users imagine performing specific movements, and cognitive tasks, where users focus on specific mental states or stimuli.

Overall, CSP is a valuable tool in EEG signal processing, enhancing discriminative features, reducing dimensionality, and improving brain state classification accuracy or analysis tasks. By leveraging the spatial patterns of brain activity, CSP empowers researchers and clinicians to gain deeper insights into brain function and associated cognitive processes. For a detailed explanation of the CSP algorithm, refer to Ref. [20].

4.2 Filter-based approaches

The BCI system offers a direct means for brain communication with the external world. CSP is a widely adopted method for effectively extracting features in MI-BCI systems. However, the performance of CSP-based BCI systems heavily relies on the choice of filter parameters. The EEG signal can be decomposed into five different frequency bands, namely delta (1–4 Hz), theta (4–8 Hz), alpha (8–13 Hz), beta (13–30 Hz), and gamma (30–80 Hz). However, these frequency bands vary between subjects due to individual differences in age, skin thickness, skull size, electrode placement, and cognitive strategies. If the CSP method is applied to unfiltered EEG signals or filtered EEG signals with poorly selected frequency bands, it can lead to low accuracy in recognizing brain states.

To determine the optimal frequency band for each subject, an exhaustive search combined with manual adjustments is often necessary, making the process time-consuming and requiring meticulous effort. Moreover, the lack of standardization in the manual selection process leads to varying performance among researchers using the same CSP algorithm independently. To address this issue, filter-based approaches for the classification of MI EEG signals in BCI applications continue to be explored. We present some state-of-the-art filter-based approaches widely utilized for EEG signal processing in this subsection.

4.2.1 Filter bank CSP (FBCSP)

The EEG signal can be typically classified into five main bands, namely alpha, beta, gamma, theta, and delta. However, the frequency range of these bands varies between subjects and appears to be lower in infants and children compared to adults. In other words, the most responsive frequency band for each subject varies, and thus, using these typical frequency bands usually does not produce optimal information and performance. To overcome this issue, initially, the subband CSP (SBCSP) [21] method was proposed. In this method, the EEG signal is decomposed into 24 subbands using Gabor filters, with each subband having a bandwidth of 4 Hz (no overlap). Subsequently, the CSP algorithm is used to extract features from EEG signals obtained from each of the subbands. To reduce the feature dimensionality, linear discriminant analysis (LDA) is then applied to the CSP variance-based features obtained from each of these subbands. This is then followed by feature fusion. Finally, an SVM classifier is trained using the selected features. To further improve the performance of the SBCSP approach, FBCSP [22] was proposed (Fig. 13.2). The main difference between the SBCSP and FBCSP approaches is that instead of performing dimensionality reduction for features obtained from each individual subband, the CSP variance-based features from all the subbands are fused together, and feature selection is performed. The selected features are then used to train the classifier. Several feature selection algorithms and classifiers were evaluated for this purpose.

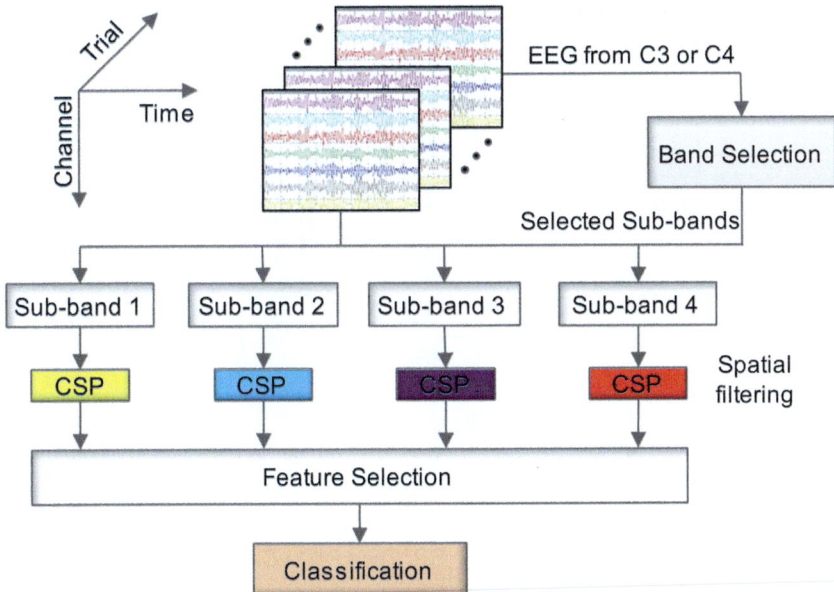

FIGURE 13.2 General framework of the FBCSP approach.

4.2.2 *Improved discriminative FBCSP (iDFBCSP)*

DFBCSP introduced a Fisher's ratio-based approach [23] to select the four subbands containing crucial information for EEG signal recognition, improving performance compared to its predecessors. However, DFBCSP uses only single-channel information for subband selection, making it susceptible to noise. To further enhance the performance, an improved DFBCSP (iDFBCSP) was introduced.

The iDFBCSP method (shown in Fig. 13.3) [15] proposes a frequency band selection approach based on mutual information, leveraging information from all available channels to effectively identify the four most discriminative subbands. To achieve this, CSP features are extracted from multiple overlapping subbands. Additionally, a wide frequency band (7–30 Hz) is introduced, and two types of features are extracted from this wide band using CSP and CSSP techniques. Mutual information is then computed from the extracted features of each subband, including the wide band, and the top four subbands are chosen for further processing. LDA is then applied for feature dimensionality reduction, and the resulting scores are fused together to train an SVM classifier. Selecting the top four subbands has been recommended as it produces the lowest average error rate.

A binary particle swarm optimization (BPSO) [25] approach has also been proposed for selecting the top subband(s) containing the most discriminative information required for EEG signal recognition. However, it requires higher computation time and computational power for selecting the subbands in comparison to the iDFBCSP approach. Moreover, the iDFBCSP outperforms the BPSO approach for subband

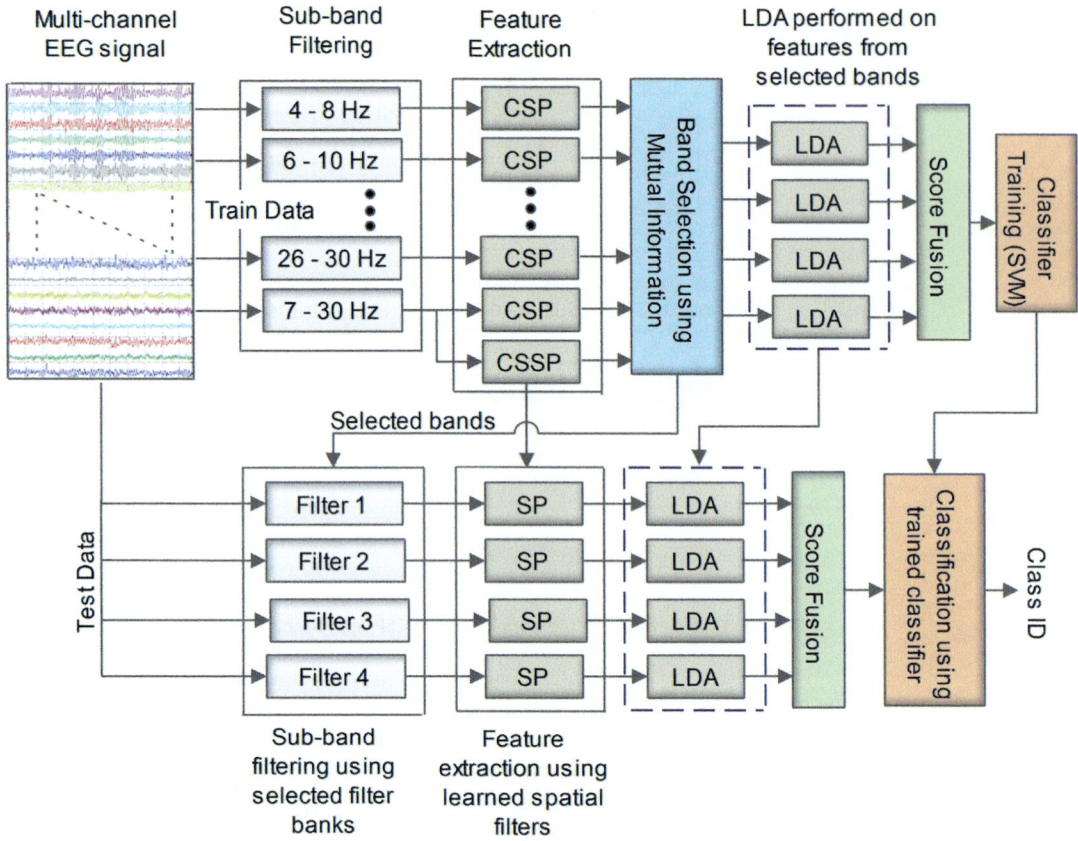

FIGURE 13.3 General framework of the iDFBCSP approach [15,24].

selection. Future works may include further fine tuning the parameters of the selected subbands and employing an adaptive method for the selection of the subbands.

4.2.3 Optimizing temporal filter parameters

The range of the frequency band of the EEG signal that contains the most discriminative information varies across different subjects and is also influenced by their varying psychological states. The complex nature of the EEG signal and its nonstationarity add to the problem and confine the performance of the system due to the filtering of the EEG signal in the time domain. Using an EEG signal that is not filtered or is filtered using inappropriate frequency band parameters will result in suboptimal results. It is of paramount importance that appropriate frequency band or subband parameters are used. This will ensure that redundant information is filtered out. Manual tweaking of the filter parameters is a tiresome and time-consuming task. To overcome this challenge, a temporal filter parameter optimization (TFPO) [7] approach has been proposed (Fig. 13.4). The fundamental parameters of a filter are the cutoff frequencies and the

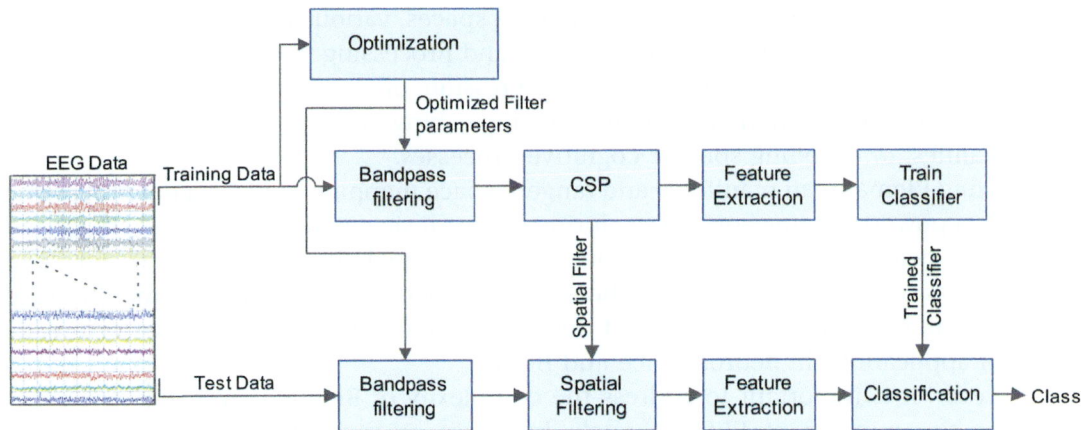

FIGURE 13.4 The general framework for the TFPO approach.

order of the filter. Thus, the TFPO approach aims to optimize these parameters. A single frequency band has been considered to keep the computational complexity of the system low. Increasing the number of frequency bands will increase the computational complexity of the system. The original TFPO approach considered particle swarm optimization (PSO), GA, and artificial bee colony algorithms for the optimization of Butterworth bandpass filter parameters. These optimization algorithms were slightly modified to suit the given TFPO challenge. It is recommended to use GA for optimization as it performs well in terms of the time taken to find the optimal parameters. However, any appropriate filter and optimization algorithm can be utilized for the TFPO approach. Moreover, the current TFPO approach utilizes the CSP algorithm for feature extraction; however, any other feature or feature extraction algorithm can also be used. The TFPO approach outperformed all the other filter-based approaches under the same experimental conditions, utilizing only a single fine-tuned frequency band. A detailed explanation of the TFPO approach can be found here [7].

4.3   Riemannian manifolds and tangent space mapping

Riemannian manifolds and tangent space mapping offer valuable tools for analyzing and understanding the underlying structure of EEG signals. By representing the space of covariance matrices derived from EEG signals on a Riemannian manifold, we can exploit its geometric properties to analyze EEG data.

The tangent space mapping is crucial in this context as it allows us to map the covariance matrices onto the tangent spaces associated with each point on the manifold. These tangent spaces represent the space of all possible variations or directions that can be taken from that point on the manifold, effectively transforming high-dimensional EEG data into a lower-dimensional space while preserving essential information.

Once the EEG data is mapped to the tangent spaces, various mathematical tools and algorithms can be applied for further analysis and processing. For instance, techniques such as tangent space-based classification, regression, or clustering can be employed to perform pattern recognition tasks such as distinguishing different brain states, detecting abnormalities, or decoding specific cognitive processes.

Utilizing Riemannian manifolds and tangent space mapping in EEG signal processing provides a powerful framework for exploring the intrinsic geometry of brain activity and extracting meaningful information from the EEG signal. It enables researchers and practitioners to leverage the rich mathematical properties of Riemannian manifolds to gain insights into brain dynamics and develop advanced analysis methods for a wide range of applications in neuroscience and BCIs.

However, it is important to address the complexity of Riemannian geometry, especially as the number of EEG channels increases. A mechanism that reduces the dimension of the input signal without losing crucial information is recommended to minimize the computation burden. A preprocessing method like CSP or CSP-TSM can be employed before using Riemannian manifolds and tangent space mapping, resulting in further improvement in the overall system's performance. This will ensure that the computational complexity of Riemannian manifolds and tangent space mapping is kept to a minimum. Fusion of the CSP variance-based features with the features obtained using Riemannian manifold and tangent space mapping results in further improvement in the performance of the overall system compared to the individual performance of the CSP approach and the Riemannian manifold and tangent space mapping approach. A detailed explanation of the implementation of this approach can be found here [26].

To enhance the performance of the CSP-TSM approach, a novel spatial-frequency-temporal feature extraction method (called the SPECTRA predictor) [27] incorporating the principles of the CSP-TSM approach has been developed. The fundamental conceptual framework of the SPECTRA predictor is illustrated in Fig. 13.5. This approach employs multiple temporal delayed windows, denoted as n, in two distinct configurations. Firstly, we compute CSP variance-based features and TSM features separately for each of the $n = 3$ windows. The number of windows used has been selected via experiments carried out using different values of n; however, any suitable number of windows can be used depending on the specific applications and limitations.

Secondly, we incorporate the CSSP (CSP with spatial shifting) method to extract additional information. The CSSP method involves the insertion of a temporal delayed window into the original signal and then performing CSP on this modified signal. This method was proposed to improve the overall performance of CSP. The selection of the time delay value τ has a significant impact on the system's performance, and careful consideration should be given to the selection of a suitable value for τ. An evaluation of four distinct feature selection algorithms, namely Lasso, sparse Bayesian learning, mutual information, and F-score-based feature selection algorithms, has been carried out to identify the best-performing feature selection algorithm. The F-score method consistently yielded the lowest misclassification rates, indicating its robustness and

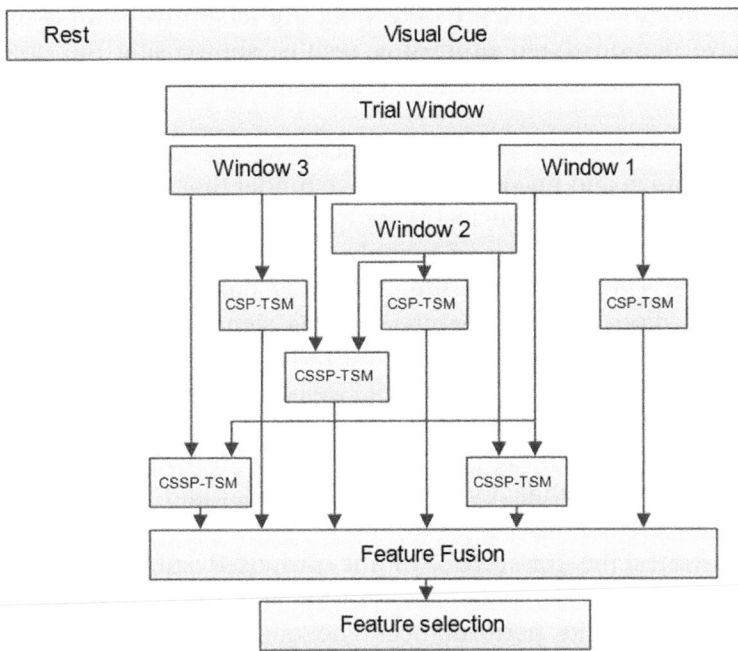

FIGURE 13.5 The general framework for the SPECTRA predictor [24,27].

reliability as a feature selection technique. However, the flexibility of the SPECTRA approach allows for the incorporation of various feature selection methods to suit specific applications and datasets.

In conclusion, Riemannian manifolds and tangent space mapping provide a powerful and flexible approach for EEG signal analysis, and their integration with CSP and novel feature extraction methods like the SPECTRA predictor shows promising results for enhancing the performance and efficiency of EEG-based applications.

4.4   Deep learning approaches for EEG signal processing

Over the years, traditional signal processing techniques have been employed to analyze EEG data and extract relevant features. However, with the advent of deep learning, there has been a paradigm shift in EEG signal processing. Deep learning approaches, particularly ANNs, have demonstrated remarkable success in a wide range of tasks across various domains.

Deep learning models possess the ability to automatically learn hierarchical representations from raw EEG data, removing the need for manual feature engineering. CNNs, RNNs, and their variants have been adapted and applied to EEG signal processing tasks. CNNs can capture spatial patterns and temporal dependencies in EEG signals, while RNNs excel at handling sequential information and dynamic patterns.

Deep learning approaches for EEG signal processing have led to significant advancements in tasks such as BCIs for MI recognition [5], emotion recognition [28,29], sleep stage classification [30], seizure detection, and cognitive state assessment. These applications have demonstrated promising results, showcasing the potential of deep learning to improve accuracy, robustness, and adaptability in EEG-based systems.

4.4.1 Optimized CSP and LSTM based predictor (OPTICAL)

Optical is an innovative and powerful predictive model that brings together two distinct methodologies: The CSP and the LSTM neural networks. This fusion aims to tackle complex challenges in the analysis of spatial and sequential data with remarkable precision and efficacy. The CSP algorithm, originally employed in BCIs and EEG data analysis, enhances discriminative features in EEG signals, allowing for effective brain activity classification. On the other hand, the LSTM network excels at processing and learning from sequential data, making it well-suited for time-series analysis. By combining the spatial filtering capabilities of CSP with LSTM's ability to capture long-range dependencies, the resulting predictor promises to revolutionize various applications, including BCIs and other domains where spatial and temporal information are crucial for accurate predictions.

Fig. 13.6 illustrates the framework of the proposed subject-dependent OPTICAL predictor [31], which combines the use of CSP and LSTM networks with Bayesian optimization to enhance its performance. The name OPTICAL is derived from this fusion. In the depicted framework, two sets of CSP spatial filters are learned by the predictor. The first set is directly acquired from the training data's trials after temporal filtering. Subsequently, CSP variance-based features are extracted from these spatially filtered data, followed by the application of LDA to obtain features in a reduced-dimensional plane. The second set of CSP spatial filters is learned from the combined data obtained after segmenting each of the trials from the training data, as represented in Fig. 13.7. This combined approach promises to yield improved predictive capabilities, enhancing the model's ability to effectively process spatial and sequential information

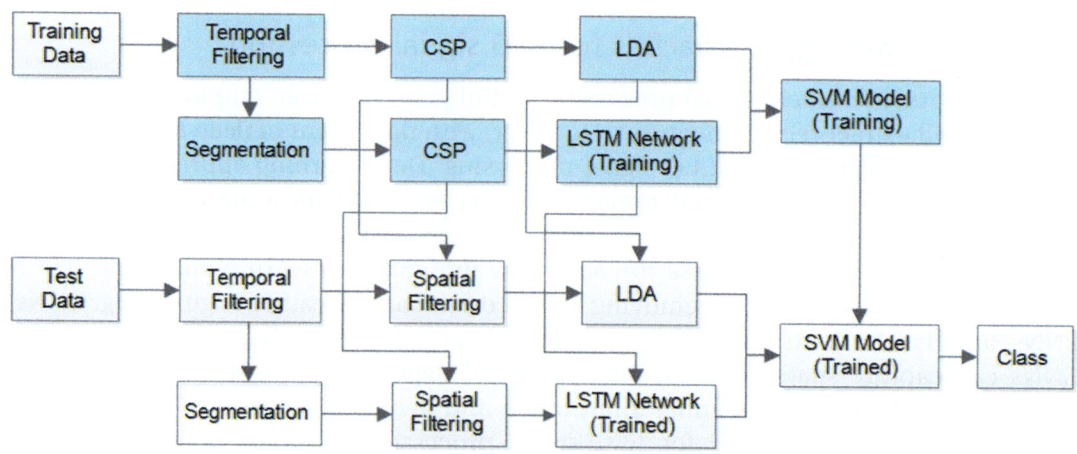

FIGURE 13.6 The framework of the OPTICAL predictor [24,31].

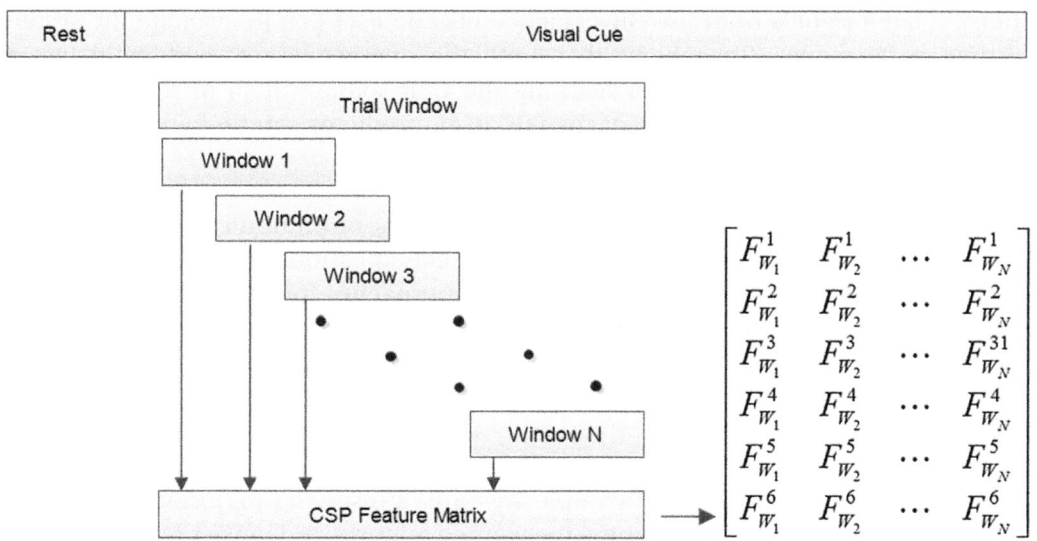

$$
\begin{bmatrix}
F_{W_1}^{1} & F_{W_2}^{1} & \cdots & F_{W_N}^{1} \\
F_{W_1}^{2} & F_{W_2}^{2} & \cdots & F_{W_N}^{2} \\
F_{W_1}^{3} & F_{W_2}^{3} & \cdots & F_{W_N}^{31} \\
F_{W_1}^{4} & F_{W_2}^{4} & \cdots & F_{W_N}^{4} \\
F_{W_1}^{5} & F_{W_2}^{5} & \cdots & F_{W_N}^{5} \\
F_{W_1}^{6} & F_{W_2}^{6} & \cdots & F_{W_N}^{6}
\end{bmatrix}
$$

FIGURE 13.7 Performing segmentation and obtaining the feature matrix [24,31].

for subject-dependent predictions. Further improvement in the performance of OPTICAL and OPTICAL + [32] has been proposed, which employs the TFPO approach.

To derive the second set of CSP spatial filters, the trial data is divided into smaller segments by employing a window of length l sample points with an overlap of t sample points between successive windows. As a result, N segments are obtained from each trial. These N segments, derived from the training trials, are utilized to learn the second set of CSP spatial filters. Subsequently, all segments undergo spatial filtering using this newly acquired set of spatial filters. From each segment in a single trial, variance-based features are computed, generating a feature matrix as depicted in Fig. 13.7. In the feature matrix, each entry represents the i-th feature obtained from the j-th windowed segment of the corresponding trial. This process is repeated for all trials, resulting in the formation of a feature matrix encompassing all trial data, which then serves as the input to the LSTM network. This sequential approach facilitates effective processing of spatial and temporal information, ultimately enhancing the predictor's performance in subject-dependent predictions.

The performance of the LSTM network is influenced by several hyper-parameters, including network size, initial learning rate, learning rate schedule (with parameters like learn rate drop factor and learn rate drop period), momentum, and L2 regularization. To determine the values for the hyper-parameters, a Bayesian optimization technique has been employed.

An essential requirement for enhancing BCI systems is the capability to implement the method effectively in real-time, which has proven to be a significant challenge. The OPTICAL approach addressed this concern by incorporating the rest-state and other nontask-related signals from the Giga Science EEG dataset [33] to evaluate the system's real-time performance. To achieve this, both the one-versus-rest and multiclass approaches have been employed for learning the CSP spatial filters in real-time implementation. A detailed explanation of the OPTICAL predictor can be found here [31].

4.4.2 Convolutional neural network approaches

CNNs have shown promising results in various fields of computer vision and image recognition, and their application to EEG signal classification [34–38] has also yielded significant advancements in BCI systems. CNN approaches for EEG signal classification have gained attention due to their ability to automatically learn relevant spatial and temporal features from EEG data, reducing the need for handcrafted feature engineering.

The key components of CNNs include convolutional layers, pooling layers, and fully connected layers. Convolutional layers apply learnable filters to extract local patterns and spatial features from the EEG signal. Pooling layers down sample the extracted features to reduce computational complexity while preserving important information. Finally, fully connected layers combine the learned features and make predictions for the given EEG signal's class label.

One common approach for EEG signal classification using CNNs is to treat the EEG data as a 2D image, where the electrodes' spatial distribution represents the image's width and height, and the EEG signal's temporal dynamics correspond to different image channels. The CNN architecture then processes this "EEG image" to extract discriminative features for classification.

Data augmentation is often used to increase the size of the training dataset and improve the generalization of CNN models. Augmentation techniques, such as random cropping, rotation, and flipping, help CNN learn more robust features and enhance its performance on unseen EEG signals.

To address the issue of limited labeled EEG data, transfer learning can be employed by pretraining the CNN on a large dataset from a related domain (e.g., natural images) and then fine-tuning it on the EEG dataset. This process leverages the learned low-level features from the pretraining to expedite convergence and enhance the performance of the CNN on EEG classification tasks.

Ensemble methods, such as combining multiple CNN models or incorporating other classifiers like SVMs with CNNs, have also been explored to boost the overall classification accuracy and improve the robustness of the EEG signal classification system.

Despite their success, CNN approaches for EEG signal classification come with some challenges. EEG data is often characterized by high dimensionality and temporal complexity, requiring careful selection of CNN architectures and hyperparameter tuning to achieve optimal results. Additionally, the interpretability of the learned features in CNNs remains a concern, as understanding how the network arrives at its decisions is crucial in many BCI applications.

In conclusion, CNN approaches for EEG signal classification offer a powerful and automated method to learn discriminative features directly from the EEG data. By addressing challenges such as data augmentation, transfer learning, and ensemble techniques, CNNs hold great potential for further advancing BCI systems and enhancing the accuracy and real-world applicability of EEG signal classification tasks.

4.5   Datasets, available packages, and performance comparison

Numerous approaches have been proposed for the classification of EEG signals for different applications. One of the drawbacks of being able to compare a newly proposed method with existing approaches is the datasets that are used for evaluation and the availability of the codes. Some approaches have been evaluated using datasets that are not publicly available. On the other hand, code for most of the methods is also not available. This makes it difficult to test the existing methods on a new publicly available dataset. Moreover, the details on the exact preprocessing methods used are not always explicitly explained, which makes it even more difficult for comparison. It is recommended that the code for the proposed methods be made available publicly. This will result in greater visibility of the work and pave the way forward for other researchers as well. Table 13.1 presents some of the commonly used datasets for EEG signals that are publicly available, while some of the methods whose codes are available online are given in Table 13.2.

In Table 13.3 we present the performance of some of the EEG signal classification techniques. Due to the large number of existing techniques and the fact that these techniques have been evaluated using different datasets, we only present selected methods that we were able to implement from scratch. These methods have been evaluated on BCI Competition IV Dataset 1 and Giga Science EEG datasets using a 10-fold cross-validation scheme. It can be seen from Table 13.2 that the DFBCSP approach performed well in comparison to the CSP approach on the BCI Competition IV Dataset 1; however, it did not generalize well when evaluated using a larger dataset, the Giga Science EEG dataset. Similarly, a lot of methods are proposed and have been evaluated using small datasets. Using a small dataset to evaluate a method is not recommended, as it does not guarantee the robustness of the system. Moreover, here we only present the accuracy of the various approaches; however, it is recommended that at least some other performance measures such as the kappa coefficient value, specificity, sensitivity, and area under the curve be also used.

4.6   Applications

The use of EEG is growing in various fields, including clinical and basic research, due to its effectiveness and affordability in studying and monitoring brain functions. Recent technological advancements, such as upgrading hardware and introducing dry electrodes, have improved data quality and reduced preparation time for EEG analysis. This has led to enhanced performance in classifying EEG signals. Additionally, combining EEG with other imaging techniques, such as neurostimulation or robotics, has sparked interest in emerging research areas. BCI has found numerous applications, and the following paragraphs provide a brief overview of some common uses.

Table 13.1   Commonly used publicly available EEG datasets.

Dataset name	No. of classes	No. of subjects	Dataset description
BCI competition III dataset IVa	2	5	Motor imagery signals Right hand and right foot Acquired using 118 EEG channels at 280 trials per subject https://www.bbci.de/competition/iii/desc_IVa.html
BCI competition IV dataset 1	2	7	Motor imagery signals 2 class from left hand, right hand and foot Acquired using 59 EEG channels at 1000 Hz 2 sessions/subject, 288 trials/session https://www.bbci.de/competition/iv/
BCI competition IV dataset 2a	4	9	Motor imagery signals Left hand, right hand, foot and tongue Acquired using 22 EEG channels at 250 Hz 2 sessions/subject, 288 trials/session https://www.bbci.de/competition/iv/
BCI competition IV dataset 2b	2	9	Motor imagery signals Left hand, right hand Acquired using 3 EEG channels at 250 Hz 2 sessions/subject, 288 trials/session https://www.bbci.de/competition/iv/
EEG datasets for MI BCI (Giga science database)	5	52	Motor imagery signals Left hand, right hand, real hand movement, rest and noise Acquired using 64 EEG channels at 512 Hz 100 or 120 trials/class (for left and right hand) http://gigadb.org/dataset/100295
Physionet MI-EEG database	2	109	Motor imagery signals Left fist/both fists, and right fist/both feet Acquired using 64 EEG channels at 160 Hz 14 experimental runs/subject https://physionet.org/content/eegmmidb/1.0.0/
DEAP dataset	4	32	Emotion analysis Valence, arousal, dominance, and liking Acquired using 32 EEG channels, 12 peripheral channels at 512 Hz 14 experimental runs/subject https://www.eecs.qmul.ac.uk/mmv/datasets/deap/readme.html

BCI applications can be broadly categorized into two main groups: those involving control of external devices, such as assistive devices, and those focused on monitoring brain activities for medical or other purposes, such as detecting diseases or emotions. Examples of BCI systems based on MI-EEG for controlling external devices include gaming [39], entertainment, wheelchair control [40] and robot control [41,42] for individuals with disabilities, as well as neurorehabilitation [43]

Table 13.2   List of EEG classification methods with codes/packages available online.

Method	Description
EEGLearn	A set of functions for supervised feature learning/classification of mental states from EEG based on "EEG images". Python code, MNE-Python Link: https://github.com/pbashivan/EEGLearn
Braindecode	Braindecode is an open-source Python toolbox for decoding raw electrophysiological brain data with deep learning models. Python code Link: https://github.com/braindecode/braindecode
EEGNet	A collection of convolutional neural network (CNN) models for EEG signal processing and classification, written in Keras and Tensorflow Python code Link: https://github.com/vlawhern/arl-eegmodels
BrainFlow	BrainFlow is a library intended to obtain, parse and analyze EEG, EMG, ECG, and other kinds of data from biosensors Python code Link: https://github.com/brainflow-dev/brainflow
TFPO-CSP	A Matlab toolbox for MI EEG signal classification based on temporal filter parameter optimization technique. Matlab code Link: https://github.com/ShiuKumar/TFPO-CSP
OPTICAL	A Matlab toolbox for MI EEG signal classification based on CSP and LSTM network. Matlab code Link: https://github.com/ShiuKumar/OPTICAL
OPTICAL+	A Matlab toolbox for frequency-based Approach using long short-term memory network (LSTM) for recognizing MI EEG signals Matlab code Link: https://github.com/ShiuKumar/OPTICAL_plus

Table 13.3   Performance comparison of different methods.

Method	Dataset	Accuracy (%)
CSP	BCI competition IV dataset 1	75.76
	Giga science EEG dataset	65.93
FBCSP	BCI competition IV dataset 1	76.83
DFBCSP	BCI competition IV dataset 1	77.14
	Giga science EEG dataset	61.54
iDFBCSP	BCI competition IV dataset 1	78.44
TFPO-CSP	BCI competition IV dataset 1	81.28
	Giga science EEG dataset	67.24
CSP-TSM	BCI competition IV dataset 1	81.06
	Giga science EEG dataset	64.98
SPECTRA	BCI competition IV dataset 1	82.10
OPTICAL	BCI competition IV dataset 1	82.52
	Giga science EEG dataset	68.19
OPTICAL+	Giga science EEG dataset	69.59

and stroke rehabilitation [44]. On the other hand, MI-EEG-based BCI systems have been utilized for medical or other reasons, such as recognizing emotions [28] or stress levels, identifying different stages of sleep [45,46], detecting and diagnosing seizures or epilepsy [19,47], and exploring biometric identification [48,49]. However, BCI systems for person recognition or identification face challenges related to temporal stability, protocol design, psychological and physiological changes, equipment performance, and commercial viability.

Epilepsy, a neurological disorder affecting around 50 million people worldwide, according to the World Health Organization, can be monitored using EEG recordings. However, manual analysis of these recordings to detect epileptic seizure activities is a time-consuming task for experts. Additionally, seizures can lead to life-threatening accidents, making it crucial to predict and reduce the risk of such events. Continuous monitoring of individuals during a seizure attack is impractical, highlighting the need for EEG signal classification methods for seizure and preseizure detection. These methods inform users in advance, allowing them to take precautions, seek help, and prevent accidents.

Another significant application of EEG-based BCI systems is sleep stage classification [45,50]. Sleep is essential for maintaining physical and mental health, but sleep-related disorders like insomnia, obstructive sleep apnea, and narcolepsy can impact individuals' quality of life and overall well-being. Accurate identification of different sleep stages aids in diagnosing these disorders. EEG signals obtained from polysomnographic recordings play a crucial role in clinical sleep stage classification, making EEG signal classification widely employed in sleep stage classification applications.

Stroke, a leading cause of disability in the United States, according to the American Stroke Association, results from blocked or ruptured blood vessels, leading to oxygen and nutrient deprivation in parts of the brain. Stroke rehabilitation helps individuals relearn lost skills caused by brain damage. EEG-based BCI systems have been increasingly investigated for stroke and neurorehabilitation, with commercial systems like Neurostyle Brain Exercise Therapy Towards Enhanced Recovery utilizing EEG signals. Furthermore, controlling wheelchairs and other assistive devices through EEG signals benefits individuals with functional brains but impaired muscle control, enabling them to live independently. Thus, the wide range of applications underscores the importance of EEG signals in various domains.

5. Conclusion and future perspectives

The remarkable advances in noninvasive EEG-based BCIs represent a significant leap forward in neurotechnology. From traditional wet electrodes to cutting-edge dry electrode technologies, signal acquisition techniques have evolved, improving user comfort and signal quality. Additionally, progress in EEG signal processing, including artifact removal, noise reduction, and feature extraction methods, has contributed to improved accuracy and reliability in decoding brain signals for BCI applications.

The emergence of machine learning algorithms and brain connectivity analysis holds great promise for further enhancing the performance of EEG-based BCIs. The fusion of EEG with other modalities in hybrid BCIs opens exciting possibilities for personalized and context-aware interactions between humans and machines.

The diverse applications of EEG-based BCIs, spanning medical rehabilitation, communication aids, and entertainment, demonstrate the versatility and potential impact of this technology in various domains. Already, these applications have significantly improved the quality of life for individuals with motor impairments, opening up new avenues for human-computer interaction.

Despite these accomplishments, challenges persist, such as intersubject variability, signal ambiguity, and the need for compact and practical BCI systems suitable for daily use. Addressing these challenges will require ongoing collaborative efforts among researchers, engineers, and clinicians, drawing on interdisciplinary expertise to drive the field forward.

Looking ahead, the future of noninvasive EEG-based BCIs is promising. As technology continues to advance, we can expect more robust, adaptive, and user-friendly BCI systems that seamlessly integrate into our daily lives. Ongoing research exploring novel signal acquisition methods, advanced signal processing algorithms, and more sophisticated machine learning models will play a pivotal role in realizing the full potential of EEG-based BCIs.

In conclusion, the continuous progress in noninvasive EEG-based BCIs presents exciting possibilities for the future, where direct communication between the human brain and external devices becomes more seamless, natural, and widespread. With the potential to revolutionize healthcare, communication, and human-machine interaction, EEG-based BCIs hold the key to empowering individuals and transforming society in ways we are only beginning to imagine. By striving for innovation, collaboration, and cross-disciplinary research, we can drive this promising technology toward a future of remarkable advancements and unprecedented possibilities.

References

[1] Mao WL, et al. EEG dataset classification using CNN method. J Phys Conf Series 2020;1456(1): 012017.

[2] Gangapuram H, Manian V. A sparse multiclass motor imagery EEG classification using 1D-ConvResNet. Signals 2023;4(1):235–50.

[3] Liu S, et al. Subject adaptation convolutional neural network for EEG-based motor imagery classification. J Neural Eng 2022;19(6):066003.

[4] Chakravarthi B, et al. EEG-based emotion recognition using hybrid CNN and LSTM classification, vol 16; 2022.

[5] Wahdow M, et al. Multi frequency band fusion method for EEG signal classification. Sig Imag Video Process 2023;17(5):1883–7.

[6] Kumar S, Sharma A, Tsunoda T. Subject-specific-frequency-band for motor imagery EEG signal recognition based on common spatial spectral pattern. In: Lecture notes in artificial intelligence: sub-series of lecture notes in computer science; 2019.

[7] Kumar S, Sharma A. A new parameter tuning approach for enhanced motor imagery EEG signal classification. Med Biol Eng Comput 2018;56(10):1861–74.

[8] Hasan MJ, et al. Sleep state classification using power spectral density and residual neural network with multichannel EEG signals. Appl Sci 2020;10(21):7639.

[9] Ali Mohammad A, Shoroq Q, Yusra MO. A novel moving window-based power spectrum features for single-channel EEG classification using machine learning. Acta Sci Technol 2022;45(1).

[10] Aliyu I, Lim CG. Selection of optimal wavelet features for epileptic EEG signal classification with LSTM. Neural Comput Appl 2023;35(2):1077–97.

[11] Furman Ł, et al. Short-time Fourier transform and embedding method for recurrence quantification analysis of EEG time series. Eur Phys J Spec Top 2023;232(1):135–49.

[12] Kant P, et al. CWT based transfer learning for motor imagery classification for brain computer interfaces. J Neurosci Methods 2020;345:108886.

[13] Lu X, et al. Study on characteristic of epileptic multi-electroencephalograph base on Hilbert-Huang transform and brain network dynamics, vol 17; 2023.

[14] Vicchietti ML, et al. Computational methods of EEG signals analysis for Alzheimer's disease classification. Sci Rep 2023;13(1):8184.

[15] Kumar S, Sharma A, Tsunoda T. An improved discriminative filter bank selection approach for motor imagery EEG signal classification using mutual information. BMC Bioinform 2017;18(16):545.

[16] Pawan, Dhiman R. Electroencephalogram channel selection based on pearson correlation coefficient for motor imagery-brain-computer interface. Meas Sensor 2023;25:100616.

[17] Zhang W, Yin Z. EEG feature selection for emotion recognition based on cross-subject recursive feature elimination. In: 2020 39th Chinese control conference (CCC); 2020.

[18] Wang M, et al. Motor imagery classification method based on relative wavelet packet entropy brain network and improved lasso, vol 17; 2023.

[19] Zeng W, et al. Epileptic seizure detection with deep EEG features by convolutional neural network and shallow classifiers, vol 17; 2023.

[20] Kumar S, et al. Decimation filter with common spatial pattern and Fishers discriminant analysis for motor imagery classification. In: 2016 international joint conference on neural networks (IJCNN); 2016 [Vancouver, Canada].

[21] Novi Q, et al. Sub-band common spatial pattern (SBCSP) for brain-computer interface. In: 3rd international IEEE/EMBS conference on neural engineering; 2007.

[22] Ang KK, et al. Filter Bank common spatial pattern (FBCSP) in brain-computer interface. In: IEEE international joint conference on neural networks. Hong Kong: IEEE World Congress on Computational Intelligence; 2008.

[23] Thomas KP, et al. A new discriminative common spatial pattern method for motor imagery brain computer interfaces. IEEE Trans Biomed Eng 2009;56(11):2730–3.

[24] Attribution 4.0 International (CC BY 4.0). [cited 2023 31 July]; Available from: https://creativecommons.org/licenses/by/4.0/.

[25] Wei Q, Wei Z. Binary particle swarm optimization for frequency band selection in motor imagery based brain-computer interfaces. Bio-Med Mat Eng 2015;26(s1):S1523–32.

[26] Kumar S, Mamun K, Sharma A. CSP-TSM: optimizing the performance of Riemannian tangent space mapping using common spatial pattern for MI-BCI. Comput Biol Med 2017;91(Suppl. C): 231−42.

[27] Kumar S, Tsunoda T, Sharma A. SPECTRA: a tool for enhanced brain wave signal recognition. BMC Bioinform 2021;22(6):195.

[28] Tang Y, et al. STILN: a novel spatial-temporal information learning network for EEG-based emotion recognition. Biomed Signal Process Cont 2023;85:104999.

[29] Yin Z, et al. Locally robust EEG feature selection for individual-independent emotion recognition. Exp Syst Appl 2020;162:113768.

[30] Xu Z, et al. Sleep stage classification using time-frequency spectra from consecutive multi-time points. Front Neurosci 2020;14(14).

[31] Kumar S, Sharma A, Tsunoda T. Brain wave classification using long short-term memory network based OPTICAL predictor. Scie Rep 2019;9(1):9153.

[32] Kumar S, Sharma R, Sharma A. OPTICAL+: a frequency-based deep learning scheme for recognizing brain wave signals. PeerJ Comput Sci 2021;7:e375.

[33] Cho H, et al. EEG datasets for motor imagery brain−computer interface. GigaScience 2017;6(7).

[34] Liu T, Yang D. A three-branch 3D convolutional neural network for EEG-based different hand movement stages classification. Sci Rep 2021;11(1):10758.

[35] Lun X, et al. A simplified CNN classification method for MI-EEG via the electrode pairs signals, vol 14; 2020.

[36] Alnaanah M, Wahdow M, Alrashdan M. CNN models for EEG motor imagery signal classification. Signal Imag Video Process 2023;17(3):825−30.

[37] Li H, et al. Motor imagery EEG classification algorithm based on CNN-LSTM feature fusion network. Biomed Signal Process Cont 2022;72:103342.

[38] Chen L, Yu Z, Yang J. SPD-CNN: a plain CNN-based model using the symmetric positive definite matrices for cross-subject EEG classification with meta-transfer-learning, vol 16; 2022.

[39] Amin M, et al. Leveraging brain−computer interface for implementation of a bio-sensor controlled game for attention deficit people. Comput Elec Eng 2022;102:108277.

[40] Ashok Kumar C, et al. EEG control of a robotic wheelchair. In: Ramana V, editor. Human-robot interaction. Rijeka: IntechOpen; 2023. Ch. 2.

[41] Guo R, et al. A robotic arm control system with simultaneous and sequential modes combining eye-tracking with steady-state visual evoked potential in virtual reality environment, vol 17; 2023.

[42] Korovesis N, et al. Robot motion control via an EEG-based brain−computer interface by using neural networks and alpha brainwaves 2019;8(12):1387.

[43] Karácsony T, et al. Brain computer interface for neuro-rehabilitation with deep learning classification and virtual reality feedback. In: Proceedings of the 10th augmented human international conference 2019. Reims, France: Association for Computing Machinery; 2019. Article 22.

[44] Al-Qazzaz NK, et al. EEG signal complexity measurements to enhance BCI-based stroke patients' rehabilitation, vol 23; 2023. p. 3889.

[45] Al-Salman W, et al. Sleep stage classification in EEG signals using the clustering approach based probability distribution features coupled with classification algorithms. Neurosci Res 2023;188: 51−67.

[46] Hasan MN, Koo I. Mixed-input deep learning approach to sleep/wake state classification by using EEG signals 2023;13(14):2358.

[47] Chen W, et al. An automated detection of epileptic seizures EEG using CNN classifier based on feature fusion with high accuracy. BMC Med Inform Decis Making 2023;23(1):96.

[48] Sayel NA, Albermany S, Sabbar BM. Use multichannel EEG-based biometrics authentication signal in real time using neural network. In: New trends in information and communications technology applications. Cham: Springer Nature Switzerland; 2023.

[49] Maiorana E. Deep learning for EEG-based biometric recognition. Neurocomputing 2020;410: 374−86.

[50] Zhang Z, et al. Sle-CNN: a novel convolutional neural network for sleep stage classification. Neural Comput Appl 2023;35(23):17201−16.

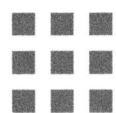

Index

'*Note:* Page numbers followed by "f" indicate figures and "t" indicate tables.'

A

Action potentials, 94
Active hand prosthesis, 256–259
ADANN algorithm, 265–266
Adaptive filtering techniques, 287
Adaptive Moment Optimization Algorithm, 9
Adaptive signal decomposition techniques, 137–138
  empirical mode decomposition (EMD), 138
  univariate, 153–154
  wavelet transform, 137–138
Affine transformations, 76–77
AIIMS Patna Department of Neurology, 180
Alpha waves, 96
Alternating direction method of multipliers (ADMMs) algorithm, 150
Alzheimer's disease neuroimaging initiative (ADNI) dataset, 181
American Epilepsy Society's epileptic seizure detection challenge, 179–180
Ant colony optimization (ACO), 120–123
Arousal symptoms, posttraumatic stress disorder (PTSD), 38–39
Artifact subspace reconstruction (ASR), 287–288
Artificial neural networks (ANNs), 71–72
  activation functions, 101
  backpropagation, 102
  brain signals decoding, 291
  hidden layers, 101
  input layer, 100
  loss functions, 101–102
  output layer, 101
  structure, 100f
Attention deficit hyperactivity disorder (ADHD)
  feature extraction
    data epochs, 111–112
    electroencephalography (EEG) signals, 111
  hyperactivity, 109
  impulsivity, 109
  inattention, 109
  model architecture
    convolutional neural network (CNN), 112
    modified ResNet neural network, 112–113, 113f–114f
  study population and dataset, 110–111
Attention-guided convolutional neural network (CNN) (AG-CNN), brain tumor (BT) recognition, 120–123
Attention level data analysis, disabled people
  normalized gamma band signals, 205–206, 206f
  normalized high beta band signals, 205, 206f
  normalized low beta band signals, 205, 205f
  raw electroencephalography (EEG) bands, 204–205, 204f
Attention U-Net model, 17
Averaging and epoching techniques, 287

B

Backpropagation, 102
Bern dataset, 179
Beta waves, 96
Binary particle swarm optimization (BPSO) approach, 295–296
Bionic hand, 257–258
Blind source separation (BSS), 220

Blinking data analysis
  double-blink detection algorithm, 208
  frequency bands, 208, 208f
  raw electroencephalography (EEG) bands,
    207—208, 207f
Bluetooth, 197
Bluetooth low energy (BLE), 197
Bonn University EEG database, 179
Brain—computer interface (BCI), 71,
    155—156, 262—263
  application interface, 195
  applications, 195, 304—306
  decision tree, 195—196
  external sensors, 195—196
  extraction techniques, 195—196
  feedback mechanism, 195
  neural signal processing, 92
  signal acquisition and processing, 195
BrainRobotics, 257—258
Brain tumors (BTs)
  clinical manifestations, 1
  deep learning-based techniques, 120—123
  genetic conditions/factors, 1—2
  genetic symptoms, 2
  grades, 119t
  imaging methods, 2
  learning and optimization experiments,
    22—27
  magnetic resonance imaging (MRI),
    119—120
  metastatic, 1—2
  prevalence, 1, 2f
  primary, 1—2
  principal component analysis (PCA),
    120—123
  random hyperparameters experiments,
    21—22
  risk factors, 1—2
  segmentation and detection, 3
    automated methods, 3, 5, 7
    convolutional neural network (CNN),
      7—8
    data augmentation, 9—10
    datasets acquisition, 14—15
    datasets preprocessing, 15—16

    deep learning (DL) method, 3
    experiments and results, 21, 21t
    feature extraction, 4
    fuzzy cognitive maps, 4
    machine learning (ML), 3—5
    manual segmentation, 6
    metaheuristic optimization, 10—12
    MRI image features, 5
    parameters optimization, 9
    performance metrics, 12—14
    phase, 16—17
    semiautomatic methods, 6
    transfer learning (TL), 9
  semisupervised generative adversarial
    network (SSGAN)
    dataset, 128, 129f
    performance evaluation, 130—131
    preprocessing, 125—127, 129
  specific symptoms, 2
  suggested technique, 132t
  transfer learning (TL), 120

C
Canonical correlation analysis (CCA), 54
Children's Hospital BostoneMassachusetts
    Institute of Technology (CHBMIT)
    dataset, 178—179
Closed-loop deep brain stimulation, seizure
    control, 189
Cognitive disorder. *See also* Attention
    deficit hyperactivity disorder (ADHD)
  deep learning, 115
  and neural signals, 115
Cognitive systems, posttraumatic stress
    disorder (PTSD), 37—38
Common spatial pattern (CSP), 289,
    292—293
Computer vision and camera sensors, 273
Computer vision techniques, 265
Connectivity analysis, neural signals, 97—98
Connectivity metrics, 286
Convolutional neural network (CNN), 3,
    76—77, 265—266
  artifact recognition
    architecture, 59—60, 60f

computational performance, 64
data filtering and augmentation, 60–61
data set distribution and training, 61–62, 62t
electroencephalography (EEG) artifacts, 57–59
experimental data set, 60
recognition errors, 64t
topoplots, 62–63, 62t
trustful artifact removal system, 63, 63f
attention deficit hyperactivity disorder (ADHD), 112
brain tumor (BT) recognition, 120–123
components, 302
computer vision applications, 7
convolution and pooling layers, 7–8
COVID-19 diagnosis, 102, 102f
data augmentation, 302
electroencephalography (EEG) signal classification, 302
fully connected layer, 7–8
kernels, 8
nonlinearity, 8
Convolutional neural network combined with bidirectional long short-term memory layers (CNNbiLSTM), 58
CyberHand, 261

D
Data-adaptive signal decomposition technique, 138
Decision trees, 185
Deep brain stimulation (DBS), posttraumatic stress disorder (PTSD), 41
Deep convolutional neural networks, 105–106
Deep learning
  algorithms, 92
  electroencephalography (EEG) signal processing
    convolutional neural network (CNN) approaches, 302–303
    Optimized CSP and LSTM based predictor (OPTICAL), 300–302, 300f

epilepsy analysis, 189–190
models, 292
techniques, 265–266
Deep learning-aided medical diagnosis
architectures, 102–103
artificial neural networks (ANNs), 99–102
attention deficit hyperactivity disorder (ADHD), 108–115
cardiology, 93–94
clinical application, 107–108
computer vision, 105
histopathology, 93
natural language processing, 105
neural signals
  deep convolutional neural networks, 105–106
  deep reinforcement learning (DRL) techniques, 107
  feature extraction techniques, 98–99
  generative adversarial networks (GANs), 107
  processing and analytics, 97–98
  recurrent neural networks, 106
  types, 94–97
optimization algorithms, 103
regularization techniques, 105
signal processing, 105
speech recognition, 105
X-ray images segmentation, 93, 93f
Deep reinforcement learning (DRL) techniques, 107
Delta waves, 95
Dice coefficient, 13
Discriminative models, 7
Dravet syndrome, 177
Dry-electrode technology, 284

E
EADC-ADNI HarP dataset, 181
EEGLAB, 57
EEGNet, 227–228
  best ICA algorithms per subject, 242–244, 242t, 243f
  training convergence speed, 244–246, 244t
Electrocardiography (ECG) artifacts, 49

Electroencephalogram (EEG), 94–95
Electroencephalography (EEG)
    applications, 285
    brain signals decoding, 291–292
    classification methods, 305t
    feature extraction, 288–289
    feature selection techniques, 289–290
        filter methods, 289–290
        wrapper methods, 290
    sensors, 271–272
    signal pre-processing methods, 286–288
    signal processing
        brain connectivity and network
            dynamics analysis, 286
        coherence analysis, 286
        connectivity metrics, 286
        feature extraction, 285–286
        graph theory, 286
        noise reduction, 285
Electroencephalography (EEG)-based brain-
        computer interface (BCI)
    advantages, 281–282
    applications, 197
    automated unsupervised algorithm, 196
    brainwave and head motion control
        android application development,
            200–201, 204f
        appliances control, 210–212
        attention level data analysis, 204–207
        blinking data analysis, 207–208,
            207f–208f
        control circuit development, 201
        electroencephalography (EEG) signal
            extraction, 198–200
        electroencephalography (EEG) signal
            processing, 202
        EMOTIV Insight EEG headset, 198
        motion data analysis, 209–210, 209f
        subsystems integration, 203, 203f
        user authentication system, 197
        wheelchair system, 197
        wireless technologies, 197
    common spatial pattern (CSP), 292–293
    deep learning approaches, 299–303
    filter-based approaches, 294–297

framework, 282f
performance comparison, 305t
publicly available electroencephalography
        (EEG) datasets, 303, 304t
Riemannian manifolds, 297–299
signal acquisition, 282
    active electrodes, 284
    dry electrodes, 284
    electrode placement methods, 284–285
    passive electrodes, 284
    traditional wet electrodes, 283
signal processing, 196
sleep stage classification, 306
stroke rehabilitation, 306
tangent space mapping, 297–299
temporal filter parameters, 296–297
Electromyogram (EMG) artifacts, 49
Electrooculogram (EOG), 96–97
Electrooculography (EOG) artifacts, 49
Electro-tactile feedback, 268
Empirical mode decomposition (EMD), 138
    adaptability, 141–142
    algorithm, 142
    intrinsic mode functions (IMFs) extraction,
        142
    local extrema points, 142
    nonlinear and nonstationary signals, 141
Empirical wavelet transform (EWT)
    empirical wavelets, 143
    multivariate EWT (MEWT), 143–145
        arbitrary function, 145
        mean Fourier spectrum, 145
        mean spectrum magnitude, 145
        wavelet and scaling functions, 145–146
    univariate, 143–145
Ensemble methods, 302
Ensemble of deep neural networks
        (ensemble DNN), 58
Epilepsy analysis
    deep learning techniques, 176
    electroencephalography (EEG) datasets,
        178–180, 178t
    feature extraction and selection, 182–183
    financial and emotional burden, 175–176
    highly prevalent condition, 175

machine learning techniques
  algorithm bias and fairness, 187–188
  clinical relevance and validity, 187
  closed-loop deep brain stimulation, 189
  collaboration and domain expertise, 187
  data augmentation and synthetic data
    generation, 189–190
  data collection and integration, 190
  data quality and quantity, 186–187
  deep learning architectures, 189–190
  deep learning-based brain lesion
    segmentation, 189
  drug responsiveness prediction, 188–189
  ethical and privacy concerns, 187
  ethical considerations, 191
  interoperability and data
    standardization, 187
  model interpretability, 187
  overfitting and generalization, 187
  patient-centric approaches, 191
  predictive and classification modeling,
    184–186
  real-time processing, 187
  real-time seizure prediction, 188
  regulatory approval and adoption, 187
  transfer learning, 189–190
  treatment response modeling, 188
model performance, 176
neuroimaging dataset, 181–182, 181f
performance evaluation, 184
predictive model, 183
preprocessing, 176, 182
working pipeline, 176f
Epileptic encephalopathies, 177
Esper Bionics, 257–258
European epilepsy database (EPILEPSIAE),
  180
Event-related desynchronization or
  synchronization (ERD/ERS), 223–224
Externally powered prostheses, 256
Extremely randomized trees, 227

F
False negative rate (FNR), 13
False positive rate (FPR), 13

Federated learning (FL), 163–164
Feed-forward neural network (FFNN), 166,
  167f
FieldTrip, 57
Filter bank CSP (FBCSP), 294, 295f
Fnite-state machine (FSM) control strategy,
  261
Focal (partial) epilepsy, 177
Force and pressure sensors, 272
Four-classes dataset experiments, 22–27,
  24t, 26t
Fourier-Bessel series expansion-based
  empirical wavelet transform (FBSE-
  EWT), 146–148
Fourier transform, 137–138
Fractal analysis, 289
Freiburg electroencephalography (EEG)
  database, 179
Frequency-domain analysis, 97
Frontal lobe epilepsy, 177
F-score, 169–173
Fuzzy inference system (FIS), 124–125

G
Gamma waves, 96
Gaussian Naive Bayes, 226
Generalized epilepsy, 177
Generative adversarial networks (GANs),
  107
Generative models, 7
Genetic algorithms (GAs), 290
Glioma, 1–2
Graphical user interface (GUI), 200
Graph theory, 286
Gray level cooccurrence matrix (GLCM)
  descriptors, 120–123

H
Hand prosthesis
  active, 256–259
  hierarchical control approach
    brain-computer interfaces (BCI),
      262–263
    collaborative control and shared
      autonomy, 263–264

Hand prosthesis (*Continued*)
   human-centric approach, 259
   myoelectric control strategy, 261–262
   sensory feedback, 259, 266–273
   user-friendly prosthesis, 259
   user's intentions (UIs), 260
  passive, 256
  signal processing techniques, 264–266
Hidden Markov models (HMM), 292
Hippocampus, fear conditioning and
   extinction, 37
Human activity recognition (HAR)
  federated learning (FL), 163–164
  split learning (SL), 163–164
   data and preprocessing, 165, 165f
   evaluation setup, 168–169
   feature extraction, 166
   F-Score, 170–173, 171f–173f
   generic architecture, 168f
   metrics and scenarios, 169–170
   model architecture, 166–167
   privacy benefits, 163–164
   setup, 167–168
   system-level aspects, 163–164
   wearable devices, 164–165
  wearable devices, 163
Hybrid prosthetic hand, 257
Hyvarinen's Fixed Point algorithm
   (FastICA), 223

I

Improved discriminative FBCSP (iDFBCSP),
   295–296, 296f
Impulsivity, 109
Independent component analysis (ICA), 287
Independent component analysis (ICA)-
   based artifact removal, 50
  algorithms, 53–54
  artifact recognition
   automatic ICA-based strategies, 57–59
   and elimination, 65f
   software tools, 57
  canonical correlation analysis (CCA), 54
  independent vector snalysis (IVA), 54
  inverse problem, 53

morphological component analysis (MCA),
   54
  multiple observed/underlying signals, 52
  preprocessing pipeline, 51–52, 52f
  recorded signals, 52
  topoplots, 54–56, 62t
Independent vector snalysis (IVA), 54
Infomax, 222
Intersection over Union (IoU), 13
Intrinsic mode functions (IMFs), 141–142

J

Jaccard index (JAC), 13
Joint approximate diagonalization of
   eigenmatrices (JADE), 222
Juvenile myoclonic epilepsy, 177

K

K-nearest neighbors (k-NN) algorithm, 185,
   291

L

Learning-based methods, 7
Lennox-Gastaut syndrome, 177
Linear classifiers, 224–225
Linear discriminant analysis (LDA), 227
Linear transformation, 76–77
Local field potentials (LFPs), 94
Logistic regression, 185, 225–226, 292

M

Machine learning techniques, 4. *See also*
   Semisupervised generative
   adversarial network (SSGAN)
  brain tumor (BT) detection, 3–5
  epilepsy analysis
   clinical trials, 186–189
   closed-loop deep brain stimulation, 189
   deep learning-based brain lesion
    segmentation, 189
   drug responsiveness prediction, 188–189
   predictive and classification modeling,
    184–186
   real-time seizure prediction, 188
   treatment response modeling, 188

Meso-corticolimbic reward pathway, 37
Metaheuristic optimization, 10—12
    algorithms, 10
    hyperparameter, 10
    minimization/maximization, 10—11
    sparrow search algorithm, 11—12
Microstructured tactile sensors, 272—273
Modular Prosthetic Limb (MPL), 259
Morphological component analysis (MCA),
        54
Motion data analysis, disabled people,
        209—210
Motor imagery-based brain-computer
        interfaces (MI-based BCI)
    BCI Competition IV dataset 2a, 219—220,
        229—237
        best classifier per ICA, 233, 233t
        best classifier per subject, 232, 232t
        best combination per subject, 231—232,
            231t
        best ICA and the best classifier, 234—237,
            235f—236f
        best ICA per classifier, 234, 234t
        best ICA per subject, 232, 232t
    classification, 224—227
    EEGNet, 227—228, 242—246
    feature extraction and selection, 223—224
    independent components (ICs), 218
    open-access datasets, 219
    OpenBMI dataset, 220, 237—241
    source extraction
        blind source separation (BSS), 220
        independent component analysis (ICA),
            221—223, 221f
        statistical independence, 220—221
    stages, 218f
Multilayer perceptron classifier (MLP), 226
Multivariate adaptive signal decomposition
        techniques
    amplitude-frequency modulated
        components, 140f
    electroencephalography (EEG) signal
        processing
        brain-computer interface (BCI), 155—156
        neurological disease diagnosis, 156—157

empirical mode decomposition (EMD),
        141—143
empirical wavelet transform (EWT),
        143—146
equal number of modes, 141
four-channel electroencephalography
        (EEG) signals, 141f
Fourier-Bessel series, 146—148
mode alignment, 140
multivariate iterative filtering (MIF),
        151—153
multivariate time series, 138—140
variational mode decomposition (VMD),
        148—151
Multivariate iterative filtering (MIF),
        151—153
Multivariate pattern analysis (MVPA),
        neural signals, 97
Myoelectric control strategy, hand
        prosthesis, 261—262
Myoelectric prosthetic hand, 256—257

N
Naive Bayes, 185, 291
Natural language processing (NLP), 105,
        265
Negative valence system, 35—37
Nested cross-validation strategy (nCV), 80
Neural networks, epilepsy analysis, 186
Neural signals, 91
Neurofeedback (nFb), posttraumatic stress
        disorder (PTSD), 39—40
Neurological disabilities, 91
Neuroprosthetic hand, 258—259
Non-invasive neural speech decoding
    data acquisition and signal processing,
        74—76
    linguistic categories, 73
    multimodal neural network
        batch normalization and dropout, 77—78,
            78f
        convolution, 76—77
        decoding performance, 83
        hyperparameter optimization, 80—81
        implementation, 78—80

Non-invasive neural speech decoding
    (*Continued*)
    non-linear activation, 77
    training procedure, 80
  study population, 73
  unimodal neural network, 80–83
Nonlinear classifiers, 224–225

O
Online recursive ICA (ORICA), 53–54, 223
Open Access Series of Imaging Studies
    (OASIS) dataset, 181–182
Open Bionics Hero Arm, 257–258
OpenBMI dataset
  best classifier per ICA, 237, 238t
  best ICA and best classifier, 238–241,
    239f–240f
  best ICA per classifier, 237–238, 238t
OpenfMRI dataset, epilepsy research, 182
Optimized CSP and LSTM based predictor
    (OPTICAL), 300–302, 300f

P
Parametric Flatten-p Mish (PFpM),
    120–123
Parkinson's disease, 91
Passive hand prosthesis, 256
Peak signal-to-noise ratio (PSNR), 127
Picard, 222
Picard-O, 223
Position and inertial sensors, 273
Positive predictive value (PPV), 13
Positive valence systems, 37
Posttraumatic stress disorder (PTSD)
  basolateral amygdala (BLA) process, 34
  chronic, 33–34
  economic challenge, 33–34
  neurocircuitry, 34, 35f
  neurotechnological strategies, 39–41
  pathophysiology, 35f
    arousal symptoms, 38–39
    cognitive systems, 37–38
    negative valence system, 35–37
    positive valence systems, 37
    social process, 38

Research Domain Criteria (RDoC), 34
  risk factors, 33
Power spectrum analysis, 288
Principal component analysis (PCA), 290
Probabilistic-based metrics, 13–14
Progressive myoclonic epilepsies, 177

Q
queryHeadsets method, 199–200

R
Random forest, 186, 227, 291
Random hyperparameters experiments,
    21–22
Rectified Linear Unit (ReLU), 8
Recurrent neural networks (RNNs),
    102–103, 103f, 106
Recursive feature elimination or selection
    (RFE or RFS), 290
Regularization techniques, 290
Reinforcement learning (RL), 265
Research Domain Criteria (RDoC), 34
Resistive tactile sensors, 272–273
ResNet50 model, 5
ResNet network, 106
Responsive neurostimulation (RNS), 41
Riemannian manifolds, 297–299
RMSProp optimization method, 9

S
Second-order blind identification (SOBI)
    algorithm, 223
Seizures, 175
Self-powered prostheses, 256
Semisupervised generative adversarial
    network (SSGAN)
  brain MR images recognition, 124
  brain tumor recognition
    dataset, 128, 129f
    MRI image preprocessing, 125–127
    Nash equilibrium, 127
    preprocessing, 129, 130f
    semisupervised loss function,
      127–128
    testing accuracy, 131t

training and validation performance, 130, 130t
GAN architectural expansion, 124
image augmentation and brightness enhancement, 124
Sensor fusion, 265
Sensory feedback, hand prosthesis control
block diagram, 267f
electro-tactile feedback, 268
natural and reliable sensory feedback, 266–267
sensor technologies
computer vision and camera sensors, 273
electroencephalography (EEG) sensors, 271–272
electromyography (EMG) sensors, 271
force and pressure sensors, 272
position and inertial sensors, 273
tactile sensors, 272–273
vibration sensors, 273
types, 267–269
upper limb prostheses, 266
vibrotactile feedback, 268–269
visual feedback, 269
Short-time Fourier transform (STFT), 137–138
Siamese Neural Network (SNN), 120–123
Smart home control, disabled people
brain-computer interface (BCI), 195
electroencephalography (EEG)-based brain-computer interface (BCI) applications, 197
automated unsupervised algorithm, 196
brainwave and head motion control, 198–212
signal processing, 196
Software development kit (SDK), 200
Southampton Adaptive Manipulation Scheme (SAMS), 261–262
Sparrow search algorithm, 11–12
Spatial filtering methods, 287
Spectral analysis, neural signals, 97
Spectral entropy, 288

SPECTRA predictor, 299, 299f
Split learning (SL)
generic architecture, 168f
human activity recognition (HAR). See Human activity recognition (HAR)
system-level aspects, 163–164
training approach, 167
Steady-state visual evoked potential (SSVEP), 197
Stochastic gradient descent (SGD), 103
Subband CSP (SBCSP) method, 294
Support vector machine (SVM), 185–186, 225, 291
Synaptic potentials, 94

T
Tactile sensors, 272–273
Temple University Hospital EEG Epilepsy Corpus (TUH EEG Corpus), 180
Temporal lobe epilepsy, 177
Temporal filter parameter optimization (TFPO) approach, 296–297, 297f
Theta (q) waves, 95–96
Time-domain analysis, 97
Time-frequency analysis methods, 288
Topoplots, 57–59
Transcranial direct current stimulation (tDCS), 40–41
Transcranial magnetic stimulation (TMS), 40
Transcutaneous Electrical Nerve Stimulation (TENS), 268
Transfer learning (TL), 9, 120
Transformer model architecture, 104f
Transformer networks, 103
True negative rate (TNR), 12–13
True positive rate (TPR), 12–13
Two-classes dataset experiments, 22, 23t, 25t

U
U-Net model, 16
U-Net++ model, 16

V

Variational mode decomposition (VMD)
    modes, 148
        multivariate VMD (MVMD),
            149–150
        optimization problem,
            148–149
Vibration sensors, 273
Vibrotactile feedback, 268–269
Visual feedback, 269
V-Net model, 17

W

West syndrome (infantile spasms),
    177
Wireless Fidelity (Wi-Fi), 197
Wrapper methods, 290

Z

Zenodo electroencephalography (EEG)
    dataset, 180
ZigBee, 197
Z-Wave, 197

Printed and bound by CPI Group (UK) Ltd, Croydon, CR0 4YY

18/11/2024

01790588-0006